Lost Crops of Africa

volume I
Grains

Board on Science and Technology
for International Development

National Research Council

NATIONAL ACADEMY PRESS
Washington, D.C. 1996

NOTICE: The project that is the subject of this report was approved by the Governing Board of the National Research Council, whose members are drawn from the councils of the National Academy of Sciences, the National Academy of Engineering, and the Institute of Medicine. The members of the committee responsible for the report were chosen for their special competence and with regard for appropriate balance.

The report was reviewed by a group other than the authors according to procedures approved by a Report Review Committee consisting of members of the National Academy of Sciences, the National Academy of Engineering, and the Institute of Medicine.

This report was prepared by an ad hoc advisory panel of the Board on Science and Technology for International Development, Office of International Affairs, National Research Council. Staff support was funded by the Bureau for Africa, Bureau for Research and Development, Office of Nutrition, and Office of Research, Agency for International Development, under Grant No. DPE-5545-A-00-8068-00.

Library of Congress Catalog Card Number: 93-86876
ISBN 0-309-04990-3

Printed in the United States of America

A Note from the Sponsors

For two decades, the U.S. Agency for International Development (AID) has supported various reports from BOSTID's Innovation Program. This current one, on the underexploited cereals of Africa, is particularly timely. Africa's nutrition situation is deteriorating, and this is a serious concern. Much of the population is more vulnerable to malnutrition and starvation than ever before. Clearly, the problem needs tangible and sustained support from the international community, but it also needs a host of fresh ideas.

This book offers many such ideas and is part of a commitment AID made at the International Conference on Nutrition (ICN) in December 1992. There, member countries, nongovernmental organizations, and the international community pledged to eliminate or substantially reduce starvation, widespread undernutrition, and micronutrient malnutrition within this decade.

By highlighting the broad potential for Africa's own native boidiversity to reduce the vulnerability of seriously at-risk people to food shortages, the book could become a major contributor to the ICN objectives. The so-called "lost crops" obviously can help provide food security in their native areas, which include many parts of Africa threatened with hunger. At the same time, however, maintaining the diversity of these ancient crops will protect options for the rest of the world to use.

For these and other reasons, we are pleased to have been this project's major sponsors. We hope the wealth of information in the following pages will stimulate much interest and many subsequent activities. If that occurs, the now largely overlooked resources described herein should contribute substantially toward achieving the goal of eliminating hunger and malnutrition by decade's end.

David A. Oot
Office of Health and Nutrition

John Hicks
Bureau for Africa

Nan Borton
Office of Foreign Disaster Assistance

The National Academy of Sciences is a private, nonprofit, self-perpetuating society of distinguished scholars engaged in scientific and engineering research, dedicated to the furtherance of science and technology and to their use for the general welfare. Upon the authority of the charter granted to it by the Congress in 1863, the Academy has a mandate that requires it to advise the federal government on scientific and technical matters. Dr. Bruce M. Alberts is president of the National Academy of Sciences.

The National Academy of Engineering was established in 1964, under the charter of the National Academy of Sciences, as a parallel organization of outstanding engineers. It is autonomous in its administration and in the selection of its members, sharing with the National Academy of Sciences the responsibility for advising the federal government. The National Academy of Engineering also sponsors engineering programs aimed at meeting national needs, encourages education and research, and recognizes the superior achievements of engineers. Dr. Harold Liebowitz is president of the National Academy of Engineering.

The Institute of Medicine was established in 1970 by the National Academy of Sciences to secure the services of eminent members of appropriate professions in the examination of policy matters pertaining to the health of the public. The Institute acts under the responsibility given to the National Academy of Sciences by its congressional charter to be an adviser to the federal government and, upon its own initiative, to identify issues of medical care, research, and education. Dr. Kenneth I. Shine is president of the Institute of Medicine.

The National Research Council was organized by the National Academy of Sciences in 1916 to associate the broad community of science and technology with the Academy's purposes of furthering knowledge and advising the federal government. Functioning in accordance with general policies determined by the Academy, the Council has become the principal operating agency of both the National Academy of Sciences and the National Academy of Engineering in providing services to the government, the public, and the scientific and engineering communities. The Council is administered jointly by both Academies and the Institute of Medicine. Dr. Bruce M. Alberts and Dr. Harold Liebowitz are chairman and vice chairman, respectively, of the National Research Council.

The Board on Science and Technology for International Development (BOSTID) of the Office of International Affairs addresses a range of issues arising from the ways in which science and technology in developing countries can stimulate and complement the complex processes of social and economic development. It oversees a broad program of bilateral workshops with scientific organizations in developing countries and conducts special studies. BOSTID's Advisory Committee on Technology Innovation publishes topical reviews of technical processes and biological resources of potential importance to developing countries.

PANEL

NORMAN E. BORLAUG, Centro Internacional de Mejoramiento de Maíz y Trigo, Mexico City, Mexico, *Chairman*

JOHN AXTELL, Department of Agronomy, Purdue University, West Lafayette, Indiana

GLENN W. BURTON, Georgia Coastal Plain Experiment Station, Agricultural Research Service, U.S. Department of Agriculture, Tifton, Georgia

JACK R. HARLAN, Department of Agronomy, University of Illinois (retired), New Orleans, Louisiana

KENNETH O. RACHIE, Winrock International (retired), Pensacola, Florida

* * *

NOEL D. VIETMEYER, *Senior Program Officer,* Board on Science and Technology for International Development, *Africa Crops Study Director* and *Scientific Editor*

STAFF

F.R. RUSKIN, *BOSTID Editor*
MARK R. DAFFORN, *Staff Associate*
ELIZABETH MOUZON, *Senior Secretary*
BRENT SIMPSON, *MUCIA Intern*
DONALD OSBORN, *MUCIA Intern*

MICHAEL MCD. DOW, *Acting Director, BOSTID*

CONTRIBUTORS

By 1993, more than 1,000 people had participated in BOSTID's overall study of the lost crops of Africa. Most had participated by nominating species of grains, fruits, nuts, vegetables, legumes, oilseeds, spices, sweeteners, and beverage plants worthy of inclusion. In a sense, all these people were contributors to this, the first product from the study. However, the following list includes only those who provided technical details that became incorporated into various chapters of this particular book. To all the contributors, both listed and unlisted, we are truly grateful.

AFRICA

SAMUEL AGBOIRE, National Cereals Research Institute, Bida, Niger State, Nigeria

OLUPOMI AJAYI, ICRISAT-WASIP, Kano, Nigeria

O.C. AWORH, Department of Food Technology, University of Ibadan, Ibadan, Nigeria

FORSON K. AYENSU, Plant Genetic Resources Unit, Crops Research Institute, Bunso, Ghana

JACOB A. AYUK-TAKEM, Institut de la Recherche Agronomique, Yaoundé, Cameroon

ROBERT CUDJOE AZIAWOR, Grains Development Board, Hohoe, Volta Region, Ghana

PAUL BECKMAN, Eden Foundation, Zinder, Niger

M.A. BENHURA, Department of Biochemistry, University of Zimbabwe, Harare, Zimbabwe

JACQUES BEYO, Institut de Recherche Agronomique, Maroua, Cameroon

STEPHEN CARR, Zomba, Malawi

CARL W. CASTLETON, International Section, Animal and Plant Health Inspection Service, U.S. Department of Agriculture, Abidjan, Ivory Coast

ABEBE DEMISSIE, Plant Germplasm Exploration and Collection, Plant Genetic Resources Centre/Ethiopia, Addis Ababa, Ethiopia

SUSAN BURNELL EDWARDS, The National Herbarium, Addis Ababa University, Addis Ababa, Ethiopia

TEWOLDE BERHAN G/EGZIABHER, The National Herbarium, Addis Ababa University, Addis Ababa, Ethiopia

SAHR N. FOMBA, Mangrove Swamp Rice Research Station, West Africa Rice Development Association, Freetown, Sierra Leone

WALTER FRÖLICH, Sorghum and Millet Section, Nyankpala Agricultural Experiment Station, Crops Research Institute, Tamale, Ghana

KIFLE GOZEGUZE, Regional Soil and Water Conservation Department, Ministry of Agriculture, Addis Ababa, Ethiopia

S.C. GUPTA, Regional Sorghum and Millets Improvement Program, International Crops Research Institute for the Semi-Arid Tropics, Bulawayo, Zimbabwe

LELAND R. HOUSE, Regional Sorghum and Millets Improvement Program, Crops Research Institute for the Semi-Arid Tropics, Bulawayo, Zimbabwe

ISRAEL AFAM JIDEANI, School of Science and Science Education, Abubakar Tafawa-Balewa University, Bauchi, Nigeria

TANTIGEGN KEREDE KASSA, Zonal Team in Soil Conservation, Bahrder, Ethiopia

HILDA KIGUTHA, Department of Home Economics, Egerton University, Njoro, Kenya

ABEBE KIRUB, Information Services, Institute of Agricultural Research, Addis Ababa, Ethiopia

J. MAUD KORDYLAS, Arkloyd's Food Laboratory, Douala, Cameroon

HELMUT KREIENSIEK, Agriculture and Soil Conservation, German AgroAction -FSAP, Maseru, Lesotho

K. ANAND KUMAR, Pearl Millet Program, Sahelian Centre, International Crops Research Institute for the Semi-Arid Tropics, Niamey, Niger

PIUS MICHAEL KYESMU, Department of Botany, University of Jos, Jos, Plateau State, Nigeria

JOYCE LOWE, Department of Botany and Microbiology, University of Ibadan, Ibadan, Nigeria

GUEYE MAMADOU, West African Microbiological Research Centre, Centre National de Recherches Agronomiques, Bambey, Senegal

FERNANDO A.B. MARCELINO, Instituto de Investigação Agronómica, Huambo, Angola

P.C.J. MAREE, Department of Agronomy and Pastures, University of Stellenbosch, Stellenbosch, Cape Province, South Africa

MATEOS MEGISO, Soil Conservation Department, Ministry of Agriculture, Addis Ababa, Ethiopia

GEBRU TEKA MEHERETA, Natural Resources Department, Ministry of Agriculture, Addis Ababa, Ethiopia

I.M. MHARAPARA, Research and Specialist Services, Chiredzi Research Station, Chiredzi, Zimbabwe

KOUAMÉ MIEZAN, West Africa Rice Development Association, Bouake, Ivory Coast

GETACHEW BEYENE MISKER, Community Forestry Department, Ministry of Agriculture, Addis Ababa, Ethiopia

HELEN MOSS, International Plant Genetic Resources Institute, University of Zimbabwe, Harare, Zimbabwe

S.C. NANA-SINKAM, Joint ECA/FAO Agriculture Division, United Nations Economic Commission for Africa, Addis Ababa, Ethiopia
NLANDU NE NSAKU, Direction des Services Généraux Techniques, Gitega, Burundi
J.C. OBIEFUNA, Department of Crop Production, Federal University of Technology, Owerri, Imo State, Nigeria
NORMAN F.G. RETHMAN, Department of Plant Production, University of Pretoria, Pretoria, South Africa
GREGORY SAXON, Tete, Mozambique
A. SHAKOOR, The Dryland Farming Research and Development Project, Ministry of Agriculture, Katumani, Machakos, Kenya
P. SOMAN, Pearl Millet Program, Sahelian Centre, International Crops Research Institute for the Semi-Arid Tropics, Niamey, Niger
P.S. STEYN, Division of Food Science and Technology, National Food Research Institute, Pretoria, South Africa
JOHN R.N. TAYLOR, Department of Food Science, University of Pretoria, Pretoria, South Africa
JANE TOLL, International Crops Research Institute for the Semi-Arid Tropics, Niamey, Niger
JENS VON BARGEN, Nyankpala Agricultural Experiment Station, Crops Research Institute, Tamale, Ghana
ADEBACHO WATCHISO, Community Forestry Department, Ministry of Agriculture, Addis Ababa, Ethiopia
G.K. WEBER, International Institute of Tropical Agriculture, Ibadan, Nigeria
J.H. WILLIAMS, Pearl Millet Program, Sahelian Centre, International Crops Research Institute for the Semi-Arid Tropics, Niamey, Niger

OTHER REGIONS

DAVID J. ANDREWS, Department of Agronomy, University of Nebraska, Lincoln, Nebraska, USA
DJIBRIL AW, Resident Mission, Silver Spring, Maryland, USA
JACQUES BARRAU, Laboratoire d'Ethnobotanique-Biogéographie, Muséum National d'Histoire Naturelle, Paris, France
J.P. BAUDOIN, Phytotechnie des Régions Chaudes, Faculté des Sciences Agronomiques de Gembloux, Gembloux, Belgium
DONALD F. BEECH, Commonwealth Scientific and Industrial Research Organization, Brisbane, Queensland, Australia
GILLES BEZANÇON, Institut Français de Recherche Scientifique pour le Développement en Coopération de Montpellier, Montpellier, France
PAULA BRAMEL-COX, Department of Agronomy, Kansas State University, Manhattan, Kansas, USA

FRANK BREYER, Bludenz, Vorarlberg, Austria
RICHARD L. BRUGGERS, International Programs Research Section, U.S. Department of Agriculture, Denver Wildlife Research Center, Denver, Colorado, USA
LYNNE BRYDON, Department of Sociology, University of Liverpool, Liverpool, England
WAYNE CARLSON, Maskal Forages, Inc., Caldwell, Idaho, USA
GEOFFREY P. CHAPMAN, Wye College, University of London, Wye, Kent, England
W. DEREK CLAYTON, Royal Botanic Gardens, Kew, Richmond, Surrey, England
MAX D. CLEGG, Department of Agronomy, University of Nebraska, Lincoln, Nebraska, USA
ELIZABETH COLSON, Department of Anthropology, University of California, Berkeley, El Cerrito, California, USA
WILLIAM CRITCHLEY, Centre for Development Cooperation Services, Free University Amsterdam, Amsterdam, The Netherlands
RONNY R. DUNCAN, Department of Agronomy, University of Georgia, Georgia Experiment Station, Griffin, Georgia, USA
ROBERT P. EAGLESFIELD, International Crops Research Institute for the Semi-Arid Tropics, Andhra Pradesh, India
JOHANNES M.M. ENGELS, International Plant Genetic Resources Institute, New Delhi, India
CONRAD L. EVANS, Office of International Programs, Oklahoma State University, Stillwater, Oklahoma, USA
CHARLES A. FRANCIS, Department of Agronomy, University of Nebraska, Lincoln, Nebraska, USA
DONALD FRYREAR, Big Spring Experiment Station, U.S. Department of Agriculture, Big Spring, Texas, USA
ZEWDIE WOLDE GEBRIEL, Department of Human Nutrition, Wageningen Agricultural University, Wageningen, Netherlands
P. GEERVANI, College of Home Science, Andhra Pradesh Agricultural University, Hyderabad, Andhra Pradesh, India
DAVID GIBBON, School of Development Studies, University of East Anglia, Norwich, Norfolk, England
HEINER E. GOLDBACH, Abta Agrarökologie, Institut für Geowissenschaften der Universität Bayreuth, Bayreuth, Germany
PAMELA M. GOODE, Environmental Resources Unit, University of Salford, Salford, England
DAVID O. HALL, Center for Energy and Environment, Princeton University, Princeton, New Jersey, USA
WAYNE W. HANNA, Georgia Coastal Plain Experiment Station, U.S. Department of Agriculture, Tifton, Georgia, USA
NAZMUL HAQ, International Centre for Under-Utilized Crops, King's College London, London, England

G. HARINARAYANA, All India Coordinated Pearl Millet Improvement Project, College of Agriculture, Shivajinagar, Pune, India
DALE D. HARPSTEAD, Department of Crop and Soil Sciences, Michigan State University, East Lansing, Michigan, USA
FRANK NIGEL HEPPER, The Herbarium, Royal Botanic Gardens, Kew, Richmond, Surrey, England
KHIDIR W. HILU, Department of Biology, Virginia Polytechnic Institute and State University, Blacksburg, Virginia, USA
R.C. HOSENEY, Department of Agronomy, Kansas State University, Manhattan, Kansas, USA
CARL S. HOVELAND, Department of Agronomy, University of Georgia, Athens, Georgia, USA
CATHERINE HOWARTH, Welsh Plant Breeding Station, University College of Wales, Old College, Aberystwyth, Dyfed, Wales
CLARISSA T. KIMBER, Department of Geography, Texas A&M University, College Station, Texas, USA
ART KLATT, Division of Agriculture, Oklahoma State University, Stillwater, Oklahoma, USA
A. DE KOCHKO, Department of Biology, Washington University, St. Louis, Missouri, USA
WENDY KRAMER, Administrative Librarian, U.S. Department of Agriculture, Philadelphia, Pennsylvania, USA
J.M. LOCK, Royal Botanic Gardens, Kew, Richmond, Surrey, England
DAVID G. LYNN, Department of Chemistry, University of Chicago, Chicago, Illinois, USA
CLARE MADGE, School of Geography, The University of Birmingham, Egbaston, Birmingham, England
JAMES D. MAGUIRE, Department of Agronomy and Soils, Washington State University, Pullman, Washington, USA
A. BRUCE MAUNDER, DeKalb Plant Genetics, Lubbock, Texas, USA
DAN H. MECKENSTOCK, INTSORMIL/Programa Internacional de Sorgo y Mijo, c/o Escuela Agrícola Panamericana, Tegucigalpa, Honduras
ALEMU MENGISTU, Department of Plant Pathology, University of Wisconsin, Madison, Wisconsin, USA
GIOVANNI MIGNONI, Risanamento, Agro Industriale Zuccheri, Roma, Inc., Rome, Italy
FRED R. MILLER, Department of Soil and Crop Sciences, Texas A&M University, College Station, Texas, USA
NOBUO MURATA, Eco-Physiology Research Division, Tropical Agriculture Research Center, Tsukuba, Ibaraki, Japan
K.V. RAMAIAH, International Crops Research Institute for the Semi-Arid Tropics, Andhra Pradesh, India

K.C. REDDY, Agency for International Development - Niamey, U.S. Department of State, Washington, D.C., USA
PAUL RICHARDS, Department of Anthropology, University College London, London, England
K.W. RILEY, IDRC/National Hill Crops Improvement Program, Kathmandu, Nepal
JAMES L. RIOPEL, Department of Biology, University of Virginia, Charlottesville, Virginia, USA
LLOYD W. ROONEY, Department of Soil and Crop Sciences, Texas A&M University, College Station, Texas, USA
RACHEL SAFMAN, Agriculture and Natural Resources, CARE, New York, New York, USA
DAVID J. SAMMONS, Department of Agronomy, University of Maryland, College Park, Maryland, USA
SHAO QIQUAN, Department of Crop Science and Plant Ecology, University of Saskatchewan, Saskatoon, Canada
BLUEBELL R. STANDAL, Department of Food Science and Nutrition, University of Hawaii, Honolulu, Hawaii, USA
MARGARET STEENTOFT, Petersfield, Hampshire, England
MICHAEL STOCKING, Soils and Land Use Development Studies, University of East Anglia, Norwich, Norfolk, England
ROBERT J. THEODORATUS, Department of Anthropology, Colorado State University, Fort Collins, Colorado, USA
H.D. TINDALL, Ampthill, Bedford, England
J.H. TOPPS, Division of Agricultural Chemistry and Biochemistry, University of Aberdeen, Aberdeen, Scotland
RICK J. VAN DEN BELDT, Winrock International Institute for Agricultural Development, Bangkok, Thailand
DAT VAN TRAN, Plant Production and Protection Division, Food and Agriculture Organization of the United Nations, Rome, Italy
PARESH VERMA, Institute of Agriculture and Natural Resources, Department of Agronomy, University of Nebraska, Lincoln, Nebraska, USA
REMKO B. VONK, Agriculture and Natural Resources, CARE, New York, New York, USA
C.E. WEST, Department of Human Nutrition, Wageningen Agricultural University, Wageningen, Netherlands
JOSIEN M.C. WESTPHAL-STEVELS, Department of Plant Taxonomy, Wageningen Agricultural University, Wageningen, Netherlands
ERICA F. WHEELER, Centre for Human Nutrition, The London School of Hygiene and Tropical Medicine, London, England
G.E. WICKENS, Royal Botanic Gardens, Kew, Richmond, Surrey, England
REBECCA WOOD, Crestone, Colorado, USA

Preface

The purpose of this report is to draw worldwide attention to traditional African cereals and especially to their potential for expanding and diversifying African and world food supplies. Africa is seen by many observers as a basket case—a vast region incorporating more than 40 nations that appears unlikely to be able to feed its burgeoning population in the coming years. To many observers, there seem to be no ready solutions. Some have given up hope that anything can be done.

What has been almost entirely overlooked, however, is that throughout that vast continent can be found more than 2,000 native grains, roots, fruits, and other food plants. These have been feeding people for thousands of years but most are being given no attention whatever today. We have called them the "lost crops of Africa."

Among the 2,000 lost foods are more than 100 native grasses whose seeds are (or have been) eaten. These can be found from Mauritania to Madagascar. Only a handful are currently receiving concerted research and development, and even those few are grossly underappreciated. Our goal is to demonstrate the potential inherent in these overlooked traditional cereals. Our hope is thereby to stimulate actions to increase the support for, and use of, the best of them so as to increase food supplies, improve nutrition, and raise economic conditions.

It should be understood that most of the plants described are not truly lost; indeed, a few are well known worldwide. It is to the mainstream of international science and to people outside the rural regions that they are "lost." It should also be understood that it is not just for Africa that the grains hold promise. Several of Africa's now-neglected cereals could become major contributors to the welfare of nations around the world. This potential is often emphasized in the following chapters in hopes of stimulating the world community into serious and self-interested support for these species that now languish.

This study began in 1989 when the staff officers mailed questionnaires to about 1,000 scientists and organizations worldwide. The questionnaire requested nominations of little-known African food plants for possible inclusion. It contained a list of 77 native African grains, roots and tubers, vegetables, fruits, legumes, oilseeds, nuts, spices, sweeteners, and beverage plants. We anticipated that perhaps 30 of these species would prove to have outstanding merit and that the report would focus on those. What actually occurred, however, was very different.

Within a few weeks of mailing the questionnaire, replies started flooding back in numbers far greater than anticipated; many recipients photocopied their questionnaire and sent the copies (as many as 50 in several cases) on to their colleagues; requests came pouring in from people we had never heard of. The staff could barely keep up with the hundreds of requests, replies, suggestions, scientific papers, and unsolicited writings that began to appear in the mail. Within 4 months, over 100 additional species had been nominated as "write-in candidates." Within a year, at least 100 more were recommended. By then it was clear that the power of this project was far greater than anyone had foreseen. It was decided, therefore, to divide it into sections dealing individually with the different types of foods.

This report on the lost grains of Africa is the first in this series. From the flood of suggestions and information on the native African cereals was fashioned a first draft. Each of its chapters was mailed back to the original nominators as well as to other experts identified by the staff. As a result, hundreds of suggestions for corrections and additions were received, and each was evaluated and integrated into what, after editing and review, became the current text.

The report is intended as a tool for economic development rather than a textbook or survey of African botany or agriculture. It has been written for dissemination particularly to administrators, entrepreneurs, and researchers in Africa as well as other parts of the world. Its purpose is to provide a brief introduction to the plants selected and to stimulate actions that explore and exploit them. The ultimate aim is to get the most promising native African grains into greater production so as to raise nutritional levels, diversify agriculture, and create economic opportunities.

Because the book is written for audiences both lay and professional, each chapter is organized in increasing levels of detail. The lead paragraphs and prospects sections are intended primarily for nonspecialists. Subsequent sections contain background information from which specialists can better assess a plant's potential for their regions or research programs. These sections also include a brief overview of "next steps" that could help the plant to reach its full promise. Finally, appendixes at the back of the book provide the following information:

- The addresses of researchers who know the individual plants well;
- Information on potential sources of germplasm; and
- Lists of carefully selected papers that provide more detail than can be presented here.

Because most of these plants are so little studied, the literature on them is often old, difficult to find, or available only locally. This is unfortunate, and we hope that this book will stimulate monographs, newsletters, articles, and papers on all of the species. One of the most effective actions that plant scientists and plant lovers can take is to collect, collate, and communicate the Africa-wide observations and experiences with these crops in such publications. They might also create seed supplies and distribute seeds of appropriate varieties. All this could stimulate pan-African cooperation and international endeavors to ensure that these crops are lost no more.

This book has been produced under the auspices of the Board on Science and Technology for International Development (BOSTID), National Research Council. It is a product of a special BOSTID program that is mandated to assess innovative scientific and technological advances, particularly emphasizing those appropriate for developing countries. Since its inception in 1970, this small program has produced 40 reports identifying unconventional scientific subjects of promise for developing countries. These have covered subjects as diverse as the water buffalo, butterfly farming, fast-growing trees, and techniques to provide more water for arid lands (see BOSTID Innovation Program, page 373).

Among these reports, the following provide information that directly complements the present report:

- *More Water for Arid Lands* (1974)
- *Triticale: A Promising Addition to the World's Cereal Grains* (1989)
- *Quality-Protein Maize* (1988)
- *Amaranth: Modern Prospects for an Ancient Crop* (1983)
- *Applications of Biotechnology to Traditional Fermented Foods* (1992)
- *Ferrocement: Applications in Developing Countries* (1973)
- *Neem: A Tree for Solving Global Problems* (1992)
- *Vetiver: A Thin Green Line Against Erosion* (1993).

Program and staff costs for this study were provided by the U.S. Agency for International Development. Specifically, these were provided by the Office of Nutrition and the Office of the Science Advisor (both of the Bureau for Science and Technology), as well as the Bureau for Africa. The panel would like to acknowledge the special contribution of Norge W. Jerome, Director of the Office of Nutrition, 1988-1991, without whose initiative the project would not have been launched. Other AID personnel who made this work possible include Calvin Martin, Tim Resch, Dwight Walker, John Daly, Frances Davidson, and Ray Meyer.

General support for printing, publishing, and distributing the report has been provided by the Kellogg Endowment Fund of the National Academy of Sciences and the Institute of Medicine as well as from the Wallace Genetic Foundation. We especially want to thank Jean W. Douglas, a foundation director, for her trust and perserverence during this project's long gestation and difficult birth.

The contributions from all these sources are gratefully acknowledged.

How to cite this report:
National Research Council. 1996. *Lost Crops of Africa. Volume 1: Grains*. National Academy Press, Washington, D.C.

NOTE ON TERMS

Throughout this book the word "Africa" always refers to Africa south of the Sahara. (The plants of North Africa are, biogenetically, part of the Mediterranean-Near East complex of plants, and so are mostly not native to the rest of Africa.) We have preferred to use English common names where possible, except in a few cases where they imply the plant pertains only to one country (for example, Egyptian lupin). Finally, because this book will be read and used in many regions beyond Africa, we have used the internationally accepted name "cassava" rather than its more common African name, "manioc," and "peanut" for "groundnut."

Nutritional values are in most cases presented on a dry weight basis to eliminate moisture differences between samples.

Contents

APPENDIXES

Lost Crops of Africa

volume I
Grains

Foreword

Africa has more native cereals than any other continent. It has its own species of rice, as well as finger millet, fonio, pearl millet, sorghum, tef, guinea millet, and several dozen wild cereals whose grains are eaten from time to time.

This is a food heritage that has fed people for generation after generation stretching back to the origins of mankind. It is also a local legacy of genetic wealth upon which a sound food future might be built. But, strangely, it has largely been bypassed in modern times.

Centuries ago, dhows introduced rice from Asia. In the 1500s, Portuguese colonists imported maize from the Americas. In the last few decades wheat has arrived, courtesy of farmers in the temperate zones. Faced with these wondrous foreign foods, the continent has slowly tilted away from its own ancient cereal wealth and embraced the new-found grains from across the seas.

Lacking the interest and support of the authorities (most of them non-African colonial authorities, missionaries, and agricultural researchers), the local grains could not keep pace with the up-to-the-minute foreign cereals, which were made especially convenient to consumers by the use of mills and processing. The old grains languished and remained principally as the foods of the poor and the rural areas. Eventually, they took on a stigma of being second-rate. Myths arose—that the local grains were not as nutritious, not as high yielding, not as flavorful, nor as easy to handle. As a result, the native grains were driven into internal exile. In their place, maize, a grain from across the Atlantic, became the main food from Senegal to South Africa.

But now, forward-thinking scientists are starting to look at the old cereal heritage with unbiased eyes. Peering past the myths, they see waiting in the shadows a storehouse of resources whose qualities offer promise not just to Africa, but to the world.

Already, sorghum is a booming new food crop in Central America. Pearl millet is showing such utility that it is probably the most promising new crop for the United States. Nutritionists in a dozen or more countries see finger millet and some sorghums as the key—finally—to solving Africa's malnutrition problem. Food technologists are finding vast new possibilities in processes that can open up vibrant consumer markets for new and tasty products made from Africa's own grains. And engineers are showing how the old grains can be produced and processed locally without the spirit-crushing drudgery that raises the resentment of millions who have to grind grain every day.

That, then, is the underlying message of this book. It should not be seen as an indictment of wheat, maize, or rice. Those are the world's three biggest crops, they have become vital to Africa, and they deserve even more research and support than they are now getting. But this book, we hope, will open everyone's eyes to the long-lost promise inherent in the grains that are the gifts of ancient generations. Dedicated effort will open a second front in the war on hunger, malnutrition, poverty, and environmental degradation. It will save from extinction the foods of the forebears. And it just might bring Africa the food-secure future that everyone hopes for but few can now foresee.

Noel D. Vietmeyer
Study Director

Introduction

Africa's savannas are probably the oldest grasslands on earth and have changed little during the last 14 million years. Humans have lived there longer than anywhere else, perhaps more than 100,000 years. Grass seeds have sustained them throughout.

Indeed, gathering Africa's wild-cereal grains is probably the oldest tradition in organized food production to be found anywhere in the world. And the operation was not small. In fact, seeds of about 60 species of wild grasses are still gathered for food in Africa.[1] In earlier eras, many were ranked as staples. At least 10 of the wild grasses were domesticated and eventually produced by farmers in their fields.

In modern times, however, this wealth of native grains has been neglected and sometimes even scorned. For this reason, we have called them Africa's "lost" grains.

Despite the neglect, these native grains are not unworthy. For the past, for the places they were grown and for the level of support they received, they may have been appropriately judged less useful than wheat, rice, or maize. But for the time that is fast coming upon us, Africa's sorghum, millets, native rice, and other indigenous cereals seem likely to become crucial for helping to keep the world fed.

INTERNATIONAL PROMISE

The present century has seen near miraculous advances in the productivity of wheat, rice, and maize. Those top three cereals have buffered much of humanity from the disasters of overpopulation. However, the next century—when human population is expected to double—cannot be built on the expectation of redoubling the production of those three.

After the year 2000, it could well be advances in today's "second

[1] Jardin, 1967.

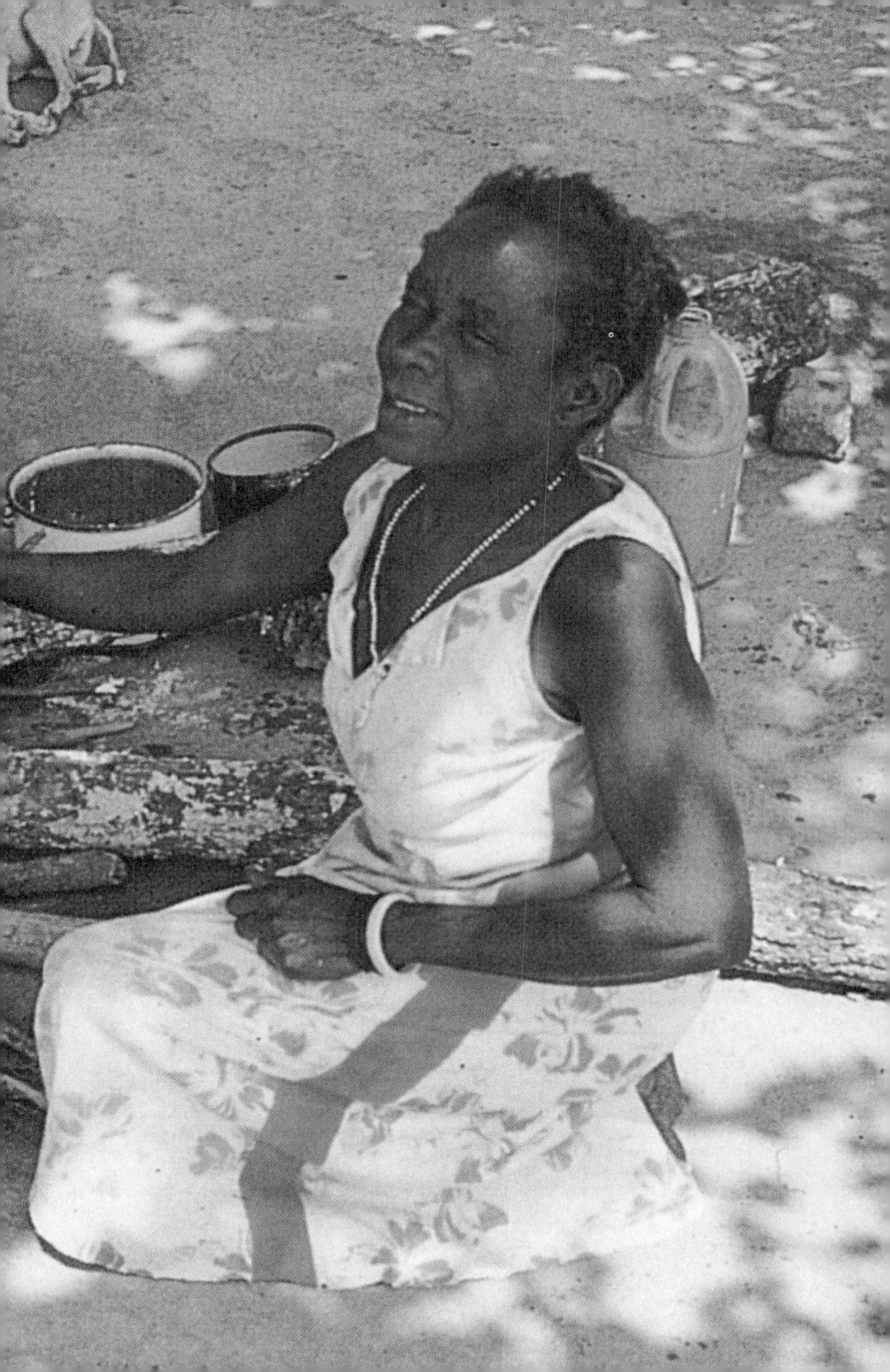

tier" cereals that are the buffers against famine. It is they that have the greatest amount of untapped potential. Among them, Africa's native grains predominate. Sorghum and pearl millet, for instance, are the fifth and sixth most important cereals in the world, and finger millet is probably the eighth.[2] Generally, they are crops of the poorest countries, which means that their improvement could directly benefit the people in greatest need.

By comparison with modern wheat, rice, and maize—respectively from the Middle East, Asia, and Central America—the grains of Africa still retain much of the hardy, tolerant self-reliance of their wild savanna ancestors. For the future, such resilient crops will be vital for extending cereal production onto the ever-more-marginal lands that will have to be pressed into service to feed the several billion new arrivals. And if global warming occurs, they could even become vital for keeping today's best arable lands in production.

Forged in the searing savannas and the Sahara, sorghum and pearl millet in particular have the merits to become crops for the shifting and uncertain conditions of an overpopulated "greenhouse age."

LOCAL PROMISE

In the last few centuries in Africa, the local grains have been superseded by foreign cereals introduced and promoted by outsiders such as missionaries, colonial powers, or researchers. In recent decades, the production of native grains has plunged even further as millions of tons of imports—particularly wheat and rice—have been sold at subsidized prices.

Despite its long history, Africa's cereal production is now low. The Green Revolution that transformed the tropics and subtropics, from the Indian subcontinent to South America, passed Africa by. In fact, per-capita production of cereals has decreased nearly 20 percent (present annual output being only about 50 million tons or a mere 11

[2] Barley, native to the Middle East, is the fourth and rye is the seventh.

Previous page. In rural Africa, the traditional cereals, such as sorghum and millets, are normally boiled into porridges (thick) or gruels (thin). Scenes like that shown have led outsiders to conclude that the crop itself is primitive. But wheat, rice, and maize were also treated this way throughout most of their 5,000-year history as crops. Only in recent times have they been commercially processed and packaged and sold in convenient forms. Given the same treatment, grains like sorghum being cooked here can also enter modern production. (ICRISAT)

kg per person). It has been estimated that Africa now needs 14 million tons more grain each year than it is producing. With the population growing at 3 percent per year and agricultural production increasing by only 2 percent, that shortfall will reach 50 million tons by 2000.[3]

Obviously a crisis is impending in Africa's food supply. Improving cereals for Africa should be a great international agricultural endeavor. Maize, rice, and wheat have much to offer and deserve greatly increased support. A crucial objective, though, must be to extend cereal production into areas where environmental stresses and plant diseases currently limit their growth. For these now-marginal lands, Africa's own grains offer outstanding promise. They are tools for helping build a new and stronger food-production framework—one of inestimable value for the hungriest and most destitute nations.

THE SPECIES

This promise (and much more) is described in the body of this book. There, the following species are covered in detail.

African Rice

Most people think of rice as an exclusively Asian crop, but farmers have grown a native rice (*Oryza glaberrima*) in parts of West Africa for at least 1,500 years. This crop comes in a wealth of different types that are planted, managed, prepared, and eaten in different ways. Some mature extremely quickly and will fit into seasons and situations where other cereals fail. The grain is much like common rice, although the husk around it is usually red. This plant not only has promise in its own right, its genes might also eventually benefit the production of common rice worldwide. (See chapter 1, page 17.)

[3] Spore, June 1995.

Following page: Pearl millet in Mali, as seen in the vicinity of Mopti. Crops such as this are a mainstay of life in the vast rural regions of Africa. Despite their vital importance to millions like this village farmer, traditional cereals receive only minuscule support from science and the world community. They are the "forgotten end" of agricultural development. Yet, as can be seen here, they are a part of the heritage and daily living of hard-working Africans. Outsiders may scorn pearl millet, but this woman's face shows the pride she feels in her harvest. (H.S. Duggal, courtesy ICRISAT)

Finger Millet

In parts of East and Central Africa (not to mention India), millions of people have lived off finger millet (*Eleusine coracana*) for centuries. One of the most nutritious of the major cereals, it is rich in methionine, an amino acid critically lacking in the diets of hundreds of millions of the world's poor. The plant yields satisfactorily on marginal lands, and its tasty grain is remarkable for its long storage life. The fact that certain Africans thrive on just one meal a day is attributed to the nutritive value and "filling" nature of this grain. (See chapter 2, page 39.)

Fonio (Acha)

An indigenous West African crop, fonio (comprising two species, *Digitaria exilis* and *Digitaria iburua*) is grown mainly on small farms for home consumption. It is probably the world's fastest maturing cereal and is particularly important as a safety net for producing when other foods are in short supply or market prices are too high for poor people to afford. But fonio is much more than just a fallback food; it is also a gourmet grain. People enjoy it as a porridge, in soups, or as couscous with fish or meat. The plant grows well on poor, sandy soils. It, too, is rich in the amino acid methionine. It also has a high level of cystine, a feature that is an even rarer find in a cereal. With its appealing taste and high nutritional value, this could become a widespread gourmet grain for savanna regions, perhaps throughout much of Africa or even much of the world. It might well have a big future as a cash crop and export commodity. (See chapter 3, page 59.)

Pearl Millet

Some 4,000 years ago, pearl millet (*Pennisetum glaucum*) was domesticated from a wild grass of the southern Sahara. Today, it is the world's sixth-largest cereal crop, but it has even greater potential than most people imagine. Of the major cereals, pearl millet is the most tolerant of heat and drought; it has the power to yield reliably in regions too arid and too hot to consistently support good yields of other major grains. These happen to be the regions that will most desperately need help in the decades ahead.

Already, water is shaping up as the most limiting resource for numerous economies—even some of the most advanced. Agriculture is usually a country's biggest user of water. Thus, for nations that have never heard of it or that perhaps regard it with scorn, pearl millet might quickly rise to become a vital resource. (See chapters 4–6; pages 17, 93, and 111.)

Sorghum

Globally speaking, sorghum is the dietary staple of more than 500 million people in more than 30 countries. Only rice, wheat, maize, and potatoes surpass it in the quantity eaten. For all that, however, it produces merely a fraction of what it could. Indeed, if the twentieth century has been the century of wheat, rice, and maize, the twenty-first could become the century of sorghum (*Sorghum bicolor*).

First, sorghum is among the most photosynthetically efficient and quickest maturing food plants. Second, it thrives on many marginal sites where other cereals fail. Third, sorghum is perhaps the world's most versatile food crop. Some types of its grains are boiled like rice, cracked like oats for porridge, "malted" like barley for beer, baked like wheat into flat breads, or popped like popcorn for snacks.

The plant has many uses beyond food as well. Perhaps the most intriguing is its use for fuel. The stems of certain types yield large amounts of sugar, almost like sugarcane. Thus, sorghum is a potential source of alcohol fuels for powering vehicles or cooking evening meals. Because of the plant's adaptability, it may eventually prove a better source of alcohol fuel than sugarcane or maize, which are the only ones now being used.

Finally, sorghum is a relatively undeveloped crop with a truly remarkable array of grain types, plant types, and adaptability. Most of its genetic wealth is so far untapped and even unsorted. Indeed, sorghum probably has more undeveloped genetic potential than any other major food crop in the world. (See chapters 7–11; pages 127, 145, 159, 177, and 195.)

Tef

This staple cereal (*Eragrostis tef*) is the most esteemed grain in Ethiopia. It is ground into flour and made into pancake-like fermented bread, *injera,* that forms the basic diet of millions. Many Ethiopians eat it several times a day (when there is enough), particularly with spicy sauces, vegetables, and stews.

Tef is nutritious; the grain is about 13 percent protein, well balanced in amino acids, and rich in iron. In many ways, it seems to have ideal qualities for a grain, yet research has been scanty and intermittent, and so far the crop is all but unknown beyond Ethiopia. In the last few years, however, commercial production has started in the United States and South Africa, and an export trade in tef grain has begun. These seem likely harbingers of a new, worldwide recognition of this crop. (See chapter 12, page 215.)

Misunderstandings

It is fair to ask why Africa's grains are not better known. At least in part, the reason can be attributed to several unjustified perceptions. Some of these misperceptions that are clouding the world's vision of Africa's native grains are discussed below.

Inferiority of Displaced Crops. Introduced crops have displaced several African ones over the past few centuries. For example, in several areas maize has replaced sorghum; in West Africa, Asian rice has replaced African rice. As a result, there is a strong inclination to consider the introduced crop superior and the native crop obsolete and unworthy of further development.

This is illogical, ill-conceived, and even dangerous. All the world's agriculture is dynamic and every crop gets displaced at certain times and certain places. In much of the eastern United States, for instance, wheat was long ago displaced by soybeans; in the Southeast, peanuts replaced rice; and in the Great Plains, wheat has supplanted maize. But no one in America considers wheat, maize, or rice to be inferior, obsolete, or unworthy.

Misclassification. Africa's cereals are inadvertently discriminated against through the way they are described. People everywhere classify sorghums and millets in a different light from wheat, rice, and maize. All the categories have pejorative connotations. For instance, these grains are typically referred to as:

- "Coarse" grains (that is, not refined; fit for animal feed);
- "Minor" crops (not worthy of major status);
- "Millets" (seeds too small);
- "Famine" foods (good for eating only when starving); and
- "Feed" grains (suitable for animals only).

Poor People's Plants. Many crops are scorned as fit only for consumption by the poor. It happens everywhere. Peanuts, potatoes, and other common crops once suffered from this same discrimination. In the United States the peanut was considered to be "merely slave food" until little more than a century ago, and in the 1600s the English refused to eat potatoes because they considered them to be "Irish food." Cultural bias against peasant crops is a tragedy; the plants poor people grow are usually robust, productive, self-reliant, and useful—the very types needed to feed the hungriest mouths on the planet.

Inferior Yield. Low yield is perhaps the most frequent comment made about Africa's grains. Yet these grains are now mostly cultivated in marginal lands under less than optimal management and the yields therefore do not reflect their true potential.

Moreover, the use of yield figures can be totally misleading. Maize may be able to outyield finger millet, pearl millet, hungry rice, and tef, but only when soil fertility, moisture, and other conditions are good. Under poor conditions, African grains often outyield the best products of modern science.

Unworthy Foods. Millets are mainly used for making porridges, fermented products, couscous, and other foods that are alien and therefore somewhat suspect to non-Africans, especially Westerners. This has led outsiders, who often serve as "decision makers," to direct resources away from native grains.

Disparaging comments about African foods are not uncommon in the writings of travelers—especially in Victorian times. They are of course only personal—often highly prejudiced—opinions but, lingering in the literature, they have a pernicious influence that can last for decades or even centuries. Europeans treated the potato and tomato this way when they first arrived from the Americas. Myths about taste and safety helped block the adoption of both for two centuries.

Cost-Effectiveness. Most of Africa's grains are exclusively subsistence crops; the remainder are partially so. Farmers grow them for their own use rather than for market, and therefore there are no statistics on production or costs. A plant may be helping feed millions, but in the international figures on area sown, tonnage produced and exported, and prices paid it never shows. It is as if it doesn't exist.

This situation might be of little consequence were it not for the fact that economic-development funding these days is overwhelmingly judged on "cost-effectiveness." Thus, a crop with no baseline data is at a cruel disadvantage. Maize or wheat researchers can pull out impressive figures to justify the promise of their proposed studies. Finger-millet or fonio researchers can only come up with guesses. To the hard-pressed, cost-conscious administrator—ever fearful of accusations that public funds may be misspent—the decision on which proposal to support is inevitably biased.

Other Cultivated Grains

Some of the cereals described previously are not, strictly speaking, "lost." But there are a number of African food grains that are indeed truly overlooked by modern science. (See chapter 13, page 237.)

Guinea Millet Perhaps the world's least-known domesticated cereal, guinea millet (*Brachiaria deflexa*), is cultivated by farmers only in the Fouta Djallon Plateau, a remote region of Guinea. At present, almost nothing can be said about its potential, but it clearly deserves exploratory research and support.

Emmer This rare wheat (*Triticum dicoccum*) originated in the Near East, but it has a very ancient African heritage. It reached Ethiopia probably 5,000 years ago or more and, although it virtually disappeared elsewhere in the world, it comprises almost 7 percent of Ethiopia's entire wheat production. Moreover, far from abandoning it, Ethiopian farmers over the last 40 years have actually increased the percentage of emmer that they grow.

The plant is adapted to a wide range of environments and should be producible in many parts of the world. The fact that it is little changed from wheat eaten in the times of the Bible and the Koran could give it special consumer appeal. But it can stand on its own culinary merits. It is one of the sweetest and best-tasting cereals.

Irregular Barley Although barley is also not native to Africa, it, too, has been used in Ethiopia for thousands of years. Indeed, Ethiopian barley has been isolated so long that it has been given its own botanical name, *Hordeum irregulare,* and has developed its own genetic "personality." This ancient barley is grown mainly in Ethiopia, where it ranks fourth among crops, both in production and area. Throughout most of the upper highlands it accounts for over 60 percent of the people's total plant food. Ethiopia is perhaps unmatched with respect to barley diversity. Indeed, some scientists think it is a source of new germplasm that could possibly boost barley growing in Africa and around the world.

Ethiopian Oats In Ethiopia is found a native oats, *Avena abyssinica.* This species was domesticated in the distant past and is a largely nonshattering plant that retains its grain so people can harvest it. It has long been used in Ethiopia and is well adapted to the high elevations there. It is, however, unknown elsewhere.

Wild Grains

As noted, people in Africa have been eating wild grains for perhaps 100,000 years. In modern times, however, various writers have discounted these grains as mere "scarcity foods." This is obviously wrong: wild grains were eagerly eaten even when pearl millet, for one, was abundant.

Many modern writers also imply that the wild cereals were gathered only on a small and localized scale. This, too, is apparently false. The harvest in the Sahara, for example, was large-scale, sophisticated, commercial, and much of it was export-oriented. The wild grains were a delicacy that even the wealthy considered a luxury. Examples of such untamed cereals are drinn, golden millet, kram-kram, panic grasses, wild rices, jungle rice, wild tefs, and crowfoot grasses.

Resurrecting the grain-gathering industry of the past might be a way to help combat desertification, erosion, and other forms of land degradation across the worst afflicted areas of the Sahel and its neighboring regions. A vast and vigorous grain-gathering enterprise might perhaps provide enough economic incentive to ensure that the grass cover is kept in place and that overgrazing is controlled. That would bring environmental stability to the world's most alarming case of desertification. (See chapter 14, page 251.)

CONCLUSION

These "lost" plants have much to offer, and not just to Africa. Indeed, they represent an exceptional cluster of cereal biodiversity with particular promise for solving some of the greatest food-production problems that will arise in the twenty-first century.

This potential for utility in the future is because Africa's native grains tend to tolerate extremes. They can thrive where introduced grains produce inconsistently. Some (tef, for instance) are adapted to cold; others (pearl millet, for example) to heat; at least one sorghum to waterlogging; and many to drought. Moreover, most can grow better than other cereals on relatively infertile soils. For thousands of years they have yielded grain even where land preparation was minimal and management poor. They combine well with other crops in mixed stands. Some types mature rapidly. They tend to be nutritious. And at least one is reputed to be better tasting than most of the world's well-known grains.

1

African Rice

To most of the world, rice connotes Asia and the vast agriculture of Far Eastern river deltas. Indeed, humanity's second major crop *is* from Asia, and 90 percent of it—the main source of calories for 2.7 billion people—is grown there.

But rice is also African. A different species has been cultivated in West Africa for at least 1,500 years. Some West African countries have, since ancient times, been just as rice-oriented as any Asian one. For all that, however, almost no one else has ever heard of their species.[1]

Asia's rice is so advanced, so productive, and so well known that its rustic relative has been relegated to obscurity even in Africa itself. Today, most of the rice cultivated in Africa is of the Asian species. In fact, the "great red rice of the hook of the Niger" is declining so rapidly in importance and area that in most locations it lingers only as a weed in fields of its foreign relative. Soon it may be gone.

This should not be allowed to happen. The rice of Africa (*Oryza glaberrima*) has a long and noteworthy history. It was selected and established in West Africa centuries before any organized expeditions could have introduced its Asian cousin (*Oryza sativa*). It probably arose in the flood basin of the central Niger and prehistoric Africans carried it westward to Senegal, southward to the Guinea coast, and eastward as far as Lake Chad. In these new homes, diligent people developed it further.

Like their counterparts in the Far East, Africa's ancient rice farmers selected a remarkable range of cultivars suited to many types of habitats. They produced "floating" varieties (for growing in deep

[1] There are rice relatives in other parts of the world, too. The genus *Oryza* is among the most ancient grasses and was able to spread to every continent before they drifted too far apart. The result is that different *Oryza* species are strung out over the tropical regions of the globe, including South America and Australia. Only one species in Asia and one in Africa were domesticated, however.

Rice growers in Sierra Leone harvesting their crop. For millennia, farmers in this area have grown African rice. Possibly it was slaves from here who introduced rice growing to the United States. For a century or more, the colony and state of South Carolina was a main rice grower. Whether African rice reached there along with the Asian species is not known. However, given the West African attachment to it, it seems likely. (M. Steber)

water), weakly and strongly photoperiod-sensitive types (for growing in different latitudes and seasons), swamp and upland cultivars (for growing under irrigated and rainfed conditions, respectively), and early and late-maturing types. And, for all of these, they selected forms with various seed characteristics.

Although modern efforts to expand rice production in Africa have largely ignored this indigenous heritage, African rice is still cultivated in West Africa—especially in remote districts. There, until recently, much of it was reserved as a special luxury food for chiefs and religious rituals. Today, however, farms that grow substantial stands of African rice are few. The area of most intense cultivation is the "floating fields" on the Sokoto fadamas (flood plains) of Nigeria and the Niger River's inland delta in Mali. However, the crop is also widely, if thinly, spread in Sierra Leone (see box, page 28) and neighboring areas, as well as in the hills that straddle the Ghana-Togo border.

From one point of view, there seem to be good reasons for abandoning this food of the forebears. In most locations farmers prefer the foreign rice because it yields better and scatters less of its seed on the ground. Millers prefer it because its grain is less brittle and therefore easier to mill. Shippers prefer it as well. For them, African rice is hardly worth a minute's consideration because it is not a trade commodity and most types are red-skinned and therefore unsuitable for mixing with conventional rice in bulk handling.

But these are concerns almost entirely of commercial farming. The situation is quite different where rice is grown strictly for localized, subsistence, or specialty use. There, yield, brittleness, color, or international interest can be unimportant. Indeed, small-scale farmers often prefer African rice. They like the grain's taste and aroma, and even its reddish appearance. They find the plant easy to produce: its rambunctious growth and spreading canopy help suppress weeds and it generally resists local diseases and pests by itself. Also, to some people traditional rituals are meaningless unless the ancient grain is employed.

Moreover, these are not the only advantages. Compared to its Asian cousin, African rice is better at tolerating fluctuating water depths, excessive iron, low levels of management, infertile soils, harsh climates, and late planting (a valued feature because in West Africa's erratic

Opposite: For hundreds, if not thousands, of years, "floating" versions of African rice have been cultivated beside the Niger River, especially here between Timbuktu and Gao. Farmers along this 600-km stretch count on the Niger to overflow its banks and flood the lowlands where they've sowed their seeds. The rice plants can survive in rising floodwater up to several meters deep. When drought reduced the Niger's flow in the 1970s, the full crop could not be planted and a million lives were put at risk. (J. Gallais, courtesy Flammarion et cie, Paris)

climate the rains are often tardy). Also, there are some types that mature much more quickly than common rice. Planted out in emergencies when food stocks are getting low, these can save lives.

PROSPECTS

What actually happens in the future to this interesting African crop will depend on individual initiatives, most of them within Africa itself. Part of the problem is its lack of prestige. Everywhere, consumers have fallen in love with processed Asian rice. If someone now makes a processed (that is, parboiled) product out of African rice, that alone may return it to high favor. Indeed, it may rise to become a gourmet food of particular interest because of its ancient and historic heritage.

Part of the problem, also, is lack of supply. Thus, if such specialty markets develop, it seems likely that African rice will survive as a commercial crop. Then, with selection and breeding, its various cultivars can almost certainly be made to compete with Asian rice in most African locations. There is evidence, for example, that certain types already match the productivity of Asian rice, and in the yield figures there is considerable overlap between the best African and the poorer Asian ones. This is remarkable considering the 5,000 years of intense effort that has been invested in improving Asian rice.

Even if the local rice never thrives as a commercial crop, it will likely continue as a subsistence crop in West Africa. However, whether this is a lingering decline for a few more decades or a robust return to massive use depends on the responses of scientists, administrators, and others. Even in its current neglected form the plant has something to offer, but just a small amount of support, promotion, and practical research seems likely to bring dramatic improvements.

The problems of shattering and brittle grain can undoubtedly be overcome by careful scrutiny of the types already spread across West Africa. A small cash prize might well produce appropriate genotypes almost overnight. The same could happen for white-skin types, which many people would find more appealing than the main type of today. Even now, not all the varieties are red-skinned. In Guinea, Senegal, and the Gambia, for example, white types are said to be already available.

Africa

Although no one can be certain of what will happen with this crop in the coming decades, the prospects for doubling its production and overcoming its various technical limitations are good. Technical

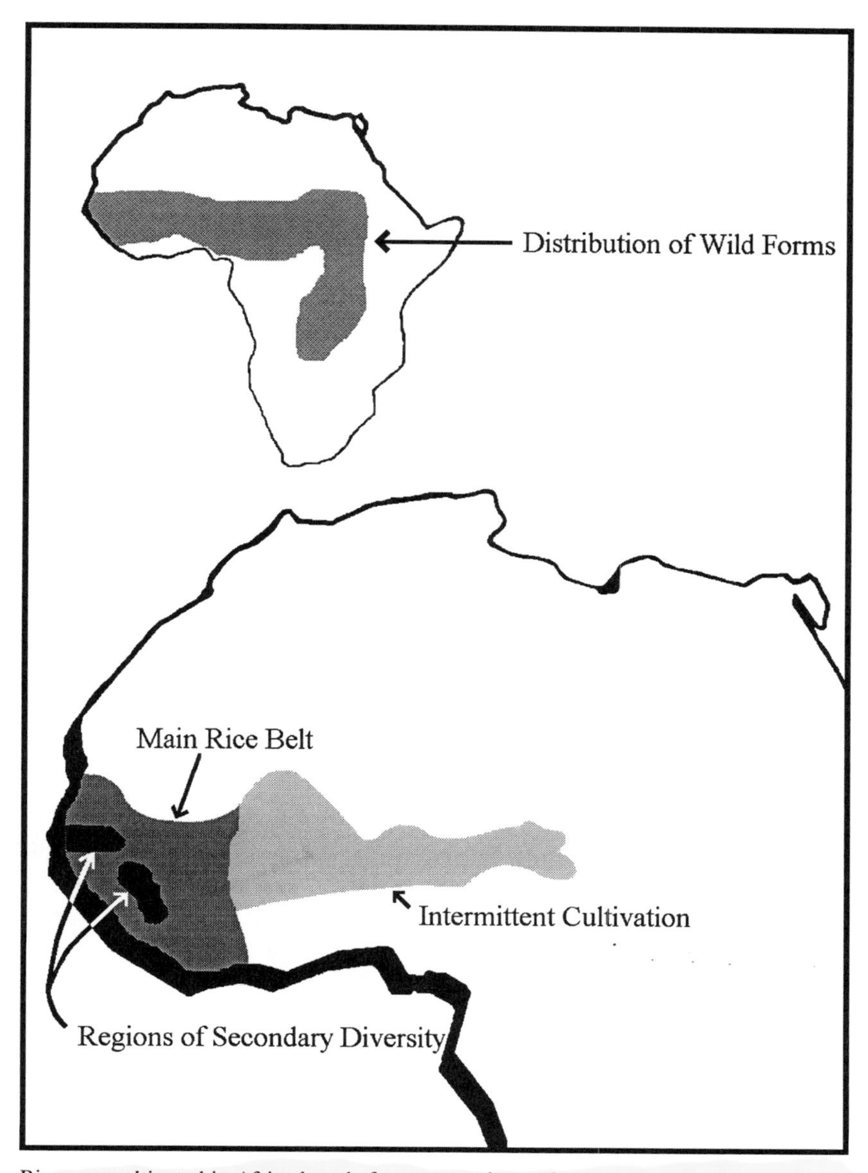

Rice was cultivated in Africa long before any navigator from Java or Arabia could have introduced their kind of rice to Madagascar or the East African coast. The native rice was grown first in the central Niger delta, and later in the Gambia, Casamance, and Sokoto basins. African rice is now utilized particularly in the central Niger floodplain, the coastal zone between Senegal and Sierra Leone, and the mountainous areas of Guinea and the Ghana/Togo border.

The primary center (small map) shows the distribution of the wild form. The secondary centers are where notable arrays of cultivated types occur. The main rice belt is the zone where African rice is cultivated the most.

improvements, such as those just mentioned, could give it a solid future. It is now known only in West Africa, but eventually it might also find a place elsewhere. Although only a few African countries grow even Asian rice in a major way, it is the continent's fourth biggest cereal (after maize, pearl millet, and sorghum) in terms of area planted. And demand is ever rising as population, standards of living, urbanization, international travel (with its exposure to new cuisines), and the search for easy-to-prepare foods increase. At present, West Africa absorbs a quarter of the world's rice exports.

Humid Areas On the face of it, African rice is at its biggest disadvantage in the humid lowlands. This is prime country for growing Asian paddy rice, whose current competitive edge makes it clearly the crop of choice. In addition, in this zone farmers and governments often invest in irrigation facilities, and to recoup their vast expenditures they must grow the highest yielding, highest selling crop. As a result, it is in this zone that African rice has suffered its most precipitous decline.

On the other hand, even here there seems to be a small but vital place for African rice. A recent survey in southern Sierra Leone, for example, found that even where Asian rice predominates farmers still retain one or two ultra-quick traditional types as "hunger-breakers." And, faced with a worsening hungry season caused by economic recession or other factors, many farmers say they would revert to the short-duration African-rice varieties, if only they could find sources of seed.[2]

Dry Areas For the truly arid zones African rice is not a suitable crop, but on moderately watered sites (for example, where annual rainfall is at least 760 mm) or seasonally flooded sites its prospects seem good. The fact that some varieties mature 10–20 days before their principal Asian-rice rivals is significant in drylands where precipitation is often erratic. In northern Sierra Leone, for example, the rainy season in recent decades has been terminating early and with unusual abruptness. For this reason alone, farmers are cultivating African rice on at least some portion of their land.[3] With it, they are assured of a harvest.

Upland Areas In West Africa's highlands[4] where this type of rice is still an important grain producer, it will continue to be important as

[2] Information from P. Richards. For additional details, see box, page 28. Two Sierra Leone varieties, *pende* and *mala,* ripen within 90–110 days: *pende* is a strongly tillering variety valued also for its ability to smother weeds.
[3] Information from P. Richards.
[4] By most standards, these lands are not very high—only about 1,000 m above sea level.

a subsistence crop. The upland varieties are notably useful in shifting agriculture. They also have a place in crop rotations because their root systems and susceptibility to soilborne diseases differ from those of the major crops. Planting them for a season or so tends to "sanitize" the site.

Other Regions

For lands beyond Africa, prospects are slight. There, African rice offers few benefits over the Asian species and may not adapt well.[5] Although it might have a future as a small specialty crop, more likely it will become an accursed weed, especially in rice fields.

USES

African rice can be used for all the same purposes as Asian rice. It is thus extremely versatile. There are, however, some specialized local uses. West Africa's Mandingo and Susu people, for instance, use rice flour and honey to make a sweet-tasting bread, so special that it is the centerpiece of ceremonial rituals. Rice beer is popular throughout West Africa, and in Nigeria a special beer (called *betso* or *buza*) is made from rice and honey. Also, in Ivory Coast there is a project to use African rice as a component of baby foods.

NUTRITION

Both rices are principally carbohydrate sources. However, in practice African rice's nutritional quality is greater than that of Asian rice.[6] This seems to be not because of any inherent difference but because it is more difficult to polish. Asian rice is invariably polished to a greater degree, and therefore more of its nutrients (especially the important vitamin, thiamine) are lost.

AGRONOMY

As with Asian rice, African rice is grown in three major ways: dryland (or upland), paddy, and "floating."

[5] For instance, one reason why African rice is not better known internationally is that it grows poorly in the Philippines, where the world's major rice-research facility is located. This is not a measure of inferiority—just a lack of adaptation to the local conditions, especially to viral diseases.

[6] Information from the Food and Nutrition Board, Food Research Institute, Ghana.

Dryland About 40 percent of the rice production in Africa's 15 major rice-producing countries relies on rain as the only source of water. Almost all of that area employs the Asian species, but West Africa still grows a small but significant amount of dryland African rice. Indeed, in certain parts of Ghana and Togo it is the chief staple.

The dryland form thrives in light soils wherever there is a rainy season of at least 4 months and minimum rainfall of 760 mm. It is often interplanted with millets, maize, sorghum, beniseed, roselle, cowpea, cassava, or cotton. Today's varieties mature in 90–170 days. Yields average 450–900 kg per hectare, but can go as high as 1,680 kg per hectare.

Paddy Only about one-sixth of Africa's rice is produced using irrigation and 60 percent of that is in just one country—Madagascar. Swamp rice, however, is being increasingly cultivated in former mangrove areas of the Gambia, Guinea-Bissau, Guinea, and Sierra Leone. Essentially all of it at present is the Asian species.

African rice can also be grown in the same way. It can be seeded into damp soil or transplanted to fields under water. These types mature in 140–220 days. The yield ranges from 1,000 to 3,000 kg per hectare.[7]

Floating In the River Niger's inland delta in Mali, farmers grow various forms of floating African rice. These plants lengthen prodigiously to keep their heads at the surface of the floodwaters, where they flower and set seed. One type (*songhai tomo*) can grow in water more than 3 m deep.

Floating varieties can utilize deeply inundated basins where nothing else can be raised. They are often harvested from canoes. They ripen in 180–250 days. Yields range from 1,000 to 3,000 kg per hectare, depending on the amount of rainfall early in the growing season and on the eventual depth of the subsequent floods.

HARVESTING AND HANDLING

African rice is handled like its more famous Asian cousin, but (as noted) its grains tend to split, and so greater care must be taken. Also, it is more difficult to hull.

As is to be expected with such a neglected crop, yields are variable and uncertain. However, there are hints that they are not as low as

[7] In the region of Timbuktu a very promising, nonfloating, dwarf type called *riz kobé* is grown on runoff water in the rainy season. (Information from W. Schreurs.)

NUTRITIONAL PROMISE

Main Components		Essential Amino Acids	
Moisture (g)	5	Cystine	2.6
Food energy (Kc)	358	Isoleucine	4.7
Protein (g)	7.6	Leucine	8.8
Carbohydrate (g)	81	Lysine	4.1
Fat (g)	1.9	Methionine	3.1
Fiber (g)	0.5	Phenylalanine	5.1
Ash (g)	3.8	Threonine	3.7
Thiamin (mg)	0.39	Tyrosine	4.6
Niacin (mg)	5.0	Valine	6.4
Calcium (mg)	25		
Iron (mg)	2.0		
Phosphorus (mg)	263		

COMPARATIVE QUALITY

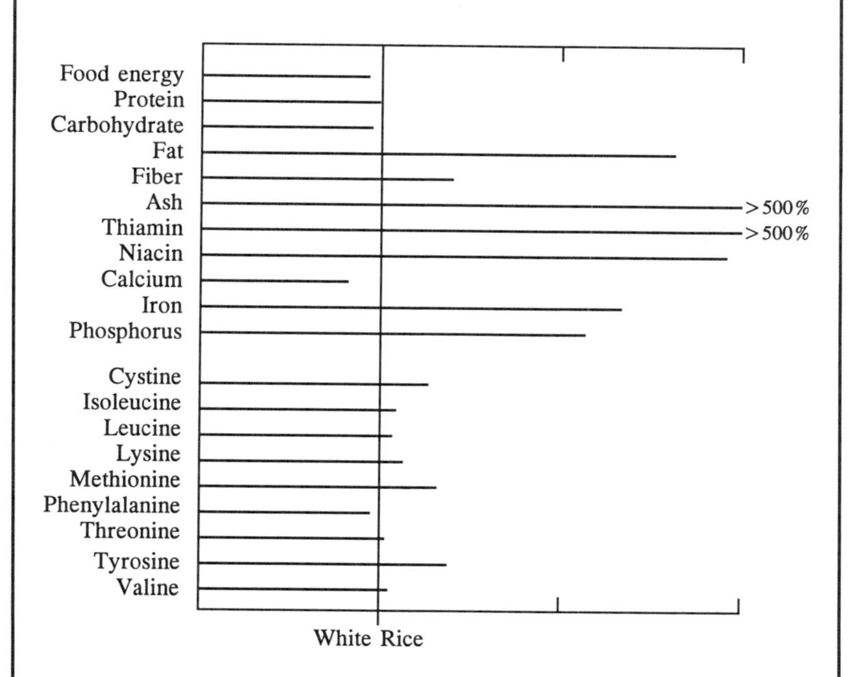

A glance at this chart shows that whole-grain African rice is at least as rich as white (i.e., Asian) rice in most nutrients. In some vitamins and minerals it is far superior.

Rice in Sierra Leone

Recently, researchers surveyed the distribution and use of rice in Sierra Leone. Following is their account of their findings. It is probably indicative of the situation throughout much of West Africa.

In visits to just over 500 farmers in all parts of Sierra Leone, we found that 245 types ("varieties") of rice were in use. Of these, 24 were African rice.

Although it generally yields less than Asian rice, African rice survives—and may even be making a modest comeback in some areas, especially in the drier northwest. There are a number of reasons for this. Compared with Asian rice, African rice:

- Seems to manage better on extremely impoverished soils.
- Competes better with weeds. Indeed, farmers pressed by labor shortages leave the crop to fend for itself. African rice will yield something even where Asian rice is choked out of existence by weeds. This is important because for small-scale rice farmers labor shortage is *the* most pressing constraint.
- Matures quicker. Nearly all the samples we collected matured in 100–125 days and are therefore among the quickest ripening rice cultivars in the country. (The average for dryland Asian rice in our sample was 130–140 days, and for wetland, 160–170 days.)
- Is preferred by many of the people. Several informants believed that African rice is nutritionally superior. They say that it is "heavy in the stomach" and keeps hunger at bay far longer

commonly claimed. For example, five years of experiments at two sites in Ivory Coast found that 16 populations of African rice (selected for their productivity) compared favorably with three top varieties of Asian rice. Despite their natural lodging and spontaneous shattering, the best African rice varieties (BG 141 and BG 187) gave average and remarkably stable yields of 1,500–1,800 kg per hectare (depending on the site) as did their Asian counterpart (Moroberekan), the traditional upland variety promoted in Ivory Coast.[8]

[8] Clement and Goli, 1987. In most cases, African-rice varieties from Guinea-Bissau proved higher yielding than those of other origin.

than the average Asian rice. Also, they often told us that it tastes "sweeter." And they said it keeps well after cooking. This is particularly important because many people prepare food only once a day, but members of the family drop by to eat at any time.

In northwestern Sierra Leone, however, Asian rice is preferred. People in this area complained that African rice is difficult to husk and that cleaning off its tough red bran takes a lot of work. Women in particular complained of the extra workload it imposes.

On the other hand, in other parts of the country redness was an important advantage. For example, Mende people in the south and east look on the red tinge (found on incompletely milled grains) as a guarantee that the sample is not a foreign rice. Rice soaked in palm oil plays a major part in their rituals, and it is unthinkable for them to use an Asian wetland variety.

In their fields, Sierra Leone farmers draw no distinction between Asian and African rices. Both species go by the same name: *mba* (Mende) or *pa* (Temne). The fields are very mixed from a genetic point of view. The farmers prefer it that way and, seemingly, they deliberately foster diversity because most of them know how to rogue out undesirable types and would do so if they wanted to.

We noticed that the African and Asian species appear to have hybridized in many places. A number of the most popular Temne rices, for example, are in fact intermediate types (judged by ligule form, grain shape, and panicle type). Certain named landraces seemed to be neither Asian nor African rice and may be assigned to either or both species.

Paul Richards, Serrie Kamara,
Osman Bah, Joseph Amara, Malcolm Jusu

LIMITATIONS

In its present state, African rice certainly has limitations, including those listed below:

- Lodging. The plants tend to have weak stalks, and late-season windstorms can sometimes topple them.
- Shattering. Today's plants tend to drop the seed as it matures.
- Splitting. The seed tends to break in half if handled roughly.
- Color. Although the grain itself is always white, most types have red husks.
- Processing. To remove the husk is laborious.

- Weediness. In West Africa, extensive genetic interaction occurs between African rice's wild and cultivated races. The mixed populations that build up can be extremely complex. The weedy results infest the rice fields and can be serious pests.[9]
- Diseases. Compared to Asian rice, it can be more susceptible to numerous fungi as well as to the parasitic plant striga and to a brown spot of unknown cause.

Although these limitations collectively add up to a fearsome combination, they mainly reflect the neglect this crop suffers from. All are now circumvented by people who grow and use African rice; research can undoubtedly reduce their severity if not overcome them entirely. Moreover, several of these limitations are also characteristic of competing grains.

NEXT STEPS

African rice must be kept from dying out as a crop. It deserves research, development, greater promotion, and support. At the very least it has genes of potential value to its near relation, the world's second biggest food crop. Actions to be taken include the following:

Friends of African Rice A good start could be made by an organization of volunteers—both professionals and amateurs—who join together in a cooperative spirit to explore, protect, promote, and provide samples of this millennia-old resource. They might also collect the legends that come with the various types before they, too, die.

Information Exchange Researchers are now working on rice in Senegal, Mali, Ghana, Ivory Coast, Burkina Faso, Cameroon, Liberia, Nigeria, Sierra Leone, and other countries. An international center, the West African Rice Development Association, specializes in the crop. And two French institutes, Office de la Recherche Scientifique et Technique Outre-Mer (ORSTOM) and Institut de Recherches Agronomiques Tropicales et des Cultures Vivrières-Centre de Coopération Internationale en Recherche Agronomique pour le Développement (IRAT-CIRAD), also have rice programs in Africa. All but one of these organizations work almost exclusively on Asian rice, but the presence of their expertise means that there are good opportunities to advance the development of its African relative.[10] One way to stimulate

[9] Such a gene flow between wild and cultivated species is also of long-term benefit to the crop. It maintains a broad genetic base and enhances its ability to resist drought, pests, diseases, and other hazards. But to a farmer faced with the need for high yields of uniform grain, it can be a curse.
[10] Research contacts for these and other programs are listed in Appendix G.

interest within the international scientific community is to collect all available research data and publish a detailed monograph on African rice.

Food Processing As noted earlier, the availability of precooked products made from African rice might do much to halt its decline and, indeed, to turn it around. Innovation, ingenuity, and marketing skill could be employed to return this food to prominence. It might well start out as a specialty product, selling at a premium to hotels for tourists and to those people dedicated to African traditions.

Seed Supply In many areas the amount of seed in circulation is so low as to render the species nonviable. It is important to keep up a supply of seed. Then, at least, the farmers who want to keep growing African rice won't be excluded as is now apparently happening in Sierra Leone.

Germplasm Samples of African rice have been gathered by various organizations, notably the International Plant Genetic Resources Institute (IPGRI), ORSTOM, and IRAT-CIRAD. This has been stored for purposes of conservation and possible plant breeding.[11]

For all that, however, many interesting types undoubtedly remain to be collected across the vastness of West Africa.

Agronomic Studies Since little hard data on this crop exists, it would be useful for students of agronomy to take up the many challenges of "filling in the map." Examples include the following:

- Selecting nonshattering genotypes or developing techniques to overcome shattering.
- Testing strains for salt tolerance.
- Locating types for drought avoidance.
- Measuring cell sap osmotic adjustment.
- Testing the plant's storage capacity and dormancy requirements.
- Reducing broken grains. Certain strains of Asian rice also suffer this problem and recent research has shown that providing adequate nitrogen fertilizer largely overcomes it.[12]
- Research in deep-water rice is vital and long overdue. The resources available—climate, water, and growing area—along with

[11] A collection of about 4,000 samples of seeds of wild and cultivated African rice, as well as Asian rice landraces that have been cultivated in Africa for a long time, is held at ORSTOM and IRAT-CIRAD. It results from 14 collecting missions in 12 African countries. Eighty-five percent are cultivated landraces (both African and Asian); 15 percent are wild African species.

[12] This research was done by Robert H. Dilday, USDA-ARS Rice Production and Weed Control Research Unit, P.O. Box 287, Stuttgart, Arkansas 72160, USA.

proper research could perhaps triple production of deep-water rice in the Niger's inland delta. This is one area of research that can do something toward reducing hunger in one of the regions of Africa most in need of help.

Genetic Improvement Although the current African types shed grain more readily than the Asian ones, some improvements have been bred into dryland varieties. Additional research emphasizing seed shattering could make a big difference. Because the gene for nonshattering is recessive, the selection of nonshattering types should be rapid, and true breeding should be immediate. Other improvements might include selection for resistance to disease. This resistance exists in the various genotypes, and the major problem is not to lose these local types as Asian rice spreads even further. For the uplands, any form of rice must resist blast and sheath blight. All types must also resist rice yellow-mottle virus; some local cultivars already do.

For areas dependent on seasonal flooding, varieties must resist lodging and respond to fertilizer; the transplant types must tolerate widely varying periods of growth in the nursery (while farmers await the onset of the unpredictable natural flooding).

Researchers are at present "mapping" the chromosomes of both African and Asian rice, identifying the portions that control various features of the plant.[13] This powerful modern technique will "jump-start" the genetic improvement of African rice (see box, page 34). Perhaps it could also facilitate the transfer of useful genetic material between the two.

SPECIES INFORMATION

Botanical Name *Oryza glaberrima* Steudel

Synonym *Oryza barthii* ssp. *glaberrima*

Common Names

English: African rice, glaberrima rice
French: riz pluvial africain, vieux riz, riz africain, riz flottant
Cameroon: erisi (Banyong)
Guinea: Baga-malé, malé, riz des Baga
Mali: Issa-mo (river rice), mou-bér (great rice)
Sierra Leone: kebelei, mba, mbei (Mende), mala (Kissi), Kono, pa (Temne)

[13] Information from G. Second.

Description

African rice is an annual grass that grows generally between 66 and 120 cm tall. It is highly variable. The dryland types have smooth, simple culms that can form roots at the lower nodes and are simply branched up to the panicle (flower cluster). The floating types can form branches and even roots at the upper nodes. The panicles are stiff, smooth, and compact. The flowers are self-fertilizing; however, some inter- and intraspecific cross-pollination occurs.

From a distance, Asian rice and African rice are similar in appearance. However, African rice has diminutive ligules (small, thin membranes found at the base of the leaf where it joins the stem). Its compact panicles have less branching. Its spikelets lack awns. It is completely annual and dies after setting seed. Asian rice, on the other hand, continues growing so that late in the season the two can look strikingly different.

Distribution

African rice is important mainly throughout the southwestern region of West Africa, but it can be found as far east as Lake Chad, especially in the lands of the Sahel that are seasonally flooded by the Niger, Volta, and other rivers.

It has apparently been introduced to India. Also, it may have been taken to Brazil by seventeenth-century Portuguese explorers. Somehow it has also reached El Salvador and Costa Rica.[14]

Cultivated Varieties

Many cultivars of African rice have been obtained by natural crossings and inbreeding, giving forms with compact panicles and heavy grains. In particular, there are numerous swamp varieties suited to different soil and drainage conditions.[15] In northern Mali alone are found about 30 cultivars of the floating type.[16] Examples of upland varieties of African rice are ITA 208, IRAT 112, Mutant 18, IRAT 104, and ISA 6.[17]

In Upper Gambia, Guinea, and Senegal (Casamance) can be found a special group of African-rice genotypes with enhanced recessive

[14] It was collected there in the 1950s by Roland Portères, a French botanist who specialized in studying West African grasses and was a pioneer in bringing the promise of African rice to world attention.

[15] One researcher collected about 180 varieties in the inland Niger delta. P. Martin. 1976. Amélioration des conditions de production du riz flottant au Mali (période 1963–1973). *L'Agronomie Tropicale* 31(2):194–201.

[16] Information from W. Schreurs.

[17] Information from J. Ayuk-Taken.

Could African Rice Go High-Tech?

The world's rice research is overwhelmingly focused on Asian rice, but the remarkable developments now emerging from laboratories may bring big advances to African rice, on the side. Following are examples.

Gene Mapping. Molecular biologists have recently "marked" the locations on rice chromosomes where genes for certain genetic attributes are carried. These markers can be used to track the genes for those traits. The ability to determine whether a desired gene is present or absent in any sample bestows enormous power. It can, for instance, help find a desired gene in wild as well as cultivated species, it can find a "hidden" gene in a given plant where the gene's outward effects are masked, and it vastly simplifies the sorting of thousands of crossbred specimens—something that formerly could take a lifetime of tedious effort.

Gene markers based on restriction-fragment length polymorphisms (RFLPs) are being developed for both Asian and African rices. For instance, in 1988 a team at Cornell University found markers for various traits on the set of 12 chromosomes that (in both species) carries all the genetic characteristics. That first map had 135 genetic landmarks; later versions have more than 300.

A particular strength of this new work is that breeders can now work with very young seedlings. In other words, they can tell whether a certain gene is present without waiting months for the plant to mature. This can cut the time needed to breed a new variety—usually 10–12 seasons—in half.

Although the genomes (chromosome sets) of both African and Asian rice have been mapped, the rest of the effort has so far been solely on Asian rice. Nonetheless, most results from Asian rice are likely to be easily transferable. The genome is relatively small, containing only a tenth as much DNA as maize.

Test-Tube Reproduction. Although until recently no grass had been cloned using tissue culture, today Asian rice, maize, sorghum, and vetiver have succumbed. African rice has so far not been cultured in the test tube but, given the new insights, it seems a likely candidate for this powerful procedure.

Several teams have managed to regenerate fertile rice plants from protoplasts—cells from which the wall has been removed. This makes it even easier to fiddle with rice genes. Already, DNA from bacteria has been transferred into rice protoplasts. Mature plants, grown from these protoplasts, have transmitted the implanted DNA to their offspring.

High-Lysine Forms. In the early 1990s, U.S. Department of Agriculture researchers discovered Asian rice plants with both high protein quality and high protein levels. This has raised hopes that extremely nutritious varieties can be bred for the first time.

To find these new forms, Gideon W. Schaeffer, Francis T. Sharpe, Jr., and John Dudley gave small clumps of rice cells a lethal dose of lysine (an amino acid vital for good health) in a laboratory dish. Only a tiny fraction survived the treatment. Those few cells, however, could allow more lysine than normal to be made. The scientists grew them into whole rice plants and found that the resulting high-lysine plants are true genetic mutants and therefore suitable for breeding new commercial varieties. Some of the crossbreeds have succeeded in producing seed of near-normal weights and good fertility but with greatly enhanced nutritional quality.

The high-lysine trait is apparently controlled by a single recessive gene. The scientists have begun isolating this gene so as to provide it to genetic engineers for incorporation into the world's Asian-rice crop. The work would likely be easily transferable to create high-lysine forms of its African cousin.

Hybrids. Both the male and female parts on rice flowers are normally fertile, but researcher J. Neil Rutger of the U.S. Department of Agriculture has found that growing certain rice plants in 15-hour daylight makes them essentially female. The plants never develop fertile pollen. This may provide a cheap and easy way to boost rice yields to a much higher level than at present. Because the modified plants cannot pollinate themselves, they are ready-made for pollination by other plants. Any pollination, therefore, produces hybrids, which are often known to produce robust and high-yielding plants. This process has not yet been tested on African rice, but Rutger believes that it might well work.

Asaf Hybrids. Recent decades have seen several dozen research papers on the genetic and morphological results of crossing Asian rice with African rice. Most have emerged from laboratories in Japan, Taiwan, and China. The driving force behind them appears to be the attempt to raise the yield of Asian rice by forming hybrids.

At least in principle, crosses between African rice and Asian rice might improve the yield of either or both. Although the botanical literature stresses their incompatibility, the two are genetically close. Both are self-pollinating diploids ($2n=24$) and possess the same genome, which rice geneticists call AA.

characters such as white husks, spikelets persisting to maturity, and vegetative and floral organs without anthocyanins. These seem to indicate a secondary region of diversity and may be particularly valuable genetic resources.

Environmental Requirements

Daylength Varies from neutral to strongly sensitive, depending on variety. However, most dryland types now in use are sensitive to photoperiod. They flower with the advent of the dry season. On the other hand, most floating types (at least in northern Mali) show little sensitivity to daylength.

Rainfall Some upland varieties can produce adequately with precipitation as low as about 700 mm.

Altitude From sea level to 1,700 m.

Low Temperature Average temperatures below about 25°C retard growth and reduce yields. Below about 20°C these effects are pronounced.

High Temperature African rice does well at temperatures above 30°C. Above about 35°C, however, spikelet fertility drops off noticeably.

Soil Type Some cultivars apparently can outperform Asian rice on alkaline sites as well as on phosphorus-deficient sites. Not unexpectedly, however, the crop performs best on alluvial soils.

Related Species

At least two of African rice's close relatives are regularly gathered for food, often in sufficient abundance to appear in the markets.

Oryza barthii (*Oryza breviligulata*)[18] is an annual that commonly occurs in seasonally flooded areas from Mauritania to Tanzania and from the Sudan to Botswana. It is the wild progenitor of cultivated

[18] The nomenclature of wild rices in Africa has been very confused. The names *Oryza stapfii*, *Oryza breviligulata*, and *Oryza barthii* often occur with uncertain usage. Much of the older literature is rendered useless by this. It is now considered that *Oryza barthii* is the direct ancestor of *Oryza glaberrima*. The name *Oryza breviligulata* is now considered invalid, as is *Oryza stapfii*, the name formerly given to the weedy races of African rice.

African rice. It can form meadows in inundated areas. Its grain falls off so easily that it must be carefully collected by hand. (People use a basket or calabash, and sometimes they tie the stalks in knots to make harvesting easier.) It tastes good and is sometimes sold in markets. However, wherever rice is cultivated, this plant is regarded mostly as a weed to be eradicated. Certain strains of this species are immune to bacterial blight of rice (*Xanthomonas*), which could give them a valuable future as genetic resources.[19]

Oryza longistaminata is a common wild rice found throughout tropical Africa as far south as Namibia and Transvaal, as well as Madagascar. Unlike the other species, it is a perennial with rhizomes. It is tall and outcrossing. It usually grows in creeks and drainage canals and reproduces by suckers, often setting few seeds. Nonetheless, these meager grains are sought in times of shortage.

[19] S. Devadath. 1983. A strain of *Oryza barthii* an African wild rice immune to bacterial blight of rice. *Current Science* (Bangalore) 52(1): 27–28.

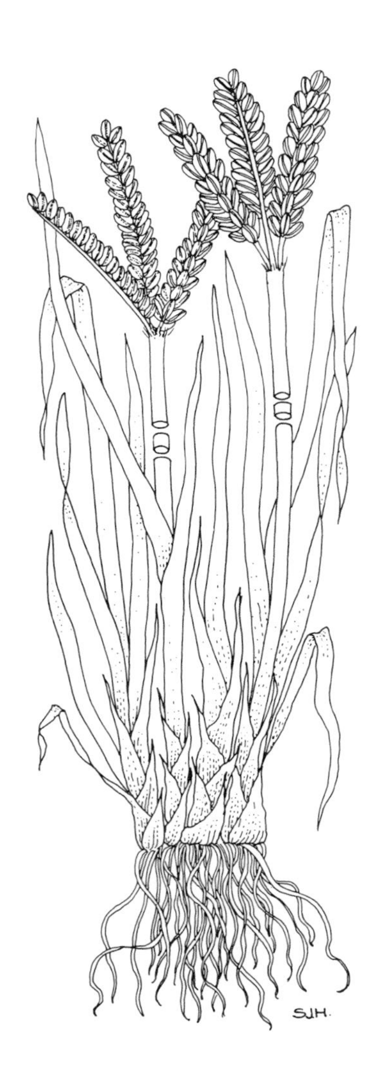
SJH.

2

Finger Millet

Finger millet (*Eleusine coracana*) is hardly "lost." Indeed, it is one of the few special species that currently support the world's food supplies. This African native probably originated in the highlands of Uganda and Ethiopia, where farmers have been growing it for thousands of years. In parts of eastern and southern Africa as well as in India, it became a staple upon which millions depend. And its annual world production is at least 4.5 million tons of grain, of which Africa produces perhaps 2 million tons.

For all its importance, however, finger millet is grossly neglected both scientifically and internationally. Compared to the research lavished on wheat, rice, and maize, for instance, it receives almost none. Most of the world has never heard of it, and even many countries that grow it have left it to languish in the limbo of a "poor person's crop," a "famine food," or, even worse, a "birdseed."[1]

Further, in recent years this neglected crop has started an ominous slide that could propel it to oblivion even in Africa. In fact, it has declined so rapidly in southern Africa, Burundi, Rwanda, and Zaire, for instance, that some people predict that in a few years it will be hard to find—even where until recently it was the predominant cereal. In those areas it clings to existence only in plots that are grown for use on feast days and other occasions demanding prestige fare.

The world's attitude towards finger millet must be reversed. Of all major cereals, this crop is one of the most nutritious. Indeed, some varieties appear to have high levels of methionine, an amino acid lacking in the diets of hundreds of millions of the poor who live on starchy foods such as cassava and plantain. Outsiders have long marveled at how people in Uganda and southern Sudan could develop such strapping physiques and work as hard as they do on just one meal a day. Finger millet seems to be the main reason.

This crop has many other advantages as well. Its grain tastes better than most; Africans who know it usually prefer finger millet over all others. The plant is also productive and thrives in a variety of

[1] This is its main use in the United States, for example.

environments and conditions. Moreover, its seeds can be stored for years without insect damage, which makes them lifesavers for famine-prone areas.

Given all these qualities, it is perhaps hard to understand why finger millet is being rejected. But the reason is simple. People are giving it up in favor of maize, sorghum, and especially cassava because producing finger millet takes a lot of work.[2]

The truth is that finger millet, as produced at present, demands a dedication to drudgery that, given a choice, few people are willing to invest. Part of the terrible toil is in weeding the fields, part in handling the harvest, and part in processing the grain.

PROSPECTS

Even though finger millet is declining in the heartland where 30 years ago it was *the* major crop of the land, all is not lost. Indeed, if immediate attention is given, the impediments causing the decline will probably be eliminated. In fact, there are already signs that the slide may be bottoming out. Prices paid for finger millet have risen dramatically in some places, and the crop is enjoying something of a resurgence—and a highly profitable one at that. In Kenya, for instance, the grain currently sells at more than twice the price of sorghum and maize.[3] In Zimbabwe, too, the government offers an attractive producer price, which has tended to slow the decline. And Uganda's most recent statistics indicate that finger millet still occupies 50 percent of its cereal area.

Africa

If this crop is given proper attention, it has the following possibilities within Africa.

Humid Areas Excellent prospects. Certain varieties are adapted to heat, humidity, and tropical conditions. (Finger millet was once the principal staple for people in southern Sudan and northern Uganda, for instance.) Given research, recognition, and sympathetic policies, production could expand dramatically.

[2] At least one reviewer speculates that abandoning this nutritious grain millet for the less nutritious ones is "likely one of the causes of increasing famine in many areas."
[3] What is more, the government-controlled price (630 shillings per quintal, or $0.29 per kilo in 1991) is only half the open-market price (1,200–1,400 shillings per quintal, or $0.60 per kilo).

Finger millet seedheads look like "hands" with the grain contained in the "digits," hence the name. Some of the hands are curled into "fists." The crop is especially appreciated by the peoples in eastern and northern Uganda. To them, it has a high social value. They traditionally hold celebrations for the new harvest, and they serve finger-millet bread to visitors and neighbors whom they want to impress. In the Buganda region, however, the people prefer finger millet in the form of hot porridge served with either sugar or banana juice.

Finger millet is grown throughout eastern and southern Africa, but especially in the subhumid uplands of Uganda, Kenya, Tanzania, Malawi, Zaire, Zambia, and Zimbabwe. The crop originated somewhere in the area that today is Uganda.

Dry Areas Fair prospects. Finger millet is not as drought tolerant as pearl millet or even sorghum, but it could play a much greater role in savanna areas that get at least moderate rainfall.

Upland Areas Excellent prospects. Certain finger millet landraces are fully adapted to highland conditions. In Africa the crop is usually grown at altitudes between 1,000 and 2,000 m and in Nepal it is grown at altitudes up to at least 2,400 m.

Other Regions

Finger millet is certainly not being abandoned in Asia. Indeed, India's national yields have increased 50 percent since 1955.[4] Moreover, in Nepal, the finger millet area is expanding at the rate of 8 percent per year.[5] Any international efforts to promote and improve the plant appear to be as beneficial to Asia as to Africa.

This high-methionine grain might also be beneficial for use in weaning foods and in many other cereal products in parts of the world (Latin America and North America, for instance) where it is now largely ignored.

USES

This is a versatile grain that can probably be used in dozens of types of foods, including many that are quite unlike its traditional ones. Its several major uses include the following:

- Porridge. The small grains—which are usually brown but occasionally white—are commonly boiled into a thick porridge.
- Bread. Some finger millet is ground into flour and used for bread and various other baked products. All are relished for their flavor and aroma.
- Malt. Malted finger millet (the sprouted seeds) is produced as a food in a few places. It is nutritious, easily digested, and is recommended particularly for infants and the elderly.
- Beverages. Much finger millet in Africa is used to make beer. Its amylase enzymes readily convert starch to sugar. Indeed, finger millet has much more of this "saccharifying" power than does sorghum or maize; only barley, the world's premier beer grain, surpasses it. In Ethiopia, finger millet is also used to make *arake*, a powerful distilled liquor.
- Fodder. Finger millet straw makes good fodder—better than that from pearl millet, wheat, or sorghum. It contains up to 61 percent total digestible nutrients.
- Popped Products. Finger millet can be popped. It is widely enjoyed in this tasty form in India (see page 298).

[4] Most of the increase occurred between 1955 and 1975 and resulted from genetic improvement of India's traditional landraces. Subsequent increases were due to crosses between those and new strains introduced from Africa.

[5] In Nepal the crop has a special niche: during monsoon rains, it continues growing well, even when the soil is almost waterlogged and where the nutrients have been leached out by the daily downpours.

NUTRITIONAL PROMISE

Main Components		Essential Amino Acids	
Edible portion (g)	95	Cystine	1.7
Moisture (g)	12	Isoleucine	4.0
Food energy (Kc)	334	Leucine	7.8
Protein (g)	7.3	Lysine	2.5
Carbohydrates (g)	74	Methionine	5.0
Fats (g)	1.3	Phenylalanine	4.1
Fiber (g)	3.2	Threonine	3.1
Ash (g)	2.6	Tryptophan	1.3
Vitamin A (RE)	6	Tyrosine	4.1
Thiamin (mg)	0.24	Valine	6.4
Riboflavin (mg)	0.11		
Niacin (mg)	1.0		
Vitamin C (mg)	1		
Calcium (mg)	358		
Chloride (mg)	84		
Copper (mg)	0.5		
Iodine (μg)	10		
Iron (mg)	9.9		
Magnesium (mg)	140		
Manganese (mg)	1.9		
Molybdenum (μg)	2		
Phosphorus (mg)	250		
Potassium (mg)	314		
Sodium (mg)	49		
Zinc (mg)	1.5		

No single set of numbers can adequately convey the nutritional promise of a grain as variable as finger millet. The numbers in these pages should be taken with caution. The dozen or so measurements that have been reported generally agree on most of the different nutrients. However, protein contents ranging from 6 to 14 percent have been claimed. The levels of fat (1-1.4 percent) and food energy (323-350 Kc) that are normally given are fairly consistent and are about the same as in maize. However, in some samples they seem to be much higher. The situation regarding iron is somewhat similar. Most analyses give the figure as about 5 mg per 100 g. But there have been two reports of iron exceeding 17 mg.

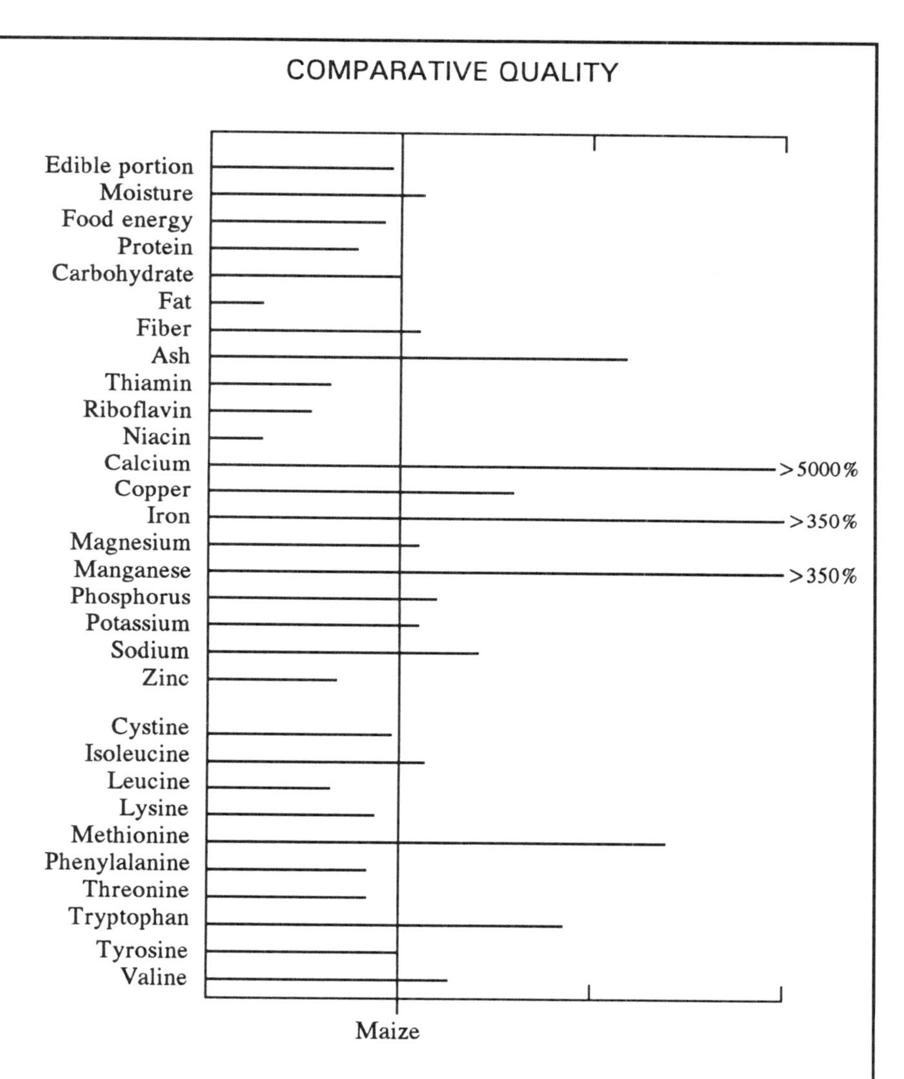

Figures reported for the essential amino acids are generally consistent, but 3 percent methionine is commonly referred to in the literature. Possibly, this was based on degerminated flour. Even that figure is outstanding for a cereal grain.

In this chart, we have compared whole-grain finger millet with the standard figures for maize. These are perhaps not fair comparisons, but they do accurately reflect the differences between the forms in which each food is normally eaten.

NUTRITION

The grain's protein content (7.4 percent) is comparable to that of rice (7.5 percent). However, it shows considerable variation, and at least one Indian cultivar contains as much as 14 percent protein.

The main protein fraction (eleusinin) has high biological value, with good amounts of tryptophan, cystine, methionine, and total aromatic amino acids.[6] All of these are crucial to human health and growth and are deficient in most cereals. For this reason alone, finger millet is an important preventative against malnutrition. The methionine level—ranging around 5 percent of protein—is of special benefit, notably for those who depend on plant foods for their protein.

Finger millet is also a rich source of minerals. Some samples contain 0.33 percent calcium, 5–30 times more than in most cereals. The phosphorus and iron content can also be high.

AGRONOMY

In Asia, finger millet is planted in rows and managed much like other cereals. But in Africa it is usually handled differently. Unlike maize, sorghum, or pearl millet—all of which are planted at individual stands in a rough seedbed—finger millet is traditionally planted in Africa by broadcasting its tiny seeds. This demands a very fine seedbed and means that the farmers must work hard and long, both to prepare the land and to weed the young plants.

Two crops a year are possible with early-maturing types.

HARVESTING AND HANDLING

In most of Africa the crop is harvested by hand. Individual heads are cut off with a knife, leaving a few centimeters of stalk attached. These are piled in heaps for a few days, which fosters a fermentation whose heat and hydrolysis makes the seeds easier to thresh.

Finger millet seeds are so small that weevils cannot squeeze inside. In fact, its unthreshed heads resist storage pests so well they can be stored for 10 years or more without insect damage. (It is said that if kept dry the seed may remain in good condition for up to 50 years!)

Yields are variable but (compared to those of other grains in the area) are generally good. In Uganda, for example, a threshed yield of 1,800 kg per hectare is regarded as average. In India, on reasonable

[6] "Total aromatic acids" is the combination of phenylalanine and tyrosine.

dryland sites, yields may run to about 1,000 kg per hectare, and on irrigated sites a normal average is more than 2,000 kg per hectare. Yields of 5,000–6,000 kg per hectare have been obtained under ideal irrigated conditions. Similar yields have been obtained in Nepal even under rainfed conditions.[7]

LIMITATIONS

As has been noted, the small size of the seeds is a serious drawback. It makes the crop difficult to handle at all stages.

Weeding is a particular problem. In Africa the dominant weed, a wild relative of the crop, looks so much like finger millet in its early stages that only skilled observers and close scrutiny can tell them apart. The problem is compounded by the practice of broadcasting seed. To weed the resulting jumbled stands, people must inspect every plant, even going through on hands and knees.

Finger millet is subject to bird predators—notably to the notorious quelea (see Appendix A).

By and large, the plant suffers little from diseases and insects, but a ferocious fungal disease called "blast" can devastate whole fields.

Finger millet is almost entirely self-pollinating and crosses between different strains can be made only with difficulty. Until recently, genetic improvement was limited to pedigree-based selection. However, in Uganda a few plants with male sterility have now been discovered. These should ease the way to breeding methods in which different lines can be crossed without trouble.

Because the seeds are so small, it takes skill and much effort to convert finger millet into flour—particularly by hand. Even hammer mills have difficulty. They must be fitted with very fine screens and run at high speed. Recently, however, a special mill for millet has been devised (see next page).

NEXT STEPS

If finger millet is ever to be rescued, now is the time. The key is to find ways to present its plight and promise to the public and politicians and to develop its markets. A few motivated individuals could do much here. Among helpful actions might be a pan-African finger millet conference, where researchers and others could compare methods used to grow it, prepare it, and sell it in the various nations. This

[7] Information from K.W. Riley.

Processing Finger Millet

Milling

Mechanical milling is of course well known; for wheat, rice, and maize, it is a major industry. But for finger millet, this primary step in the commercial processing of a food grain is essentially unknown. Machinery for rubbing the bran (embryo) off finger millet has never been available, perhaps through a lack of interest but mainly because the grain is exceptionally difficult to mill by machine. Finger millet, therefore, is usually eaten as a whole-grain flour, and the presence of oil in the embryo means that its shelf life is short and its commercial use limited.

Finger millet seed is a challenge to mill because it is very small and because its seed coat is bound tightly to the edible part (endosperm) inside. Moreover, the grain is so soft and friable that conventional milling equipment cannot remove the outside without crushing the inside. However, farmers have long known that moistening finger millet (for about 30 minutes) toughens the bran and reduces its grip enough that it can be mechanically separated without crushing the rest.

A machine for doing this has now been developed in India. This so-called "mini millet mill" consists of a water mixer, a plate grinder, and various sifter attachments. It is a versatile device in which debranning and sizing the endosperm (into either flour or semolina) take place in a single operation. It yields fairly white products. It can also be used to process wheat, maize, sorghum, and pearl millet and will even remove the outer husk from finger millet seeds if the clearance between the grinder plates is reduced.

This machine, and others like it, could initiate a new era for finger millet as a processed grain of commerce. The flour would then have a good shelf life and could be trucked to the cities and sold in stores as are wheat, rice, and maize. Commercial horizons would open up that have never before been contemplated.*

Malting

Finger millet could be the key to providing cheap and nutritious foods for solving, at last, the malnutrition that each year kills millions of babies throughout the warmer parts of the world.

As is described elsewhere (notably in appendixes C and D), the process of germinating finger millet activates enzymes that break down the complex structures of starches into sugars and other simple carbohydrates that are easy to digest. The enzymes

* For more information, contact N.G. Malleshi, Central Food Technological Research Institute (CFTRI), V.V. Mohalla PO, Mysore 570 013, India.

are of course there to benefit the seeds in which they occur—to mobilize food for the growing seedling; but long ago people found that they could use them also to break up starches from other sources. This process (usually called malting) became the first step in making beer and liquor out of starchy foods such as potatoes, maize, rice, or sorghum (see page 168).

What has been overlooked to a large extent is that malting can be used for more than just brewing. Indeed, it is probably the key to making cheap, digestible, liquid foods with little effort and no extra cooking fuel. These foods are particularly promising for children facing the life-threatening dietary switch from mother's milk to solid foods.

Adding a tiny amount of malted grain turns a bowl of hot starchy porridge into a watery liquid. The resulting food matches the viscosity of a bottled baby food, such as those sold in American supermarkets. A child who is too small or too weak to get down solids can then get a full meal—and get it out of the food its mother is preparing for the rest of the family.

The germinated grain acts as a catalyst to liquefy any of the world's major starchy foods: wheat, rice, maize, sorghum, millet, potatoes, cassava (manioc), yams, and the rest. Moreover, it does more than turn those staples into liquid form: it predigests the starches, making the food easy for a body to absorb, and (by releasing sugars) it renders even the blandest staples palatable.

The malted grain is readily available, cheap, and safe to eat. It should develop healthy bodies and fully functioning brains in the millions of children whose health and happiness is now jeopardized by malnutrition.

Of all the world's cereal grains, finger millet is second only to barley in its ability to hydrolyze starches ("malting power"). And it has the inestimable value of growing in the latitudes where malnutrition is rife. (Barley is strictly a temperate-zone resource.)

But for all its potential to benefit the malnourished, not much attention has been paid to malting internationally. Only in India and Nepal have malt-based children's foods been intensively studied. In both countries, food scientists have created malted-grain products that can overcome malnutrition. And in almost every product, malted finger millet was the prime ingredient.

The fact that malting is a cheap and widely understood process that can be easily accomplished in the home or village and requires no fuel or special equipment is a major benefit. This means that top-quality weaning foods can be made by the poor, who cannot afford to buy commercial baby-food concoctions.

meeting would provide the opportunity to exchange experiences and to begin the process of preparing papers, pamphlets, recipes, and perhaps a monograph. Another might be the establishment of a "finger-millet action program" to share seeds and research results in the future. There might even be established a pan-African finger millet "SWAT" team to provide advice and stimulus to the countries where finger millet is now declining toward economic extinction.

Rescuing this crop may be easier than now seems probable. Lifestyles and eating habits may have changed, but in much of Africa people still appreciate finger millet. Subsistence farmers like finger millet also. Every seed sown can return between 200 and 500 seeds (other grain crops seldom go above 100 even under ideal conditions). And this crop has many uses. To those whose very lives and livelihoods depend on what they grow, its flexibility is vital.[8]

Beyond Africa, finger millet should also be given a higher research priority. It is a good way to help the rural poor in parts of Asia. Much of the spectacular rise of wheat occurred in areas where irrigation could be used. Overcoming finger millet's yield constraints would, more importantly, benefit rainfed agriculture.

Research Needs

Research is needed on all aspects of this plant, which now is little known to scientists in general. ICRISAT is conducting research on it, but more effort is needed. Research operations might include those discussed below.

Trials in New Areas Entrepreneurs in the United States as well as in Australia and other countries that specialize in cereal breeding could probably do much to benefit this crop. It is already grown in a small way in the United States. It grows well, but so far is used only for birdseed. Nonetheless, it might support a small specialty grain industry for local and national food uses. And enlisting the country's outstanding cereal-science capabilities could perhaps transform this crop's potential worldwide.

Farming Methods As far as Africa is concerned, finger millet's greatest immediate needs lie not so much in plant breeding as in farming practices. Reducing the current drudgery involved with its production would bring the biggest and quickest benefits.

Surprisingly, techniques for making finger-millet production less

[8] For crops like these, perhaps we need a whole new measure of performance, one that takes into account not just the yield in the field but the all-around value to people's welfare.

laborious can probably be employed rapidly and widely. For instance, planting the seed in rows would dramatically slash the need for weeding. One or two hoeings (or perhaps a layer of mulch) would eliminate most of the weeds with little further effort. To make this practical, however, a device is needed that can deliver small seed with precision. It would have to be easy to make and simple to use. Such devices do indeed exist (see Appendix A) but have not yet been introduced to finger millet farmers.

Examples of other types of farming practices worth exploring are the following:

- Minimum tillage seeding.
- Wide rows for water capture.
- Control of birds.
- Intercropping or undersowing with legumes. (The foliage from leguminous shrubs or ground cover may be especially helpful by supplying nitrogen to the crop.)
- Sowing or transplanting with other crops. (In Nepal, for instance, it is often planted with maize.)
- Weeding using animal power and other labor-saving techniques.
- Developing ox-drawn implements for planting, cultivating, harvesting, and threshing finger millet.

Erosion Control In some parts of southern and eastern Africa finger millet has been abandoned because it "causes" severe soil erosion. In these areas, farmers typically clear forest from a hillside, burn it, and sow finger millet in the ashes. The tiny plants hold soil poorly, and it easily washes away. For such sites there is a need for alternative methods of erosion control. One example might be vetiver (see Appendix A). Another is mulching with stubble from the previous crop.

On the other hand, other parts of Africa actually employ finger millet for erosion control. In fact, when broadcast—or even line sown—across the slope it is good for reducing erosion. Data from Zambia, for example, show that the plant prevents erosion more effectively than legumes do. Farmers in Nepal also report that finger millet "holds the soil."

Plant Breeding In its genetic development as a crop, finger millet is about where wheat was in the 1890s. Many landrace types are known but have not been systematically evaluated, codified, or analyzed, Thus it is likely that the best-yielding, best-tasting, and best-handling types have not been isolated or created out of the massive gene pool. Since the 1890s, average yields of wheat have risen from about 500 kg per

Ragi

Finger millet crossed the Indian Ocean more than 1,000 years ago and since then has become extremely important in South Asia. In India, where it is generally called "ragi," this native African grain is now grown on more than 2 million hectares.

In its new home, scientists and farmers have created numerous ragi races. There are, for instance, plants that are purple; seed-heads that are short, long, "open," "curved," or "fisty"; seeds that range from almost black to orange-red; and there is also a popular type whose seeds are pure white. Some ragi varieties are dwarfs (less than 50 cm), some tiller profusely, some are slow to mature and are grown mainly under irrigation, while others mature quickly and lend themselves to dryland production.

Ragi is considered one of India's best dryland crops, and most of it is produced without supplemental water. The plant is both adaptable and resilient: it survives on lateritic soils, it withstands some salinity, and it has few serious diseases or pests. Ragi also yields well at elevations above those suitable for most other tropical cereals. In the Himalaya foothills, for example, it is cultivated up to slightly over 2,000 m above sea level.

Despite its importance in the Himalayas, about 75 percent of the ragi area lies in South India, particularly in Karnataka, Tamil Nadu, and Andhra Pradesh. In parts of this vast region farmers can get two crops a year; in Tamil Nadu and Andhra Pradesh three are not unknown. Wherever the rains at sowing time are uncertain, the farmers often transplant ragi like rice. In fact, the two crops are commonly grown in a "relay" that is good for both. For instance, in May a farmer may start out by sowing ragi seeds in the nursery; in June, he (or she) transplants the seedlings to the field and replants the nursery with rice seeds; in August, the ragi crop is harvested and the rice seedlings are put out into the just vacated fields. This process is efficient, highly productive, and a good insurance against the vagaries of the weather.

Ragi yields as much as 5,000 kg of grain per hectare. Because the seed can be stored for decades (some say 50 years), it is highly valued as a reserve against famines.

However, ragi is much more than just a famine food. In certain regions it is an everyday staple. It is, for instance, a principal cereal of the farming classes in Karnataka, Tamil Nadu, and Andhra Pradesh, as well as in the Himalaya hill tracts (including those of Nepal). The grain is mainly processed into flour, from which is made a variety of cakes, puddings, porridges, and other

Indian farmer holding ragi (ICRISAT).

tasty foods. Some, however, is malted and turned into beer as well as into easily digested foods for infants and invalids.

As in its African homeland, ragi enjoys a reputation for being both nutritious and sustaining, and Indian studies lend scientific support to this view. Certain grain types, particularly the white ones, can match the most nutritious local cereals, at least in protein content.

hectare to more than 4,000 kg per hectare; finger millet's could rise similarly and much more quickly.

Various finger-millet landraces possess genes for blast resistance, robust growth, early vigor, large panicle size, high finger number and branching, and high-density grain. Similarly, there are water-efficient types with high carbon dioxide fixation and low leaf area that could be outstanding new crops for semiarid conditions. Long-glume types with high seed weight are especially promising for increasing seed size. All of these, and more, are genetic raw materials that could transform this crop.

The grain is already nutritious, but it might be improved even more. As noted, types containing up to 14 percent protein are known. Also, it is a high-methionine protein and, of all the essential amino acids, is the most difficult to find in grain-based foods. Thus these finger millets could be a "super cereal" in nutritional terms.

White-seeded forms that make good unleavened bread and bakery products are also known, and they too are undeveloped. Today's crop in Africa is overwhelmingly the coarse, rusty-red form that is mainly useful for porridge and brewing beer.

Hybrids between Indian and African varieties seem promising as well. These high-yielding "Indaf" types are popular in India. Similar hybridization and selection for improved Indaf varieties for African conditions is now being started.[9] Hybridization, however, is difficult and mutation breeding is another approach worth exploring.

Some of finger millet's relatives have interesting traits that might be transferable. Among wild *Eleusine* species are perennials that might lend some of their enduring characteristics to finger millet. Others have genes for tolerance of heat, cold, drought, and waterlogging, as well as resistance to salinity and an ability to mobilize phosphorus and utilize nitrogen efficiently.[10]

Less dramatic but more immediately practical plant-breeding needs are the fine-tuning of today's varieties. The most important objectives are resistance to blast,[11] helminthosporium (another fungus), striga (parasitic witchweed), lodging, stressful soil and moisture conditions, and grain that can be more easily dehulled and ground. Other objectives might include fast seedling growth to compete better with weeds, shade-tolerant types for relay and intercropping, and types with anthocyanin pigmentation in the leaves (possibly obtainable through

[9] This work is beginning at the SADCC/ICRISAT Center at Bulawayo, Zimbabwe (see Research Contacts).

[10] These wild relatives are currently being collected by IPGRI, but several that could be part of the primary or secondary gene pool are not yet represented by even a single collection.

[11] Recently, a number of blast-resistant types have been selected at ICRISAT and are undergoing yield tests in different sites.

induced mutation), which could be spotted easily in the fields and would make weeding a much easier task.[12]

Post Production Research Reducing the labor to dehull and to grind grain is obviously a vital need. Less urgent needs include: (1) improvement of malting quality (important both for brewing and for making high-methionine weaning foods); and (2) new methods of processing, such as parboiling, milling, and puffing (see Appendix B).

SPECIES INFORMATION

Botanical Name *Eleusine coracana* (L.) Gaertner

Common Names

Afrikaans (and Dutch): vogel gierst
Arabic: tailabon
Bantu: bule
English: finger millet, African millet; koracan
French: petit mil, eleusine cultivée, coracan, koracan
German: Fingerhirse
Swahili: wimbi, ulezi
Ethiopia: dagussa (Amharic/Sodo), tokuso (Amharic), barankiya (Oromo)
India: ragi
Kenya: wimbi (kiswahili), mugimbi (Kikuyu)
Malawi: mawere, lipoko, usanje, khakwe, mulimbi, lupodo, malesi, mawe
Nepal: koddo
The Sudan: tailabon (Arabic), ceyut (Bari)
Tanzania: mwimbi, mbege
Uganda: bulo
Zambia: kambale, lupoko, mawele, majolothi, amale, bule
Zimbabwe: rapoko, zviyo, njera, rukweza, mazhovole, uphoko, poho

Description

Finger millet is a tufted annual growing 40–130 cm tall, taking between 2.5 and 6 months to mature. It has narrow, grasslike leaves and many tillers and branches. The head consists of a group of digitately arranged spikes.

It is a tetraploid.

[12] Information from A. Shakoor.

Distribution

Finger millet derives from the wild diploid *Eleusine africana*.[13] There is archaeological evidence that before maize was introduced it was a staple crop of the southern Africa region. Today it is found throughout eastern and southern Africa and is the principal cereal grain in Uganda, where it is planted on more than 0.4 million hectares (especially in northern and western regions), as well as in northeastern Zambia. It is also an important backup "famine food" as far south as Mozambique.

Finger millet does not appear to have been adopted in ancient Egypt, and it is said to have reached Europe only about the beginning of the Christian era. However, it arrived in India much earlier, probably more than 3,000 years ago, and now it is an important staple food in some places, particularly in the hill country in the north and the south.

Cultivated Varieties

Numerous cultivars have been recognized in India and Africa, consisting of highland and lowland forms, dryland and irrigation types, grain and beer types, and early- and late-maturing cultivars. By and large, there are highland races and lowland races—each adapted to its own climate.

Environmental Requirements

Daylength Finger millet is a short-day plant, a 12-hour photoperiod being optimum for the best-known types. It has been successfully grown in the United States as far north as Davis, California (with considerable problems of photoperiod sensitivity), and it is widely grown in the Himalayas (30°N latitude); however, it is mainly produced within 20°N and 20°S latitude. Daylength-neutral types probably exist.

Rainfall It requires a moderate rainfall (500–1,000 mm), well distributed during the growing season with an absence of prolonged droughts. Dry weather is required for drying the grain at harvest. In drier areas with unreliable rainfall, sorghum and pearl millet are better suited. In wetter climates, rice or maize is preferable.

Altitude Most of the world's finger millet is grown at intermediate elevations, between 500 and 2,400 m. Its actual altitude limits are unknown.

Low Temperature The crop tolerates a cooler climate than other millets. For an African native, this crop is surprisingly well adapted to the temperate zones.

[13] This wild ancestor has at least one genome derived from *Eleusine indica* (Hilu, 1988).

High Temperature Finger millet thrives under hot conditions. It can grow where temperatures are as high as 35°C.[14] In Uganda, the crop grows best where the average maximum temperature exceeds 27°C and the average minimum does not fall below 18°C.[15]

Soil Type The crop is grown on a variety of soils. It is frequently produced on reddish-brown lateritic soils with good drainage but reasonable water-holding capacity. It can tolerate some waterlogging.[16] It seems to have more ability to utilize rock phosphate than other cereals do.[17]

[14] Information from J.A. Ayuk-Takem.
[15] Thomas, 1970.
[16] In recent trials of nine cereal species subjected to waterlogging from seedling to heading, finger millet was the most resistant, except for rice. It resisted waterlogging much better than maize. (Kono et al., 1988.)
[17] In pot experiments, the rock phosphate mobilizing capacity increased in the order maize: pearl millet: finger millet. (Flack et al., 1987.)

3

Fonio (Acha)

Fonio (*Digitaria exilis* and *Digitaria iburua*) is probably the oldest African cereal. For thousands of years West Africans have cultivated it across the dry savannas. Indeed, it was once their major food. Even though few other people have ever heard of it, this crop still remains important in areas scattered from Cape Verde to Lake Chad. In certain regions of Mali, Burkina Faso, Guinea, and Nigeria, for instance, it is either *the* staple or a major part of the diet. Each year West African farmers devote approximately 300,000 hectares to cultivating fonio, and the crop supplies food to 3–4 million people.

Despite its ancient heritage and widespread importance, knowledge of fonio's evolution, origin, distribution, and genetic diversity remains scant even within West Africa itself. The crop has received but a fraction of the attention accorded to sorghum, pearl millet, and maize, and a mere trifle considering its importance in the rural economy and its potential for increasing the food supply. (In fact, despite its value to millions only 19 brief scientific articles have been published on fonio over the past 20 years.)

Part of the reason for this neglect is that the plant has been misunderstood by scientists and other decision makers. In English, it has usually been referred to as "hungry rice," a misleading term originated by Europeans who knew little of the crop or the lives of those who used it.[1] Unbeknownst to these outsiders, the locals were harvesting fonio not because they were hungry, but because they liked the taste. Indeed, they considered the grain exotic, and in some places they reserved it particularly for chiefs, royalty, and special occasions. It also formed part of the traditional bride price. Moreover, it is still held in such esteem that some communities continue to use it in ancestor worship.[2]

Not only does this crop deserve much greater recognition, it could have a big future. It is one of the world's best-tasting cereals. In recent

[1] Information from J. Harlan. In Nigeria it is usually called "acha."

[2] It is important this way to the Dogon, a people of Mali. To them, the whole universe emerged from a fonio seed—the smallest object in the Dogon experience—a sort of atomic cosmology. (Information from J. Harlan.)

times, some people have made side-by-side comparisons of dishes made with fonio and common rice and have greatly preferred the fonio.

Fonio is also one of the most nutritious of all grains. Its seed is rich in methionine and cystine, amino acids vital to human health and deficient in today's major cereals: wheat, rice, maize, sorghum, barley, and rye. This combination of nutrition and taste could be of outstanding future importance. Most valuable of all, however, is fonio's potential for reducing human misery during "hungry times."

Certain fonio varieties mature so quickly that they are ready to harvest long before all other grains. For a few critical months of most years these become a "grain of life." They are perhaps the world's fastest maturing cereal, producing grain just 6 or 8 weeks after they are planted. Without these special fonio types, the annual hungry season would be much more severe for West Africa. They provide food early in the growing season, when the main crops are still too immature to harvest and the previous year's production has been eaten.

Other fonio varieties mature more slowly—typically in 165–180 days. By planting a range of quick and slow types farmers can have grain available almost continually. They can also increase their chances of getting enough food to live on under even the most changeable and unreliable growing conditions.

Of the two species, white fonio (*Digitaria exilis*) is the most widely used. It can be found in farmers' fields from Senegal to Chad. It is grown particularly on the upland plateau of central Nigeria (where it is generally known as "acha") as well as in neighboring regions.

The other species, black fonio (*Digitaria iburua*), is restricted to the Jos-Bauchi Plateau of Nigeria as well as to northern regions of Togo and Benin.[3] Its restricted distribution should not be taken as a measure of relative inferiority: black fonio may eventually have as much or even greater potential than its now better-known relative.

PROSPECTS

Unlike finger millet, African rice, sorghum, and other native grains, fonio is not in serious decline. Indeed, it is well positioned for improved production. First, it is still widely cultivated and is well known. Second, it is highly esteemed. (In Nigeria's Plateau State, for example, the present 20,000-ton production is only a quarter of the projected state demand.[4]) Third, it tolerates remarkably poor soil and will grow

[3] Both have white seeds but black fonio has black or dark brown spikelets.
[4] T. Mabbett. 1991. *African Farming* Jan/Feb:25–26.

Fonio has a lacy appearance. It is often less than knee high. (Nazmul Haq)

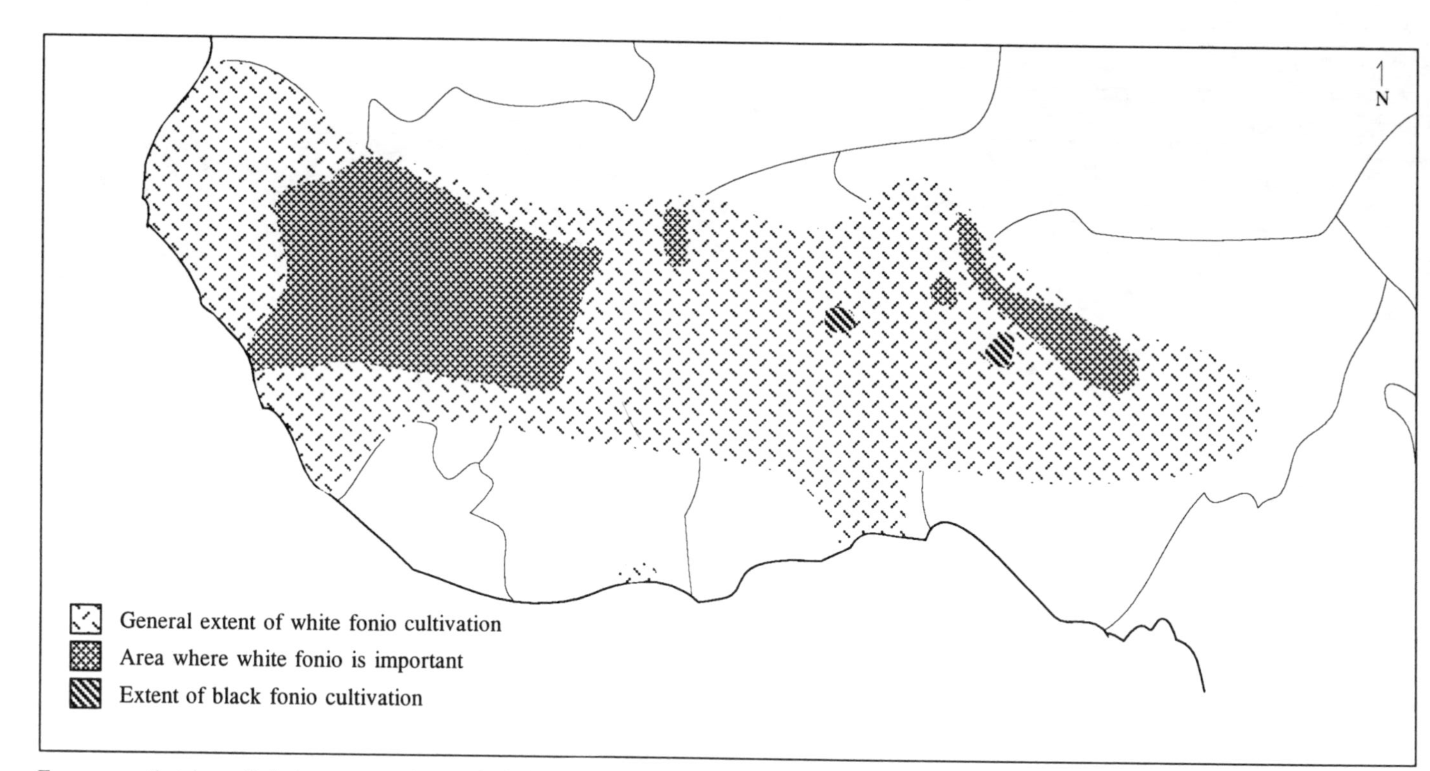

For a crop that is so little known to science, fonio is surprisingly widely grown. It is employed across a huge sweep of West Africa, from the Atlantic coast almost to the boundary with Central Africa.

where little else succeeds. These are good underpinnings for fonio's future advancement.

Africa

Humid Areas Low prospects. Fonio is mainly a plant of the savannas and is probably ill adapted to lowland humid zones. It seems likely to succumb to various fungal and bacterial diseases. However, white fonio does grow around the Gola Forest in southeastern Sierra Leone, and black fonio is reportedly cultivated in Zaire and some other equatorial locations. These special varieties (occasionally misnamed as *Digitaria nigeria*) are possibly adapted to hot and humid conditions.

Dry Areas High prospects. People in many dry areas of West Africa like fonio. They know that it originated locally, and they have long-established traditions for cultivating, storing, processing, and preserving it. During thousands of years of selection and use, they have located types well adapted to their needs and conditions. Although the plant is not as drought resistant as pearl millet, the fast-maturing types are highly suited to areas where rains are brief and unreliable.

Upland Areas Excellent prospects. Fonio is the staple of many people in the Plateau State of Nigeria and the Fouta Djallon plateau of Guinea, both areas with altitudes of about 1,000 m.

Other Regions

This plant should not be moved out of its native zones. In more equable parts of the world it might become a serious weed.[5]

USES

Fonio grain is used in a variety of ways. For instance, it is made into porridge and couscous, ground and mixed with other flours to make breads, popped,[6] and brewed for beer. It has been described as

[5] It is a relative of crabgrass, a European crop introduced to the United States in the 1800s as a possible food and now a much-reviled invader of lawns. However, white fonio is grown for forage in parts of the United States—apparently without causing problems.

[6] Little or nothing has been reported on popping this crop, but in southern Togo women put a little fonio into a metal pot and swirl it over a fire. Within a few seconds the grains begin bursting and bouncing, and the result is a light and puffy white material. Information from D. Osborn.

a good substitute for semolina—the wheat product used to make spaghetti and other pastas.

In the Hausa region of Nigeria and Benin, people prepare a couscous (*wusu-wusu*) out of both types of fonio. In northern Togo, the Lambas brew a famous beer (*tchapalo*) from white fonio. In southern Togo, the Akposso and Akebou peoples prepare fonio with beans in a dish that is reserved for special occasions.

Fonio grain is digested efficiently by cattle, sheep, goats, donkeys, and other ruminant livestock. It is a valuable feed for monogastric animals, notably pigs and poultry, because of its high methionine content.[7] The straw and chaff are also fed to animals. Both make excellent fodder and are often sold in markets for this purpose. Indeed, the crop is sometimes grown solely for hay.

The straw is commonly chopped and mixed with clay for building houses or walls. It is also burned to provide heat for cooking or ash for potash.

NUTRITION

In gross nutritional composition, fonio differs little from wheat. In one white fonio sample, the husked grain contained 8 percent protein and 1 percent fat.[8] In a sample of black fonio, a protein content of 11.8 percent was recorded.[9]

The difference lies in the amino acids it contains. In the white fonio analysis, for example, the protein contained 7.3 percent methionine plus cystine. The amino acid profile compared to that of whole-egg protein showed that except for the low score of 46 percent for lysine, the other scores were high: 72 for isoleucine; 90–100 for valine, tryptophan, threonine, and phenylalanine; 127 for leucine; 175 for total sulfur; and 189 percent for methionine.[10]

This last figure means that fonio protein contains almost twice as much methionine as egg protein contains. Thus, fonio has important potential not only as survival food, but as a complement for standard diets.

AGRONOMY

Fonio is usually grown on poor, sandy, or ironstone soils that are considered too infertile for pearl millet, sorghum, or other cereals. In

[7] Göhl, 1981.
[8] De Lumen et al., 1986.
[9] Carbiener et al., 1960.
[10] Using the FAO, A/E approach. Information from B. Standal. One analysis has reported a methionine level as high as 5.6 percent.

NUTRITIONAL PROMISE

Main Components		Essential Amino Acids	
Moisture	10	Cystine	2.5
Food energy (Kc)	367	Isoleucine	4.0
Protein (g)	9.0	Leucine	10.5
Carbohydrate (g)	75	Lysine	2.5
Fat (g)	1.8	Methionine	4.5
Fiber (g)	3.3	Phenylalanine	5.7
Ash (g)	3.4	Threonine	3.7
Thiamin (mg)	0.47	Tryptophan	1.6
Riboflavin (mg)	0.10	Tyrosine	3.5
Niacin (mg)	1.9	Valine	5.5
Calcium (mg)	44		
Iron (mg)	8.5		
Phosphorus (mg)	177		

COMPARATIVE QUALITY

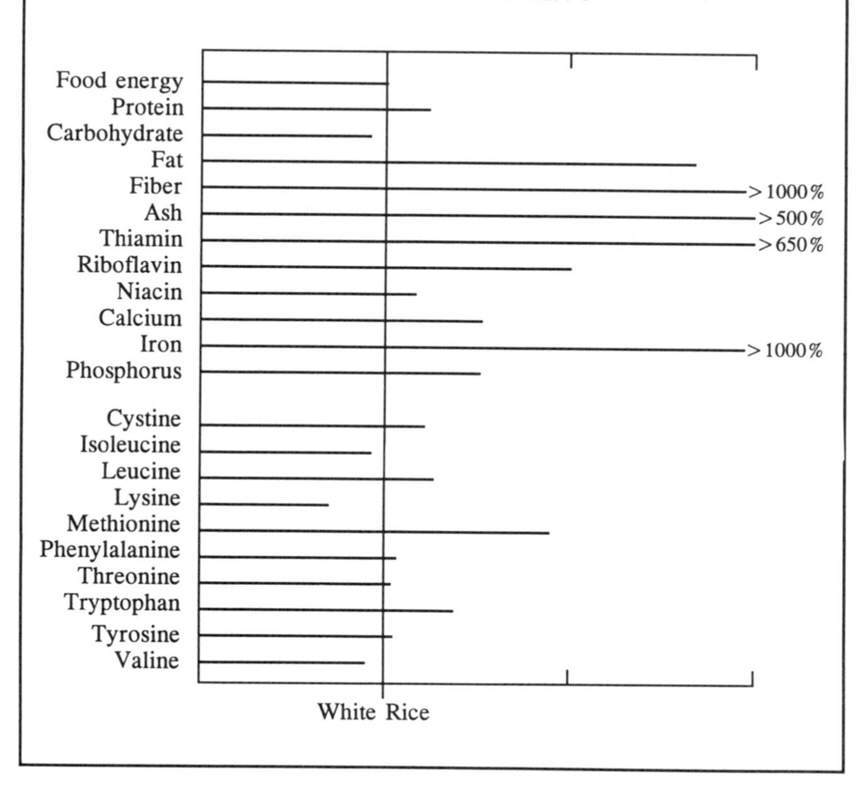

Fonio is an extremely adaptable plant that is little affected by climatic or soil conditions. Much of it is found growing in semiarid areas. In the Fouta Djallon Plateau of Guinea (shown here), it grows on acidic soils with high aluminum content that are deadly to other crops. (Nazmul Haq)

Guinea's Fouta Djallon region, where fonio is common, the soils are acidic clays with high aluminum content—a combination toxic to most food crops. It is generally grown just like upland rice, and the two are frequently produced by the same farmers. Normally, the seed is broadcast and covered by a light hoeing. It germinates in 3–4 days and grows very rapidly. This quick establishment and the heavy seeding rate (usually 10–20 kg of seed per hectare) ensures that the fields seldom need weeding. In a few cases the crop is transplanted from seedbeds to give it an even better chance at surviving the harsh conditions.

In Sierra Leone, and probably elsewhere, fonio is often grown following, or even instead of, wetland rice. This is done particularly when the season proves too dry for good paddy production and the farmers decide to give up on the rice. Fonio thus serves as an insurance against total crop failure.

In certain areas, fonio may sometimes be planted together with sorghum or pearl millet. Indeed, it is frequently the staple, while the other two are considered reserves. Commonly, farmers in Guinea sow multiple varieties of fonio and then later fill in any gaps with fast-maturing varieties of guinea millet (*Brachiaria deflexa*).[11]

HARVESTING AND HANDLING

Fonio grain is handled in traditional ways. The plants are usually cut with a knife or sickle, tied into sheaves, dried, and stored under cover. Good yields are normally 600–800 kg per hectare, but more than 1,000 kg per hectare has been recorded. In marginal areas, yields may drop to below 500 kg and on extremely poor soils may be merely 150–200 kg per hectare.[12]

Traditionally, the grain is threshed by beating or trampling, and it is dehulled in a mortar. This is difficult and time-consuming.

The seed stores well.

LIMITATIONS

Because of the lack of attention, fonio is still agronomically primitive. It suffers from small seeds, low yields, and some seed shattering.

The plant responds to fertilizers, but most types are so spindly that fertilization makes them top-heavy and they may blow over (lodge).

[11] Portères, 1976. This fonio-like grain is described in the chapter on other cultivated grains, page 237.

[12] As noted elsewhere, yield figures such as these can be very misleading. They may be low, but hungry rice produces a yield on sites or in seasons when other cereals yield nothing whatever.

Fonio: It's Not Just a Famine Food

Late in 1990, I interviewed a farmer with a largish plot of fonio. It was just a few kilometers from Bo town, in central Sierra Leone. What especially intrigued me was that this was not, as I at first supposed, a poverty-stricken woman's attempt to grow a little food for household subsistence. It was instead a commercial venture, aimed at the Bo market. There, fonio sells (cup for cup) at a better price than rice. By selling her crop she would be able to buy a larger amount of rice. To me, this was a striking confirmation of the commercial potential of this almost entirely neglected crop. To the people who know it, fonio is treasured more highly than rice!

Paul Richards

Birds may badly damage the crop in some areas; bird-scaring is usually necessary in those locations. The plants are also susceptible to smut and other fungal diseases.

It has been reported that fonio causes soil deterioration, but this appears to be a misperception. It is often sown on worn-out soils, sometimes even after cassava (the ultimate crop for degraded lands elsewhere). It is this association with poor soils that has given rise to the rumor, but the soils were in fact impoverished long before the fonio was put in.

Some groups dislike black fonio because, compared with the white form, it is more difficult to dehusk with the traditional pestle.

The seed loses its viability after two years.

Because of its small seed size, the harvest is very difficult to winnow. Sand tends to remain with the seed and produces gritty foods. It is therefore necessary to thresh fonio on a hard surface rather than on bare ground. Also, just before cooking, the grains are usually washed to rid them of any remaining sand.

NEXT STEPS

Clearly, fonio is important, has many agronomic and nutritional virtues, and could have an impressive future. This crop deserves much greater attention. Modern knowledge of cereal-crop improvement and dedicated investigations are likely (at modest cost) to make large advances and improvements. Yields can almost certainly be raised dramatically, farming methods made less laborious, and markets developed—all without affecting the plant's resilience and reliability. These results, and more, are likely to come about quickly once fonio becomes as important to the world's scientists as it is to West Africa's farmers.

Promotion General activities to raise awareness of this crop's value and potential include a monograph, a newsletter, a "friends of fonio" society, a fonio cookbook, a series of fonio cook-offs, and fonio conferences. These could be complemented by publicity, seed distributions, and experiments to test fonio's farm qualities and cultivation limits.

It should not be too difficult to generate excitement for this "lost gourmet food of the great ancestors." It might prove a good basis for recreating traditional cuisines. Even export as a highly nutritious specialty grain is a possibility.

Scientific Underpinnings Despite its importance, fonio is a crop less than halfway to its potential. There have been few, if any, attempts to optimize, on a scientific basis, the process of growing it. Its taxonomy, cultivation, nutritional value, and time to harvest are only partially documented. Varieties have neither been compared, nor their seed even collected, on a systematic basis. Little or no research has been done on postharvest deterioration, storage, or preservation methods.

Germplasm Collection An early priority should be to collect germplasm.[13] Varieties are particularly numerous in the Fouta Djallon Plateau in Guinea and around the upper basins of the Senegal and Niger Rivers.[14] Among these will certainly be found some outstanding types. This alone seems likely to lead to better cultivars that will bring marked advances in fonio production. The collection should also be screened to determine if yield is limited by viruses.[15] If so, the creation of virus-free seed might also boost yields dramatically.

Seed Size The smallness of the grain offers a special challenge to cereal scientists: can the seeds be enlarged—perhaps through selection, hybridization, or other genetic manipulation?

[13] One reviewer suggested asking village schoolmasters to collect seeds of all the different types in their areas. He reports getting outstanding assistance in this way on a project (in northern Nigeria) dealing with another widespread but little-known crop.
[14] Historically, these were the domains of the old empires of Mali, developed in the twelfth and thirteenth centuries, and it is there that fonio probably was brought to its apogee.
[15] In 1985, pangola grass (*Digitaria decumbens*), a related species that is widely planted as a tropical forage, was found to carry a stunt virus.

Yield The cause of the low yields needs investigation. Is it because of the sites, diseases and pests, poor plant architecture, inefficient root structure, lodging, poor tillering, bolting, or daylength restrictions? What are optimum conditions for maximum yields? Can fonio's productivity approach that of the better-known cereals?

Grain Quality Cereal chemists should analyze the grains. What kinds of proteins are present? What are the amino-acid profiles of the different proteins? Nutritionists should evaluate the biological effectiveness of both the grains and the products made from them. There are probably happy surprises waiting to be discovered. In particular, protein fractionation is likely to turn up fractions with methionine and cystine levels even greater than fonio's already amazingly high average.

The exceptional content of sulfur amino acids (methionine plus cystine) should make fonio an excellent complement to legumes. Feeding studies to verify this are in order. The combination could be nutritionally outstanding.

Cytogenetics As a challenge to geneticists, fonio has a special fascination. It has no obvious wild ancestor. That it appears to be a hexaploid ($2n=6x=54$) may help account for this. Does it, in fact, contain three diploid genomes of different origin? What are its likely ancestors, and might they be used to increase its seed size and yield?

Plant Architecture Lodging is a serious drawback, especially when the soil is fertile. This may be overcome by dwarfing the plant or endowing stronger stems by plant breeding. How "free-tillering" are the various types?

Other Uses Certain other *Digitaria* species are cultivated exclusively as fodder, whereas some are notable for their soil-binding properties and ability to produce an excellent turf. Is fonio also useful for such purposes? Could it, too, become a valuable all-purpose plant for many regions? Could improved fonio be "naturalized" in the northern Sahel to increase the availability of wild grain to nomadic groups?

Sociocultural Factors How is the crop currently cultivated, distributed, and processed? What roles are played by social and cultural

Opposite: Fonio is characterized by the very small size of its seeds. The tiny white grains have many uses in cooking: porridge, gruel, and couscous, for example. They are also the prime ingredient in several choice dishes for religious and traditional ceremonies. (Brent Simpson)

factors such as the division of labor, traditional beliefs, and people's expectations? (Fonio, after all, is seldom if ever grown under optimum conditions.) Its promotion will succeed best in West Africa if its development is placed within such local constraints.

Processing The processing and cooking of this crop is extremely arduous. Unless this can be relieved, fonio will probably never reach its potential.

SPECIES INFORMATION

Botanical Names *Digitaria exilis* Stapf and *Digitaria iburua* Stapf[16]

Synonyms *Paspalum exile* Kippist; *Panicum exile* (Kippist) A. Chev.; *Syntherisma exilis* (Kippist) Newbold; *Syntherisma iburua* (Stapf) Newbold (for *Digitaria iburua*)

Common Names

English: hungry rice, hungry millet, hungry koos, fonio, fundi millet
French: fonio, petit mil (a name also used for other crops)
Fulani: serémé, foinye, fonyo, fundenyo
Bambara: fini
Nigeria: acha (*Digitaria exilis,* Hausa); iburu (*Digitaria iburua,* Hausa); aburo
Senegal: eboniaye, efoleb, findi, fundi
The Gambia: findo (Mandinka)
Togo: (*Digitaria iburua*); afio-warun (Lamba); ipoga (Somba, Sampkarba); fonio ga (black fonio); ova (Akposso)
Mali: fani, feni, foundé
Burkina Faso: foni
Guinea: pende, kpendo, founié, pounié
Benin: podgi
Ivory Coast: pom, pohin

Description

As noted, there are actually two species of fonio. Both are erect, free-tillering annuals. White fonio (*Digitaria exilis*) is usually 30–75 cm tall. Its finger-shaped panicle has 2–5 slender racemes up to 15 cm

[16] Black fonio has been known to science only since 1911, when a botanist recognized that what was growing in fields with pearl millet in the Zaria region of northern Nigeria was a species new to science.

Fonio as Fast Food

As noted elsewhere (especially in Appendix C), a lack of processed products is holding back Africa's native grains. One grass-roots organization is doing something about this: it is turning fonio into a convenience food.

In southern Mali, fonio is mainly grown by women on their individual plots. Perhaps not unexpectedly, then, it is a women's group that has chosen to foster the grain's greater use. The group aims to raise fonio consumption by producing a precooked flour.

The project, backed by the Malian Association for the Promotion of the Young (AMPJ), is staffed and run entirely by women. Their goal is a fast-cooking fonio that will challenge parboiled rice and pre-packaged pasta (both of which are usually imported) in the Bamako markets.

The new "instant" fonio comes in 1-kg plastic bags and is ready for use. It requires no pounding or cleaning. It can be used to prepare all of the traditional fonio dishes. It is simple to store and handle. It is clean and free of hulls and dirt. And it requires less than 15 minutes to cook. For the user, then, it offers an enormous saving in both effort and time.

The project is currently a small one, designed to handle 6 tons of raw fonio per year. It uses local materials, traditional techniques, and household equipment: mortars, tubs, calabashes, steaming pots, sieves, matting, kitchen scales, and small utensils. The women sieve, crush, wash, and steam-cook the fonio; then they dry and seal the product in the airtight bags. The most delicate operation is a series of three washes to separate sand from the fine fonio grains.

The women have organized themselves into small working groups, formed for (1) the supply of raw materials, (2) production and packaging, and (3) marketing.

Fonio is considered a prestige food in local culinary customs. Yet, on the Bamako market this precooked product currently sells at a very competitive price: between 500 and 550 CFA Francs per kg. (By comparison, couscous sells at 650–750 CFA Francs.)

This small and homespun operation exemplifies what could and should be done with native grains throughout Africa. It is good for everyone: diversifying the diet of city folks, reducing food imports, and, above all, benefitting the local farmers by giving them a value-added product.

long. Black fonio (*Digitaria iburua*) is taller and may reach 1.4 m. It has 2–11 subdigitate racemes up to 13 cm long.

Although both species belong to the same genus, crossbreeding them seems unlikely to yield fertile hybrids, as they come from different parts of the same genus.[17]

The grains of both species range from "extraordinarily" white to fawn yellow or purplish. Black fonio's spikelets are reddish or dark brown. Both species are more-or-less nonshattering.

Distribution

Fonio is grown as a cereal throughout the savanna zone from Senegal to Cameroon. It is one of the chief foods in Guinea-Bissau, and it is also intensively cultivated and is the staple of many people in northern Nigeria. Fonio is not grown for food outside West Africa.

Cultivated Varieties

There are no formal cultivars as such, but there are a number of recognized landraces, mainly based on the speed of maturity.

Environmental Requirements

Daylength Flowering is apparently insensitive to daylength.

Rainfall Fonio is extremely tolerant of high rainfall, but not—on the whole—of excessive dryness. The limits of cultivation (depending on seasonal distribution of rainfall) are from about 250 mm up to at least 1,500 mm. The plant is mostly grown where rainfall exceeds 400 mm. By and large, the precocious varieties are cultivated in dry conditions and late varieties in wet conditions.

Altitude Although fonio is grown at sea level in, for instance, Sierra Leone, the Gambia, and Guinea-Bissau, its cultivation frequently is above 600 m elevation.

Low Temperature Unreported.

High Temperature Unreported.

Soil Type It is grown mainly on sandy, infertile soils. It can, however, grow on many poor, shallow, and even rocky soils. Most

[17] Information from G.P. Chapman.

varieties do poorly on heavy soils. However, by working with a range of varieties, one can generally adapt the crop to almost all terrains and exposures; for example, to fertile or unproductive conditions: sandy, limy, gravelly, or pebbly soils; slopes; plateaus; valleys; or riverbanks.

4
Pearl Millet

Of all the world's cereals, pearl millet (*Pennisetum glaucum*)[1] is the sixth most important. Descended from a wild West African grass, it was domesticated more than 4,000 years ago, probably in what is now the heart of the Sahara Desert (see map, page 80). Long ago it spread from its homeland to East Africa and thence to India. Both places adopted it eagerly and it became a staple.

Today, pearl millet is so important that it is planted on some 14 million hectares in Africa and 14 million hectares in Asia. Global production of its grain probably exceeds 10 million tons a year,[2] to which India contributes nearly half. At least 500 million people depend on pearl millet for their lives.

Despite its importance, however, pearl millet can be considered a "lost" crop because its untapped potential is still vast. Currently, this grain is an "orphan" among the significant cereals. It is poorly supported by both science and politics. In fact, few people outside of India and parts of Africa have ever heard of it. As a result, it lags behind sorghum and far behind the other major grains in its genetic development. For instance, its average yields are barely 600 kg per hectare and it is almost entirely a subsistence crop; perhaps for this last reason alone pearl millet has attracted little research or industrial support.

Indeed, largely due to neglect, pearl millet is actually slipping backwards. Production in West Africa during the last two decades has increased by only 0.7 percent a year—the lowest growth rate of any food crop in the region and far less than the population's growth rate. Furthermore, even this meager increase has been mainly due to expanding the area cultivated rather than to boosting yields. Elsewhere in Africa the decline has been even more dramatic. Just 50 years ago,

[1] Most taxonomists today believe that the most valid name for cultivated pearl millet is *Pennisetum glaucum* (L.) R. Br. Common synonyms are *Pennisetum typhoides* and *Pennisetum americanum*. The crop is also known as "bulrush millet" and in India it is normally called "bajra."

[2] Exact figures cannot be determined because some countries lump all the millets and sorghum together in their statistics. Also, many countries cannot provide statistics because their pearl millet does not enter organized commerce and is therefore never counted.

Kavango, Namibia. A farmer carrying millet heads to prepare the daily meal. (S. Appa Rao)

pearl millet was of almost incalculable value to millions of rural people in eastern and southern Africa. But over the decades, more and more farmers—especially in southern Africa—have abandoned it and switched to maize.

There are several reasons for this. For one thing, international research efforts have made maize more productive than pearl millet; for another, government incentives have given maize an added financial advantage; and for a third, easier processing has made maize more convenient to use. The momentum for change has now gone so far that maize is often pushed into pearl millet areas to which it is poorly suited and where it cannot perform reliably.

Now, however, a new era may be dawning. Pearl millet is supremely adapted to heat and aridity and, for all its current decline, seems likely to spring back as the world gets hotter and drier. Perhaps the best of all "life-support" grains, pearl millet thrives where habitats are harsh. Of all the major cereals, it is the one most able to tolerate extremes of heat and drought. It yields reliably in regions too hot and too dry to consistently support good yields of maize (or even sorghum). These happen to be the regions most desperately in need of help. It is there that the famines of recent decades have brought mass devastation and death. It is there that expanding deserts are destroying the productivity of perhaps 25 million hectares every year. And it is there that agricultural development could have its greatest humanitarian benefits.

These reasons alone should be sufficient to make pearl millet the target of a global initiative. But this crop has even more promise. Rising climatic temperatures are starting to concern almost all countries. And water is shaping up as the most limiting resource for dozens of the world's nations—including some of the most advanced. Agriculture is usually a country's biggest user of water, so that crops that sip, rather than gulp, moisture are likely to be in ever greater demand. Thus, even for economies that until now never heard of it, pearl millet could quickly become a vital resource.

Agronomically, there is no reason why pearl millet could not (like sorghum) become used worldwide. Indeed, recent research in the United States is showing that its prospects are much higher than most people now think. Already, the crop is showing promise for the heartland of America. It might also become widely used in the hotter and drier parts of Latin America, Central Asia, and the Middle East.[3] It could have a bright future in dry areas of Australia and other countries as well.

[3] Sorghum also has much promise here. It, too, will grow where it is dry, but is best when conditions are cool as well. Pearl millet's important characteristic is its concomitant ability to withstand both heat and low moisture.

Pearl-millet-growing areas in Africa. There are an estimated 14 million hectares of millet in this zone, making it the third most widely grown crop in sub-Saharan Africa. The plant was probably domesticated some 4,000–5,000 years ago along the southern margins of the central highlands of the Sahara. It has since become widely distributed across the semiarid tropics of Africa and Asia. Today, approximately one-third of the world's millet is grown in Africa; about 70 percent of it in West Africa. Africa's major pearl-millet producing countries include Nigeria, Niger, Burkina Faso, Chad, Mali, Mauritania, and Senegal in the west; Sudan and Uganda in the east. In southern Africa, the commercialization of agriculture has resulted in maize partially or completely displacing this traditional food crop. (ICRISAT, 1987; each dot represents 20,000 hectares)

Pearl millet is easy to grow. It suffers less from diseases than sorghum, maize, or other grains. Also, it has fewer insect pests.

The widespread impression that pearl millet grain is essentially an animal feed, unpalatable to all but the desperately hungry, is wrong. The grain is actually a superior foodstuff, containing at least 9 percent protein and a good balance of amino acids. It has more oil than maize and is a "high-energy" cereal. It has neither the tannins nor the other compounds that reduce digestibility in sorghum.

Pearl millet is also a versatile foodstuff. It is used mainly as a whole, cracked, or ground flour; a dough; or a grain like rice. These are made into unfermented breads (*roti*), fermented foods (*kisra* and *gallettes*), thin and thick porridges (*toh*), steam-cooked dishes (*couscous*); nonalcoholic beverages, and snacks.

Grain from certain cultivars is roasted whole and consumed directly. The staple food of the mountainous regions in Niger is millet flour mixed with dried dates and dried goat cheese. This nutritious mixture is taken on long journeys across the Sahara and eaten mixed with water—no cooking required.

Grain from other types is used to make traditional beer. In Nigeria, it is fermented, like maize or sorghum, to produce *ogi*—a traditional weaning food that is still common.

In future, pearl millet may be used in many more types of foods. The fact that it can be made into products resembling those normally produced from wheat or rice should make it acceptable to many more people.[4] With new technology, there seem to be possibilities of using it even to make raised breads (see Appendix C).

All this is not to say that pearl millet is perfect. Indeed, the crop has several serious problems. For one, the raw grain is difficult to process. Many consumers decorticate (dehull) the grain before grinding it into various particle sizes for use in different products. Dehulling by traditional hand pounding produces low yields of flour (around 75 percent) and the product has poor storage stability.[5]

Despite these impediments, this plant's promise is so great that we have devoted the following two chapters to its various types. The next chapter highlights its promise for subsistence farmers—the millions in Africa and Asia to whom pearl millet means life itself. The subsequent chapter highlights commercial pearl millets—the types that are increasingly grown by farmers who produce a surplus to sell.

[4] Information H.S.R. Desikachar.

[5] For a probable solution to this problem, see Appendix C. Semi-wet milling and parboiling are two techniques that have recently been shown capable of overcoming the storage stability problem. (Information from D.E. Blyth, ICRISAT).

Bajra

About 3,000 years ago pearl millet crossed the Indian Ocean and became a vital contributor to South Asia's food supplies. Today it is India's fourth most important cereal, surpassed only by rice, wheat, and sorghum. *Bajra*, as it is called, is currently grown on almost 10 percent of India's food-grain area, and it yields about 5 percent of the country's cereal food. Rajasthan, Maharashtra, Gujarat, and Uttar Pradesh account for nearly 80 percent of the 14 million hectares planted and 70 percent of the 5 million tons of pearl millet grain produced each year.

India's farmers grow some pearl millet under irrigation during the hot, dry months and routinely reap harvests as high as 3 or 4 tons per hectare. But most grow it in the arid areas, particularly where the rainfall is just insufficient for sorghum or maize. Here, the soils are usually depleted in fertility and there is no irrigation. Some plots receive as little as 150 mm of rainfall per year. But pearl millet survives and produces food.

Bajra-growing areas in the Subcontinent. (ICRISAT, 1987; each dot represents 20,000 hectares)

Indian farmer with a sampling of his bajra harvest. (The Rockefeller Foundation)

Indians commonly grind pearl millet and make the flour into cakes or unleavened bread (*chapati*). Some goes into porridges, which may be thin or thick. Much is cooked like rice. The grain is sometimes parched and eaten, the product (known as *akohi*, *bhunja*, *lahi*, or *phula*) being similar to popcorn. In some regions, the green ears are also roasted and eaten like a vegetable.

Although small quantities of the grains are used for feeding cattle and poultry, the plant is more often fed to animals as a green fodder. It is well suited for this purpose because it is quick-growing, tillers very freely, lends itself to multiple cutting, and usually has thin and succulent stems.

All in all, pearl millet is not a neglected crop in India. Authorities realize that it stabilizes the nation's food basket. Improved strains, suited to various regions, have been created and released for cultivation. Indeed, its potential is being increasingly exploited, especially as the swelling population requires increased cultivation of marginal land.

Let Them Eat Millet Bread

Millet once played a greater role in the world of cereals for many rural people in eastern and southern Africa, but it has declined in importance over the last 30–50 years because of a preference for maize.

The decline has been compounded by increased research on maize leading to greater productivity of the crop and by the incentives given to maize production through government policies. Maize has been grown, as a result, in dry conditions to which it is not adapted and it has failed too often in these conditions. Governments have come to realize this as well as the farmers themselves.

So it is now necessary to reestablish the importance of millet and sorghum in these drier areas and to do so we must make the production of these crops attractive enough so that they can compete with maize, not only in the worst and most severe droughts but in at least a majority of years. Here is work for the scientists in millet.

But in the long run, even in Africa, maize is not the problem at all. The problem is wheat, or more correctly, bread. Politicians are going to give the people bread. They have been saying this for a long, long time, and they mean it. Technocrats may decry this trend, particularly in tropical areas where wheat cannot be grown satisfactorily, but I can assure you that the protestations will be to little avail. They may slow the process down but they will not stop it. The people of the cities want bread, and the elected officials will ensure that they get it. The people are already exposed to bread and they will ask for it, they will insist upon it, and they will get it.

In many tropical countries it will be very expensive to satisfy this demand unless millet can become bread. And this, too, the politicians recognize and they will support this demand whether efforts can be made to decrease the cost of giving people the food that they demand. So here is something else for the millet scientists to do. Don't ask me how you do it. You know far better than I do. I am just telling you it's got to be done.

From an address by L.D. Swindale
Former Director-General, ICRISAT

NUTRITION

Pearl millet's average composition is given in the tables on the following pages. Some highlights are summarized below.

Carbohydrates usually make up about 70 percent of the dry grain, and they consist almost exclusively of starch. The starch itself is composed of about two-thirds amylopectin (the insoluble component that forms a paste in water at room temperature) and one-third amylose (the soluble component that forms a gel in aqueous solution).

Measurements made on several hundred types have shown that the protein ranges from 9 to 21 percent, with a mean of 16 percent. However, the varieties now used in farm practice have an average of about 11 or 12 percent. Of the different protein types, prolamine constitutes 40 percent and globulins 20 percent; the presence of an albumin has been also reported, but no gluten. The protein's biological value and digestibility coefficient have been measured as 83 percent and 89 percent, respectively.[6] The protein efficiency ratio has been found to be 1.43, which is even better than that of wheat (1.2).[7]

The grain has about 5 percent fat, roughly twice the amount found in the standard cereals. It is composed of about 75 percent unsaturated and 24 percent saturated fatty acids.

The vitamin values of pearl millet grain are generally somewhat lower than those of maize, although the level of vitamin A is quite good. The carotene value is also good—for a cereal.[8]

Of the grain's edible portion, ash comprises about 3 percent, an amount somewhat higher than in wheat, rice, or maize. The various mineral constituents, accordingly, tend to occur in greater quantities as well. Compared with maize, phosphorus (average 339 mg) is half again as much, iron (average 9.8 mg) is more than three times, and calcium (average 37 mg) is more than five times as much. Traces of barium, chromium, cobalt, copper, lead, manganese, molybdenum, nickel, silver, strontium, tin, titanium, vanadium, zinc, and iodine have also been noted.

In feeding trials, pearl millet has proved nutritionally superior to rice and wheat. A review of research in India[9] states that a diet based on pearl millet and pulses is somewhat better at promoting human growth than a similar diet based on wheat. In one trial, for instance, researchers made up vegetarian diets typical of those eaten by the

[6] These figures were determined by feeding experiments on rats at a 5-percent level of protein intake (CSIR, 1966).
[7] This was determined at the 10-percent level of protein intake (CSIR, 1966).
[8] The value reported (as vitamin A) is 22 retinol equivalents (RE), which is not outstanding in itself, but any amount of vitamin A is good for a cereal.
[9] CSIR, 1966.

NUTRITIONAL PROMISE

Main Components		Essential Amino Acids	
Moisture (g)	10	Cystine	1.8
Food energy (Kc)	353	Isoleucine	3.9
Protein (g)	11.8	Leucine	9.5
Carbohydrate (g)	70	Lysine	3.2
Fat (g)	4.8	Methionine	1.8
Fiber (g)	1.9	Phenylalanine	4.1
Ash (g)	2.3	Threonine	3.3
Vitamin A (RE)	22	Tryptophan	1.4
Thiamin (mg)	0.31	Tyrosine	3.0
Riboflavin (mg)	0.19	Valine	4.9
Niacin (mg)	2.6		
Calcium (mg)	37		
Chloride (mg)	43		
Copper (mg)	0.5		
Iron (mg)[a]	9.8		
Magnesium (mg)	114		
Manganese (mg)	0.8		
Molybdenum (μg)	190		
Phosphorus (mg)	339		
Potassium (mg)	418		
Sodium (mg)	15		
Zinc (mg)	2.0		

[a]Values range from 1.0-20.7 mg.

The pearl millet grain is nutritious. It has no husk, no tannin, contains 5–7 percent oil, and has higher protein and energy levels than maize or sorghum. The unsaturated fatty acids making up the oil are oleic (20–31 percent), linoleic (40–52 percent), and linolenic (2–5 percent). The saturated fatty acids are palmitic (18–25 percent) and stearic (2–8 percent).*

In general, pearl millet has a higher protein content than other cereals grown under similar conditions. In 180 pearl millet lines tested

* Information from L.W. Rooney.

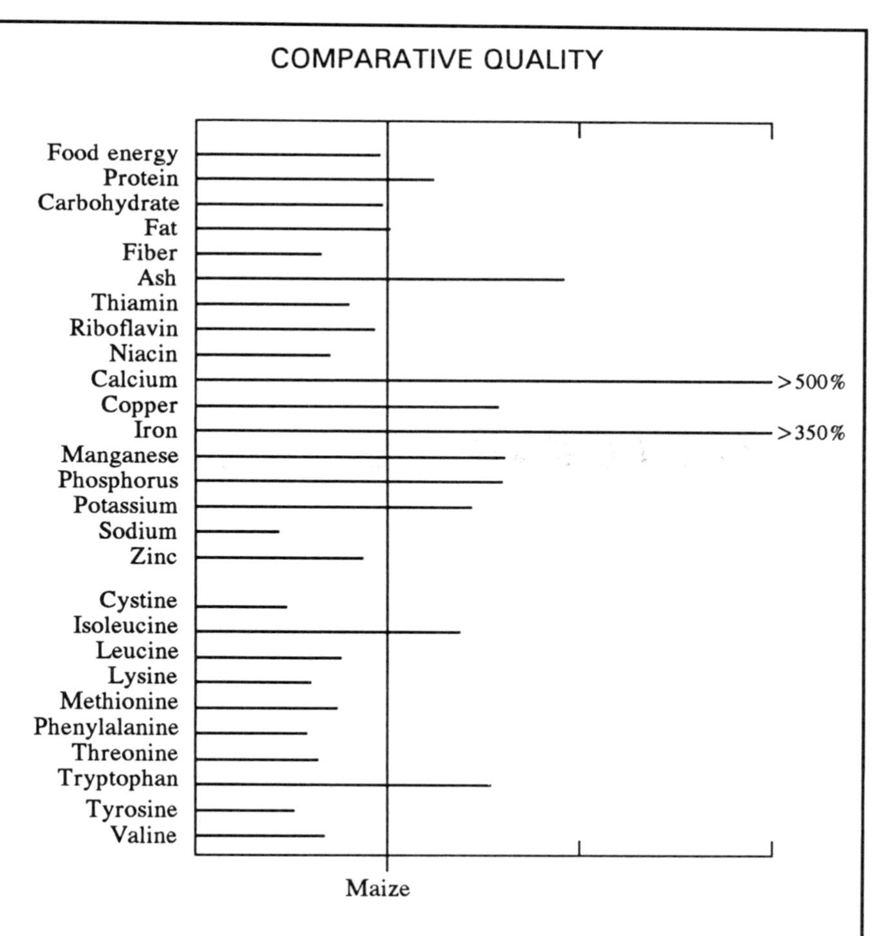

in 1972, protein contents ranged from 9 to 21 percent with a mean of 16 percent. It has an excellent amino acid profile and, depending on the variety and perhaps on the growing conditions, the levels of the various amino acids making up the protein can vary by as much as a factor of two. In general, however, the reported values show higher tryptophan, threonine, and valine and lower leucine, but otherwise similar essential amino acids in pearl millet compared with grain sorghum. What is uncertain, however, is the digestibility of pearl millet protein. It is possible that the actual amount of digestible protein is less than that of other major grains.

poor. When pearl millet partially or completely replaced rice, the nutritive value increased appreciably.

Studies conducted on children showed that all the subjects fed diets based on pearl millet maintained positive balance with respect to nitrogen, calcium, and phosphorus. The protein's apparent digestibility was about 53 percent, an amount close to that for finger millet and sorghum proteins, but less than that of rice protein (65 percent). It was also found that pearl millet could replace 25 percent of the rice in a child's diet without reducing the amount of nitrogen, calcium, or phosphorus its body absorbed.

SPECIES INFORMATION

AMABELE

Botanical Name *Pennisetum glaucum* (L.) R. Br.[10]

Synonyms *Pennisetum typhoides* (Burm.f.) Stapf and Hubbard, *P. americanum* (L.) Leeke, *P. spicatum* Roem and Schult.

Common Names

Angola: massango
Arabic: duhun, dukhon
English: pearl millet, bulrush millet, cattail millet, candle millet
Ethiopia: bultuk (Oromo), dagusa (Amharic)
French: mil du Soudan, petite mil, mil
India: bajra, bajri, cumbu, sajje
Kenya: mi/mawele, mwere (Kikuyu)
Mali: sanyò, nyò, gawri
Malawi: machewere (Ngoni), muzundi (Yao), uchewere, nyauti (Tumbuka)
Niger: hegni (Djerma), gaouri (Peul), hatchi (Haussa)
Nigeria: gero (Hausa), dauro, maiwa, emeye (Yoruba)
Shona: mhunga, mhungu
Sotho: nyalothi
Sudan: dukhon
Swahili: uwele, mawele
Swati: ntweka
Zambia: mawele, nyauti, uchewele (Nyanja), bubele, kapelembe, isansa, mpyoli (Bemba)
Zimbabwe: mhunga (Chewa), u/inyawuthi (Ndebele)
Zulu: amabele, unyaluthi, unyawoti, unyawothi

[10] The widely used name *Pennisetum americanum* is not taxonomically valid, according to most (but not all) authorities.

Description

Pearl millet is an erect annual, usually between 50 cm and 4 m tall. Tillering and branching are not uncommon and are sometimes profuse. The straw is coarse and pithy.

The numerous flowers are tucked tightly around a cylindrical spike (rachis) that can range in length from 15 to 140 cm. This inflorescence is usually greenish yellow, and it may be cylindrical throughout its length or may taper at one or both ends.

The flowers can be either cross-pollinated or self-pollinated. The female part (stigma) emerges before the male part is ready to shed its pollen.[11] As a result, cross-pollination normally occurs. However, where the timing overlaps, some self-pollination can occur.

Grain begins developing as soon as fertilization occurs and is fully developed 20–30 days later. The whole process, from fertilization to ripening, takes only about 40 days.

The seeds range in color from white to brown, blue, or almost purple. Most are slate gray. They are generally tear shaped and smaller than those of wheat. The average weight is about 8 mg. Some thresh free from glumes, while others require husking.

The seeds are quick to germinate. If conditions are favorable, they sprout in about 5 days. Freshly harvested seed may not germinate immediately; however, a dormancy of several weeks after harvesting has been reported.

Pearl millet is a diploid (2n = 14).

Distribution

The two vast areas of West and East Africa where pearl millet is prominent have already been mentioned (see page 80).

Soon after its domestication, the crop became widely distributed across the semiarid tropics of both Africa (15 million hectares) and Asia (14 million hectares). Pearl millet first became known in Europe about 1566 when plants were raised in Belgium from seed said to have been received from India. This form, sometimes known as *Pennisetum spicatum*, is still grown in Spain and North Africa. Pearl millet was introduced into the United States at least as long ago as the 1850s.

Cultivated Varieties

There are vast numbers of types, differentiated by features such as the following:

[11] The first flush of flowering (the female part) is completed in about 2 days; a day later the anthers from the male flowers emerge and a second flush of flowering (the one that produces pollen) continues for a further 2 days. A day or two after that a third flush of flowering begins. This one is from the female-sterile florets.

Double Dip

Pearl millet's extremely deep roots can reach down into soil layers untapped by other plants. Tests in the southeastern United States have revealed that it can pull up nutrients from residues that have accumulated below the root zones of the previous farm crops.

This finding, should it prove more widely true, has profound implications. Much of the fertilizer now put on crops leaches past their roots where it is not only lost but becomes a pollutant. Having an annual crop that can scavenge the lower layers gives farmers a second shot at the (expensive) fertilizer as well as a tool for cleaning the environment. They might even make a profit from it by selling the pearl millet.

- Quick maturity (about 80 days), medium maturity (100 days or so), or slow maturity (180 days or more)
- Height
- Amount of tillering
- Stem thickness and branching
- Leaf size and hairiness
- Seedhead size, shape, and "tightness"
- Number, length, rigidity, brittleness, and hairiness of bristles
- Size, shape, and color of grain
- The degree to which the glume adheres to the grain.

For pearl millet, the bulk of the systematic breeding has been done in India, but substantial contributions have also come from several African countries, France, and the United States. Most yield improvements have resulted from incorporating genes from African varieties into Indian breeder stocks. However, a breakthrough came in the late 1950s when plants carrying cytoplasmic male sterility were discovered. This genetic trait made hybrids a practical possibility. Today, single-cross pearl-millet hybrids, using male-sterile seed parents, are the basis of vigorous private and semi-public seed industries, especially in India (see chapter 6, page 111).

Environmental Requirements

Daylength Pearl millet is usually a short-day plant (see next chapter), but some varieties are daylength neutral.

Rainfall Although the crop is grown where rainfall ranges from 200 to 1,500 mm, most occurs in areas receiving 250–700 mm. The lowest rainfall areas rely mainly on early-maturing cultivars. Although very drought resistant,[12] pearl millet requires its rainfall to be evenly distributed during the growing season. (Unlike sorghum, it cannot retreat into dormancy during droughts.) On the other hand, too much rain at flowering can also cause a crop failure.

Altitude Pearl millet is seldom found above 1,200 m in Africa, but occurs at much higher altitudes elsewhere (for instance, in western North America).

Low Temperature The plant is generally sensitive to low temperatures at the seedling stage and at flowering.

High Temperature High daytime temperatures are needed for the grain to mature. In Africa's pearl millet zone, temperatures are typically above 30°C.

Soil Type Like most plants, pearl millet does best in light, well-drained loams. It performs poorly in clay soils and cannot tolerate waterlogging. It is tolerant of subsoils that are acid (even those as low as pH 4–5) and high in aluminum content.

Related Species

Pearl millet has many relatives. A number are quite troublesome. In much of Africa, for instance, wild *Pennisetum* species manage to get their pollen in, and this cross-pollination quickly reduces the crop's productive capacity. The hybrid swarms of weedy "half-breeds" (called *shibras* in West Africa) are common contaminants in the farmer's crop. Whereas the cultivated races have broad-tipped persistent spikelets and large, mostly protruding grains, the wild species have narrower, pointed spikelets. Also, their grains are smaller, entirely enclosed by husks, and prone to fall out (shatter). Luckily, the weedy species did not accompany the crop to India.

Although hybridization and introgression between the crop plants and the wild relatives is a problem for farmers, it can be a blessing for plant breeders, giving rise to new forms both of the crop and of the weed. (see page 121).

[12] The northern limit of sorghum in West Africa is around the 375-mm isohyet; that of pearl millet is further north—around the 250-mm isohyet. The crop's drought resistance comes from its rapid growth, short life cycle, high temperature tolerance and developmental plasticity.

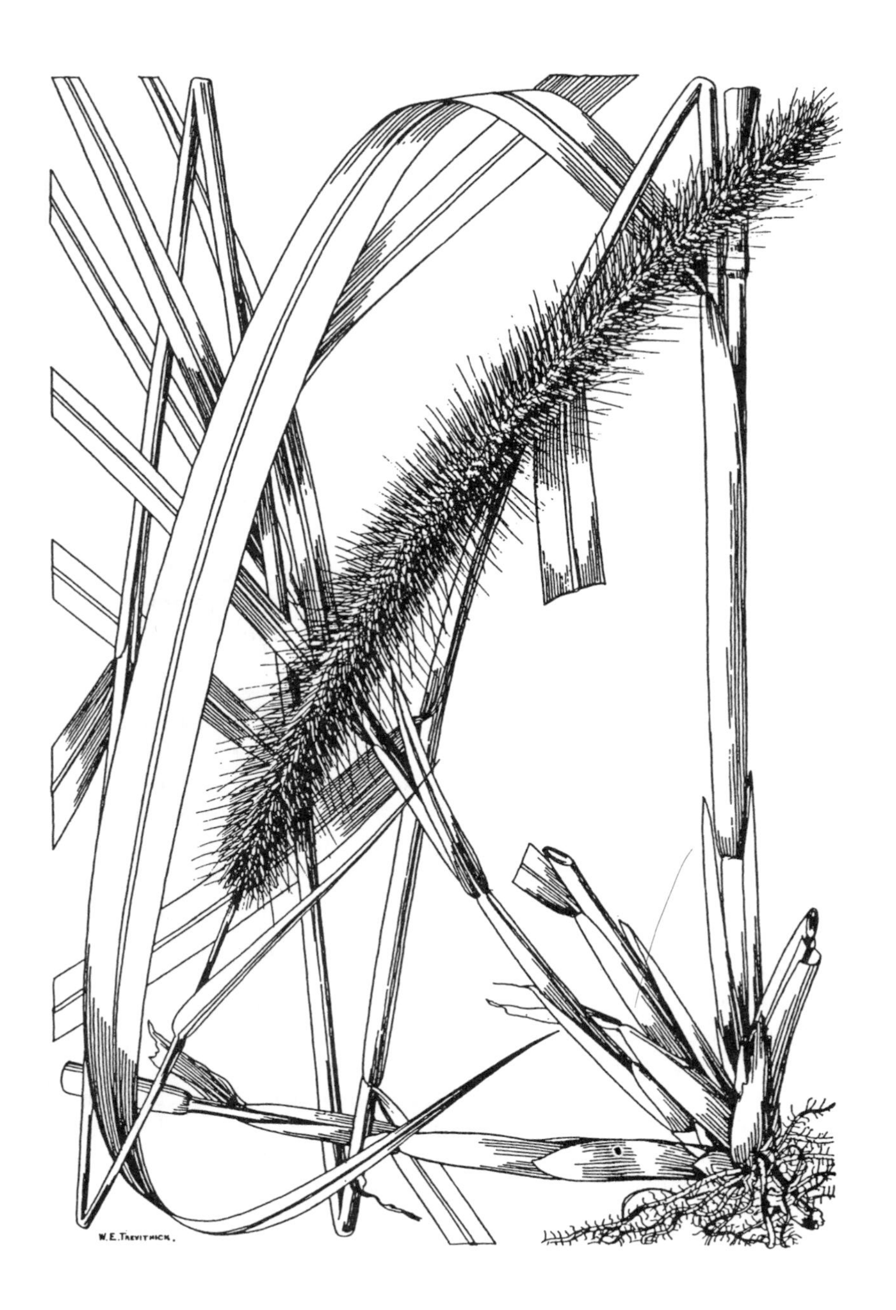
W. E. TREVITHICK.

5

Pearl Millet: Subsistence Types

Pearl millet is the staple of what is perhaps the harshest of the world's major farming areas: the arid and semiarid region stretching over 7,000 km from Senegal to Somalia (almost one-sixth of the way around the globe at that latitude). There, on the hot, dry, sandy soils, farmers produce some 40 percent of the world's pearl millet grain.

How to help these farmers—who live in the often drought-devastated zone on the edge of the world's biggest desert and who have no access to irrigation, fertilizer, pesticides, or other purchased inputs—is perhaps the greatest agricultural challenge facing the world. The answer may lie in their age-old staple, pearl millet.

Indeed, there is probably no better cereal to relieve the underlying threat of starvation in the Sahel, the Sudan, Somalia, and the other dry lands surrounding the Sahara. Millions entrust their lives to this single species every day, and, of all the peoples on the planet, they are the ones most needing help. Yet, at the moment, pearl millet suffers from neglect and misunderstanding—in part because the crop grows in some of the poorest countries and regions and in some of the least hospitable habitats for humans (including research workers). People have thus unjustly stigmatized it as a poor crop, fit only for interim support while something better is located.

This chapter's purpose is to counter that misguided notion.

SUBSISTENCE MILLETS

Most pearl millets grown in Africa are necessarily oriented toward survival under harsh conditions rather than high yields.[1] For want of a better name, we have called them "subsistence types."

To any outsider used to the robust look of wheat, rice, or maize, subsistence pearl millets may seem puny, unproductive, and downright unworthy of consideration. To an agronomist or cereal breeder, they

[1] The next chapter discusses pearl millet varieties that are adapted to commercial production and more salubrious sites. They tend to be productivity oriented.

Farming on the Fringe

Pearl millet is the last cereal crop of arable farming on the edge of the desert—beyond it there is only pasturing and open grazing. There is not a more drought-tolerant cereal crop to relieve the threat of starvation. When it fails, nothing else can be substituted. Thus, millions are forced to entrust their lives to this plant. It is not an easy bargain to make.

Most of Africa's pearl millet is grown where the danger of drought is ever present; where the landscape abruptly changes between the wet and dry seasons; where the rains are sometimes limited to only a month or two or three; and where utter aridity prevails the rest of the time.

It seems a cruel irony that the most destitute of people are forced to depend upon foods that they must produce for themselves in the harshest lands. But pearl millet has "rusticité," a French term implying that it will produce *something* no matter what. Droughts, floods, locusts, diseases, and other hazards may hurt, but the plant produces food nonetheless. All other grains, on the other hand, are more vulnerable and more subject to complete collapse.

It is remarkable that any crop can cope with the sites where pearl millet is grown. Local cloudbursts can dump the year's precipitation in a few hours. On crusted and hard soils, such deluges result in massive rushing runoff, heavy erosion, and the nearly complete loss of desperately needed moisture. Early-season rains are preceded by severe dust storms that damage, bury, and desiccate tender emerging seedlings. Scorching heat can kill an entire crop before it becomes established.

Because of problems like these, the threat of crop failure is omnipresent. Farmers must repeat their sowings, often two or three times. Most sow more area—and in widely separated sites—than they anticipate getting a harvest from. During the planting period they may scatter seeds continually wherever their herds trample the soil, and thereby give the seeds a chance to survive. To farmers elsewhere, tossing a few seeds in cow tracks may seem futile, but to those of the Sahel it can mean life itself.

Previous pages: The pain of growing crops on the desert fringe. Pearl millet underpins millions of desperate lives, including this Fulani farmer in Niger. "I will not plant again," he told the photographer after a rainstorm flattened his seventh millet planting that year. Even his resilient subsistence-type pearl millet has succumbed. Most plantings in this region are lost to drought or sandstorm; this one, ironically, to a flash flood. (Steve McCurry)

look particularly terrible. The plants perform poorly even when they are unstressed. They are tall and top-heavy; they are generally photosensitive; they exhibit low rates of fertilizer response; they have low harvest indexes; and they are localized in adaptation so that even the best of them cannot be easily moved around for use in other places. Above all, they are low yielding—averaging only around 500 kg per hectare.

In reality, though, subsistence pearl millets are some of the most remarkable food plants to be found anywhere. In the area of West Africa where pearl millet is paramount, the droughts can be fierce, the heat searing, and the rainstorms terrible. The sandstorms are even worse. Early in the growing season, the ever-present winds increase in intensity and often swirl the soil so powerfully that it literally sandblasts the tender seedlings. Then, heated by the Sahara sun, the new-blown sand may "cook" the seedlings before they can grow tall enough to shade and cool the land around their roots. Finally, as the soil dries out, its surface often hardens into a crust so impenetrable that any surviving seeds cannot break through.

Because of conditions like these, crop failure is omnipresent and Sahelian farmers must repeat their sowings, often several times. But of all food crops, subsistence pearl millets tend to survive best—they sometimes survive even in bare Sahara sand dunes.[2] They are cereals for "base-line food security" and give the farmer the best chance of staying alive.

By and large, subsistence pearl millets can:

- Germinate at high soil temperatures;
- Germinate in crusted soil;
- Tolerate some sand blasting in the seedling stage;
- Yield grain at low levels of soil fertility;
- Resist downy mildew;
- Tolerate stem borer and head caterpillar; and
- Hold up reasonably well against the parasitic weed striga.

Few of the scientists' varieties could be relied upon to produce food under conditions of such uncompromising hostility.[3] Some of the "faults" perceived by outsiders are actually of great local importance, as the following examples show.

[2] They are also found in bare dunes elsewhere: on the coastal plain of Yemen, for instance. Information from M.W. Brown.

[3] A reviewer from a large research facility in West Africa sent us the following comment: "The fact is, that after 40 years of [pearl] millet breeding, only one 'improved' line—CIVT—consistently surpasses (but not by much) local cultivars. Breeders' varieties routinely underperform local cultivars, even in on-station trials."

Northern Namibia. An Owambo farmer in front of his harvest holding large, compact heads he has selected for seed to sow next season. (S. Appa Rao)

Late Maturation

Elsewhere in the world, plant breeders have tried to speed up their cereals—to make them mature quickly so that more than one crop can be grown per year; so that weeds, pests, and diseases have less chance of causing destruction; and so that food can be produced where growing seasons are short. This is one reason why subsistence pearl millets look bad: many tend to mature very slowly.

The long growing season certainly leads to problems. Since flowering generally takes place after the rains end, even a brief early drought can hit the plants before there is any chance of forming seed and thereby bringing on total crop failure.

However, to Sahelian farmers the delay is all important. They want the grains to ripen after the rains have ceased. Although agronomically inefficient, it eliminates many drying and storage problems. (The grains can be easily dried, and they do not grow molds.) It probably also reduces problems caused by grain diseases and insects, both of which need moisture to thrive.

For the same reason, some subsistence pearl millets are "open-headed." This, too, is inefficient, and plant breeders elsewhere try to replace loose seedheads with compact ones. For the farmer in much of Africa, however, the open form eliminates many of the drying and storage problems encountered with tight-headed varieties.

The long vegetative growth phase is probably also a major adaptive advantage in this region where the soils are lacking in both moisture and fertility: it gives the roots a chance to explore larger soil volumes. For one thing, this probably contributes to the plant's drought tolerance. For another, it probably helps the plant amass the nutrients necessary to grow a good head of grain. This may take considerable time, because roots grow slowly and because in those depleted soils the release of any remaining mineral nutrients is itself often slow.

A related, subtle feature is that the traditional crop varieties usually mature at the same time. This means that only one generation of birds, insects, and diseases gets a chance to attack the flowers and seeds. Adding a mixture of types that mature successively is a disaster: it provides a "rolling nursery" that builds up multiple generations of pests and diseases that then wipe out all late-maturing types.

Daylength Sensitivity

Many of the world's wild plants (as well as most traditional landraces) are sensitive to the length of day. Modern plant breeders try to eliminate this restrictive trait so the plants they produce can be grown in different latitudes and seasons. But, for the subsistence pearl millets of West Africa, daylength sensitivity is what ensures that grain will be ready to harvest just at the right time in the dry season. It is the length of day that triggers the plant to flower, not the age of the plants. The yield may be poor if the season has been difficult, but the plant will at least flower and mature whatever grain it can.

By-Products

Traditional rustic varieties tend to be big, tall, leafy plants that perform best when spaced far apart. While these varieties produce massive amounts of greenery (6–12 tons per hectare even under the prevailing circumstances), the harvest index is often less than 20 percent. This means that less than 20 percent of the plant (above ground) is grain and more than 80 percent is stalk and leaves, as compared to 30 percent or more for improved high-yield-potential varieties.

But farmers who must produce almost every necessity right on their own land look at these cereals in totality. To them, there is no such thing as excessive stalk. For anyone who cannot buy fencing, roofing, or fuel, stalks are as valuable as grains. And for those who have a cow or some goats, the leaves are what keep the animals alive during the dry season.[4]

Consumer Preferences

To a subsistence pearl-millet farmer, the kernel characteristics—shape, color, processing qualities, and endosperm texture—can be more important than the absolute yield. A grain is almost worthless if it doesn't have the right (and often very subtle) properties for the type of foods the family eats. Subsistence growers choose among the varieties mainly on grounds of suitability for preparing such dishes as:

- *Toh*. The principal food, served at least once a day in the northern Sahel, *toh* is a stiff porridge prepared by adding pearl millet to boiling water while stirring.
- *Koko*. This is prepared by mixing pearl millet flour with water into a fine paste, which is then put aside in a warm place for a day or two to ferment. The resulting sourdough is then dropped into boiling water to form a thin porridge of creamy consistency.
- *Marsa*. This favorite snack of Ghanaians is a deep-fried pancake, prepared from the leavened batter of pearl–millet flour.

[4] This feature is restricted neither to Africa nor to this crop. Even today in parts of Turkey and Syria wheat straw sells for more per kilo than wheat grain (and wheat, of course, is a high-priced crop). Information from J. Harlan.

Genetic Diversity

Pearl millets grown under truly marginal conditions are usually heterogeneous enough to ensure stable production over seasons with widely differing weather patterns. In a sense, the African farmers for centuries have been performing "population breeding," a technique that is only now becoming popular in science. With this technique, a cluster of genotypes acts as a "cohort" able (collectively) to make the best of varying conditions. The genetically different plants in the "swarm" help create a successful harvest, no matter what hazards the season may bring. Should one type be depressed by weather, pests, disease, or mismanagement, others carry the brunt.

Advancing the qualities of a plant along a broad genetic front helps ensure a reliable—although not maximum—yield. And when your life depends on what you can grow, reliability is the most fundamental need.

WHAT TO DO?

Supporting greater production of subsistence pearl millets is one of the world's most humane endeavors. But improving the plants in this case is probably of secondary importance. Given the already remarkable qualities of these time-tested survival crops, given the infertile soils and harsh climates, and given the resources at the farmer's disposal, it would be difficult to come up with a better plant than he has already.

More important is research to make the farming methods easier, more reliable, and more effective; research to make storing and handling the harvest better and safer; and research to ease the daily drudgery of processing the raw grain into edible forms.

This book is of course designed to highlight promising plants rather than farming, storage, or processing methods. However, during the course of this study we came across some innovative ideas that may help boost the performance and reliability of subsistence pearl millets. We mention them here briefly. In the appendixes can be found ideas on potential breakthroughs in pest control, grain storage, milling, and other pertinent aspects.

REDUCING VULNERABILITY TO CLIMATE

Helping farmers to deal with the uncertainties of the early rains—not to mention the droughts, sandstorms, and high soil temperatures—are perhaps the most valuable interventions that can be made. These

The Dual Track

In this report we have given equal weight to species for both subsistence and commercial production. This is certainly an uncommon approach: in recent years polarization and even rancor have prevailed between the proponents of each viewpoint. However, in a broad sense, subsistence and commercial farming, although separate, are parallel and equally worthy—a fact not widely recognized by the public and one that sometimes befuddles even the best-intentioned scientific minds.

Subsistence farming is vital to the lives of millions, of course, and strengthening it is perhaps the most humanitarian contribution that can be made to African agriculture. But it is often operationally impossible to reach the neediest in the way they want. To create a new variety—even of a well-understood crop such as wheat—can easily take a decade of dedication and perhaps a million in money. It is therefore clearly impractical to reach, individually, the thousands of subsistence regions, each with its likes and dislikes, needs and desires, climates and conditions.

Although technical farming is not inimical to traditional farming, it is often much criticized by those most motivated to helping the neediest farmers. Everybody wants to help the most poverty stricken, of course. However, there is probably not a single subsistence farmer who doesn't dream of producing a surplus for sale. And that surplus is much more than a way to pay for a daughter's dowry or a transistor radio; it is, after all, the way out of poverty.

For this reason, then, those who are developing modern cultivars and hybrids for use in even the poorest nations are not wrongheaded or misguided. Subsistence farmers may be in the overwhelming majority, but the other farmers are the ones who, producing more than they can eat, feed the nonfarming public—the city dwellers, businessmen, doctors, teachers, tourists, and, yes, even the visiting researchers. Nor is there any reason to deny subsistence farmers a route to prosperity by withholding from them the means for producing commercially desirable varieties. Any nation, to survive and prosper, must help its farmers feed more than themselves.

Commercial farming has different requirements and goals from subsistence farming, but it poses no threat. This can be seen in many parts of the world. Throughout the Middle East, for example, farmers grow rustic and advanced wheats side by side—one for family use, the other for market day. Also, in the highlands of

Peru, Indians commonly grow traditional potatoes for themselves and modern potatoes for the cities.

Some have pointed out that the Green Revolution wheats in India and Pakistan were grown largely for sale. They conclude (rightly or wrongly) that commerce was the main motivation and that no quantum leap in food production can occur in Africa until similar commercial opportunities are available. Thus, despite the current polarized approaches, subsistence farming and commercial farming in the Third World are inextricably linked. Improvements in one can benefit the other.

Traditional Farmers Are Superb, But . . .

Subsistence farmers are to be admired and even emulated. Their techniques have been honed in the uncompromising harshness of an unforgiving climate as well as in the ever-present knowledge that failure means hunger or even death. However, no one should get carried away with the romantic notion that peasants always know best.

In the 1860s, when the United States proposed putting an agricultural university in every state, there was much opposition and many claims that American farmers needed no technical help—that professors in universities could not possibly teach the people of the soil how to farm better. But it proved otherwise—the so-called "land grant colleges" provided the engine of basic knowledge that has driven U.S. agriculture to its current heights.

It was through those universities and similar research facilities that the life cycles of many farm pests were worked out, the effects of fertilizer demonstrated, crop genetics illuminated, soil types and soil micronutrients identified, and myriad other basic facts underlying any farming operation brought to light. With this knowledge, even the most stubborn traditionalists were able to coax more from their land, with less effort and more consistency.

All in all, there are many ways in which a basic biological understanding can benefit the subsistence farmers of the hungry nations. Even the best of those farmers can, in this way, be helped to grow their crops more easily, more reliably, and with higher returns.

In the past, scientific findings were applied mostly to commercial agriculture, but that was because larger scale farmers are usually easier to reach and more susceptible to change. Knowledge is not detrimental to subsistence farming, and the polarization that now pervades rhetoric and thinking worldwide is deplorable.

would provide more secure environments early in the planting season and would do much to reduce a farmer's vulnerability to total crop failure before the crop is even started. Following are six possibilities.

Tillering

The pearl millets grown in the Sahel tend to be nontillering—each seed puts up only a single stem. This adds a major vulnerability because if that stem dies in a drought or sandstorm, for example, the plant is lost.

But certain pearl millets put up as many as five heads—not all of them at once. In this case, then, the destruction of a stem still leaves the plant alive and with a chance to rebound.

Other things being equal, adding some tillering types would dramatically reduce the severity of crop losses in the bad years and it would reduce the need to replant damaged fields. And in the good years when the rains are plentiful and timely, two or three (or perhaps more) stems would all emerge and survive, thereby doubling or tripling the yield.

Deep Planting

In the United States, researchers are studying how different types of pearl millet perform while in the seedling stage. They have found that the seedlings show large differences in the length and in the speed with which they lengthen.[5] By selecting types that produce tall seedlings and rapid elongation they have been able to plant the crop as deep as 10 cm.[6] This gives the newly germinated and highly vulnerable seedling a better chance at surviving: it can reach deeper moisture; it is less likely to be killed if the soil surface dries out; and, if it is a fast grower, it can perhaps get through to the air before the soil crusts over.

Although the tests were done in germinators and greenhouses in the United States, they successfully identified lines possessing improved stand-establishment capabilities of high potential value for the subsistence farmers facing the elements a world away.

Water Harvesting

There are many possible ways to help concentrate moisture at the base of seedlings. A companion report identifies a considerable number.[7] That these are likely to have significant value is suggested

[5] This work has been spearheaded by W.D. Stegmeier.

[6] Highly significant differences in elongation of mesocotyl (MC) and coleoptile (CL) organs were found among 1,100 entries germinated at 30°C with MC and CL lengths ranging from 14 to 130 mm and 6 to 40 mm, respectively.

[7] *More Water for Arid Lands*. For a list of BOSTID reports, see page 377.

by a recent paper on the use of soil imprinting and tied ridges.[8] Both techniques produce little "basins" around the plants where water collects.

In the trials (conducted in an area of West Africa where annual rainfall is 600–900 mm), tied ridges captured 85–100 percent of the rainfall received on the site during the season. Normal ridging or flat planting captured only 55–80 percent—the rest was lost as runoff. Tied ridging also reduced the soil's surface bulk density, maintained soil fertility (by reducing losses of soil nutrients), and improved the soil's water-holding capacity. In the case of the pearl millet crop, tied ridging increased the depth of rooting, the root density, the vegetative growth, and the yields—and it did it in both wet and dry years.

Transplanting

The use of nurseries is one of the oldest strategies to avoid water stress in the seedling stage. For centuries, Asians have transplanted rice seedlings and West Africans have transplanted sorghum seedlings (see page 184). Now farmers in parts of Asia are transplanting maize in the same way. Direct sowing is of course much easier, but wherever catastrophic failure is a probability, transplanting provides added security.[9]

In this process, the seeds are planted not in the fields, but in small irrigated nurseries; they are taken to the fields only after the rains have commenced in earnest. This technique seems particularly promising with subsistence pearl millet (not to mention other crops in this book) because the crop must be established during the least favorable season, the time available is often short, the water supply limited, and the weather unpredictable. On top of all that, the farmer feels pressure to plant early because the family needs food and because the growing season is all too brief.

Transplanting not only overcomes the hazards of the unreliable early rains, but compared with a seeded crop, the transplanted crop is in the field for a much shorter time. It also needs far less water for an equivalent yield, and its resistance to the elements is greater. Growing

[8] N.R. Hulugalle. 1990. Alleviation of soil constraints to crop growth in the upland alfisols and associated soil groups of the West African Sudan savannah by tied ridges. *Soil and Tillage Research* (Netherlands) 18(2–3):231–247.

[9] Vietnam, long familiar with transplanting rice, pioneered the technique with maize in tropical conditions. Today, the cultivation of transplanted maize is widespread in the Red River Delta. The technique has boosted the maize crop from 50,000 hectares a year in 1983–1986 to almost 250,000 hectares in 1990. North Korea also uses transplanted maize these days. In fact, it has been said that without transplanting, the area under maize would probably never have exceeded 350,000 hectares, whereas today it covers about 700,000 hectares.

the seedlings in a nursery also allows the farmer to cull diseased plants and thereby reduce the intensity of infection.

Although transplanting is so far associated mainly with other crops, there seems to be no reason why it couldn't prove most beneficial with subsistence pearl millets. Indeed, in a few parts of India and Africa this is already practiced, and with considerable success.

Mulching

As we have noted, burning-hot soil is one of the major hazards to the newly planted subsistence millets. Anything that could cool the surface of the land would help. Apparently, little or no innovation has yet been applied to this problem, although some tests using shade have resulted in a tenfold increase in survival and yield.[10]

Windbreaks

The "sand-blasting" effect can surely be overcome by various kinds of barriers around (or at least on the windward sides of) the fields. One suggestion is the use of vetiver (*Vetiveria zizanioides*) hedges. This tall, extremely rugged grass would probably be unaffected by the blasting sand as its stems are enclosed in tough sheaths. When the time for planting crops arrives—even at the end of the driest of seasons—this perennial should still be standing stiff and straight and able to battle the wind.[11]

IMPROVING CROP MANAGEMENT

Ideas on helping subsistence farmers handle their crops with less work or higher returns can be found in various books, journals, research-station reports, and PVO newsletters, for example. We have included a few ideas in the appendixes to this volume. It is thus not our intention here to belabor such fairly well-recognized issues as the use of fertilizers, optimum levels of tillage, optimum crop population size, and the use of less-laborious cultivation practices such as hoes, plows, and draft animals.

There are, however, some promising lines of research that fit in with the spirit of innovation that lies at the heart of this book. Following are three examples.

[10] Information from J.H. Williams, who wrote to us saying: "As a point of interest my personal research has shown that millet growth varied tenfold as a result of manipulation of soil surface temperature by 6°C (I used shading techniques), but the same manipulation allowed maize to grow in 40°C air temperatures as well!"

[11] More information on this most interesting grass can be found in the companion volume *Vetiver: a Thin Green Line Against Erosion*. For a list of BOSTID reports, see page 377.

Cropping Systems

Subsistence pearl millets are essential components of traditional agricultural systems. They are usually intercropped with cereals such as sorghum and maize or with legumes such as cowpea or peanut. To most farmers, the combined production is more important than the yield from either crop by itself. This mixed cropping is difficult for today's researchers to deal with, but there are some interesting developments. One is dwarfing.

To reduce the size of a cereal plant is a common strategy (see next chapter). It provides a compact plant that is more resilient, easier to handle, and higher yielding. In the case of subsistence pearl millets, however, dwarfing is done not for such a yield advantage. Researchers have found that simply reducing the plant height can contribute greatly to the associated cowpea and other low-growing legumes.[12] The millet no longer shades its shorter companion, which, with the increased photosynthesis, results in better yields. Initial results in Niger are quite encouraging. Farmers there have adopted dwarf millets eagerly.

Building Tilth

The soil under subsistence pearl millet is usually coarse textured, containing at least 65 percent sand. Such porous sites are not only poor in fertility, they are very poor at holding water. Any rain that does fall tends to drain away below the reach of the roots. Ways to keep it in the root zone would bring marked benefits, both in the crop's yield and its reliability.

It has been found, for example, that leaving crop residues in the field dramatically raises pearl millet yields in West Africa's deteriorating semiarid areas. In three recent trials, grain yields rose by 300, 450, and 550 percent, respectively. The residues not only increased the sandy soil's moisture-holding capacity, they also lowered soil temperatures and boosted fertility.[13]

Biological Fertilization

The areas where subsistence pearl millet is prevalent are usually so remote and so poverty stricken that despite the soil's barrenness commercial fertilizer can seldom, if ever, be used. But all plants, even those as robust as subsistence pearl millets, need food in the form of nitrogen, phosphorus, potassium, and a few so-called "micronutrients." How to provide plant foods under subsistence conditions is one of the

[12] Reddy et al., 1990.
[13] *INTSORMIL Bulletin*, 1990.

Pearl Millet Helps Namibia

Namibia's farming lands are among the driest and most unpredictable to be found. Perhaps for that reason, its farmers rely on *mahangu* (pearl millet) to provide the basic foods to keep their families fed. In the north of the country, where two-thirds of the population live, it is the staple.

In the past, Namibia's farmers could hope to obtain only about 300 kg of grain per hectare—a pitifully small amount. Indeed, production was so low that the country had to import maize to feed its people.

In 1986, however, the country asked ICRISAT for help, and 50 highly productive varieties were brought in and planted out for testing. In March 1987, at the new nation's first "Farmers' Field Day," approximately 100 farmers came to see the results. The variety Okashana 1 proved particularly impressive even though the rainfall that season had been only 170 mm (but well distributed).

Namibia then requested 200 kg of Okashana 1 seed for multiplication, large-scale testing, and demonstration to farmers. At the March 1988 Farmers' Field Day, 250 visitors showed up to buy Okashana seed. A year later, more than 500 farmers came, and they bought about 4 tons of the seed.

Since this new variety's arrival, Namibia's farmers have reaped bumper harvests. Even using traditional cultivation practices, they doubled their yields. But those who employed better methods obtained yields of 2.4 tons per hectare, about eight times the traditional amount.

Okashana 1 results from intensive plant breeding at ICRISAT, but it still retains its rustic resilience and is especially suited to subsistence farmers' needs. Among its characteristics are high grain yield, large seed size, early maturity, resistance to downy mildew, and ability to mature grain even when end-of-season droughts rob the plants of moisture.

According to Wolfgang Lechner, of the Mahanene Research Station at Oshkati, more than half of Namibia's pearl-millet farmers now grow the new variety. "Okashana 1 gives a light-colored flour that is highly acceptable," Lechner explains. "With this and the increased yields, within a couple of years the country may not have to rely on maize imports any more. That will save us a lot of valuable foreign exchange."

greatest of all agronomic challenges—not just for Africa and not just for pearl millet.

In certain places, deposits of rock phosphate have been located. This almost insoluble phosphorus-containing mineral has seldom been tapped for fertilizer in the past. But it is potentially a major source of phosphate for regions in extremity. Unlike standard soluble fertilizers, it doesn't provide an instant jolt of good nutrition, but it is nonetheless a most valuable source of a prime nutrient that plants need to remain healthy, robust, and high yielding. Certain parts of West Africa have deposits of rock phosphate that could be tapped for this purpose.

For providing nitrogen to a subsistence farmer's crops, probably nothing is more practical than biological sources. Nitrogen can be obtained in this way by:

- Incorporating crop residues or animal manures into the soil;
- Using leguminous food plants (such as cowpea or peanuts) in crop rotations;
- Intercropping with herbaceous soil-building legumes such as stylosanthes or macroptilium; or
- Incorporating nitrogen-fixing tree species such as *Acacia albida* into the fields.[14]

With pearl millet there is also the potential to get nitrogen directly from a beneficial microorganism that can live on its roots. Such nitrogen-fixing symbioses between a plant and a microbe are characteristic of many legumes, but of only a few grasses. Pearl millet is one of those few. It benefits from a nitrogen-fixing bacterium called azospirillum. Recent trials in Maharashtra, India, have shown that when pearl millet plants were inoculated with azospirillum, the yield of both grain and fodder was significantly increased.[15]

[14] This very interesting African tree, which can add nitrogen to the cropping system and also provide important windbreak effects, is described in the companion report *Tropical Legumes*. For a list of BOSTID reports, see page 377.

[15] A.S. Jadhav, A.A. Shaikh, A.B. Shinde, and G. Harinarayana. 1990. Effects of growth hormones, biofertilizer and micronutrients on the yield of pearl millet. *Journal of Maharashtra Agricultural Universities* 15(2):159–161.

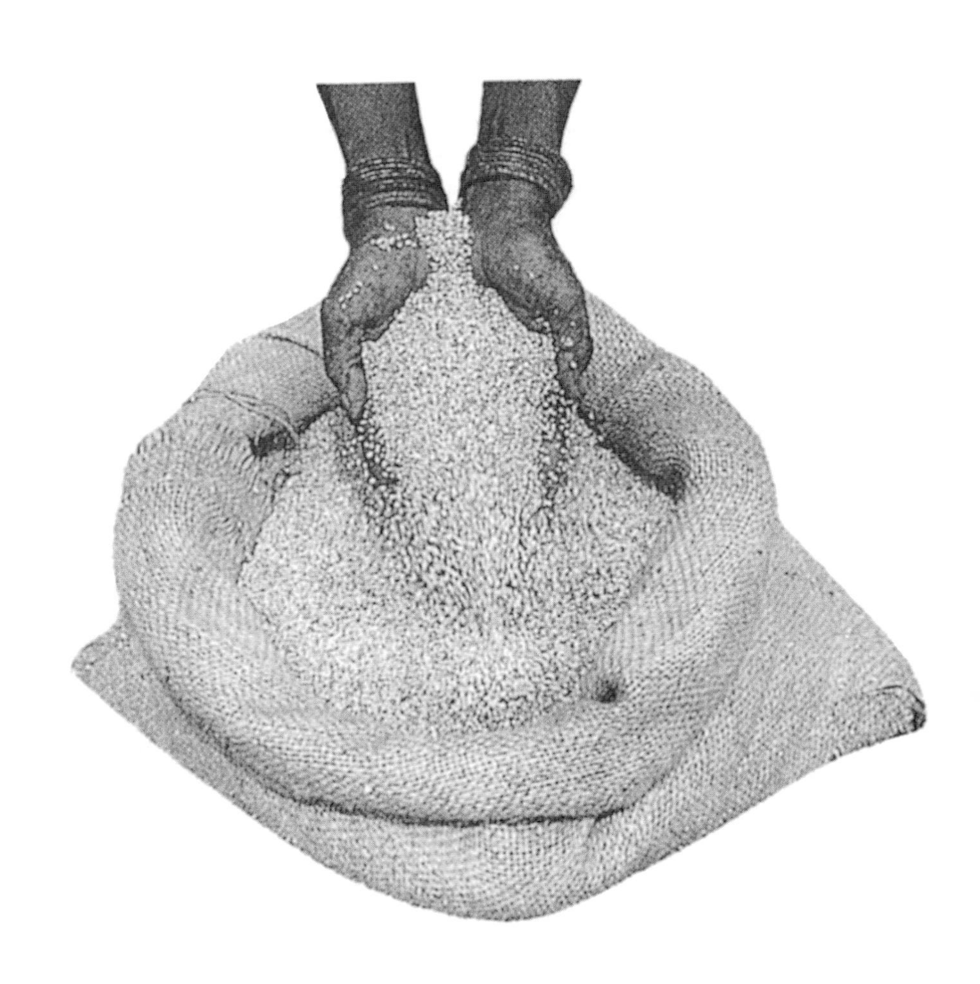

6

Pearl Millet: Commercial Types

Although it is one of the best means for sustaining life in the most desolate and difficult parts of the farming world, pearl millet also grows well under pampered conditions—under irrigation and in equable climes, for example. Because this fact is not widely known, most people dismiss pearl millet as a crop for good lands, pointing out that its low yield, low harvest index, and generally low fertilizer response mean that it cannot match the better known cereals under high-tech management.

However, it is far too early to dismiss pearl millet as a crop for regions that now grow modern maize and wheat and rice. The plant, as we have said earlier, has remarkable qualities, and some of its environmental resilience happens to be of the type that Latin America, North America, Australia, Europe, and others may soon need desperately. Moreover, pearl millet is no longer a rustic relic. Hybrids and other advanced forms are coming available for worldwide use. The old impressions no longer hold.

In fact, a new vision of this ancient crop's potential is becoming clearer from research in the United States, where pearl millet is already exciting increasing interest (see box, page 114). Indeed, fast-maturing types that ripen grain in as few as 90 days after planting and can be harvested by giant combines are now viewed as important resources for a vast belt spanning the nation from the Carolinas to Colorado.

This recognition is starting a new era in pearl millet production. For almost the first time the crop is being seriously investigated with sophisticated methods in the world's finest research facilities. Male-sterile forms, dwarfs, hybrids, and even some very unusual hybrids that produce fertile seed, have all recently been created. So far (at least in the United States), the emphasis has been on producing pearl millet as a feed grain— and with excellent reason: in U.S. Department of Agriculture trials, beef cattle, young pigs, and poultry fed pearl millet grain have grown as well as (or better than) those eating maize (see box).

More and more, however, America's pearl millet proponents are realizing that they have in their hands a potential new food grain for the nation and for the world.

There are good reasons for that assumption. Despite the current widespread notion that pearl millet is a second-rate cereal, the plant actually has a high potential growth rate—higher even than sorghum. Like maize and sorghum, it has the super-efficient C_4 photosynthesis. Some types mature very fast and can produce two or even three generations a year if conditions permit. And there are other advantages as well. Pearl millet is, for example, "a plant-breeder's dream" and can be developed quickly into numerous and widely different forms (see box, page 122). It is a cross-pollinating species on which several different breeding methods can be successfully employed. And, by a strange twist of genetic luck, it can also be easily inbred.

In terms of large-scale commercial production, therefore, this crop is poised for revolutionary advances. It stands at about the point maize did in the 1930s. Hybrids are known but are not in widespread use; yields are only a fraction of what they might be; and although the basic understanding of the crop's physiology and genetics is still rudimentary, it is beginning to become clear. Seizing the opportunity now could propel pearl millet (like maize since the 1930s) to far higher levels of productivity by using the best of modern techniques. Indeed, pearl millet might well result in a similar leap in food production in many new areas.[1]

Reasons for thinking this are not hard to find. The world's drylands are faced with an increasingly serious food crisis. Already this is becoming clear in the Middle East. For example, in 1989 Syria's parliamentary speaker announced at a meeting called to discuss Arab development and population problems that, unless the Arab world produces more food, one-third of its people will face starvation.[2] In such places the world's most drought- and heat-tolerant cereal obviously has vital promise.

All in all, then, this plant's adaptability to *both* good and bad conditions makes it a potentially outstanding food crop for vast areas of a "greenhouse-afflicted" world where climates may change wildly from decade to decade or even from year to year, and where more and more people must obtain food from hot, dry soils.

The chances for boosting pearl millet's productivity and usefulness are good, but the improvements may not come rapidly. To make the

[1] The increase in food supplies resulting from the creation of hybrid maize is considered to be a triumph second only to that of the Green Revolution (based on wheat and rice) in Asia in the 1960s and 1970s.

[2] "Danger is moving fast and if we do not . . . face it seriously and sincerely we will never be able to overcome the crisis," said Speaker Abdel-Qader Qaddoura, who noted that Arab food consumption was increasing by 7 percent per year, while production was increasing by only a little over 2 percent.

crop a modern and globally useful food resource, varieties with large, dense, spherical, light-colored kernels that taste good are needed. In addition, improved dehulling characteristics are vital if pearl millet is going to be employed in human foods on a truly wide scale.

Eventually, all of these and more seem likely to come about, as can be seen from the following promising lines of development.

HIGH-GRAIN TYPES

The worldwide cereal-breeding advances of the last 100 years have increased rice, wheat, and maize yields dramatically, but, contrary to popular perception, the plants still produce about the same amount of growth (that is, their overall dry matter is largely unchanged). Yields have risen because the plants were reconfigured to reduce the proportion of stems and leaves and increase the proportion of seeds. Usually, this meant reducing the plant height, but sometimes it also meant increasing the number of seedheads per plant.

Such rearranged plants have been the key to the remarkable jump in cereal yields that have occurred in most parts of the world. They respond well to good management; they make it possible to use fertilizer and other inputs profitably; and they create an upward spiral of yield and income that goes far beyond food production alone. For example, they help farmers to rest part of their land to restore its physical condition and fertility.

As of now, however, Africa's pearl millets are not of the rearranged type. After centuries of trying to stretch their heads above the rampant weeds, they are too tall for maximum grain production. In creating excess stalk, they are consuming energy and moisture that could be used to develop more grain.[3] Also, they cannot fully enjoy the benefits of fertilizer because it makes the plants top-heavy so that rain or wind can easily topple them into the dirt. Paradoxically, more fertilizer can mean less yield.

This was the situation of Mexico's wheats before the 1950s when genes from Japanese dwarf varieties helped create short, strong-stemmed plants that could hold their heavy heads up during lashing winds and pounding storms. Strengthening the plant's architecture allowed fertilizer to work to the fullest benefit and was a prime component of the wheats that generated the Green Revolution.

A similar transformation is now occurring with pearl millet. Strong-stemmed dwarf types are being put to use for the first time. Such types have already been developed in the United States, for example. Yields of 4,480 kg per hectare have been achieved on research stations, and demonstration plots on farms in 1991 yielded 3,024 kg per hectare.

[3] We are of course focusing here on the interests of the farmer whose main goal is to produce grain. For many subsistence farmers, the stalk is also an important resource.

Millet in the USA

Although pearl millet has long been grown in the United States, few Americans have ever heard of it. That may soon change, however. A number of pioneering researchers see this crop as a valuable grain for the nation. High-yielding cultivars are being selected and bred; even hybrids have been created (see page 119). However, owing to an oversupply of food, pearl millet is currently being developed mainly as a way to feed animals. Recent results have indicated that it has exceptional promise for the American livestock industry.

Part of the research has been done in Nebraska and Kansas, where the plant's tolerance of drought and acid soils, its resistance to pests, as well as its low requirements for nitrogen fertilizer, make it a potential boon to farmers. The experiments showed the plant could fit into multiple cropping systems for the Great Plains region.

Pearl millet might be used as a quality-protein grain for many livestock–feeding purposes. Compared with maize, it had higher crude protein and ether-extract concentrations, as well as higher concentrations of all essential amino acids. Already, it is showing promise for feeding both poultry and cattle.

Poultry

Trials in different parts of Georgia have shown that pearl millet grain can fully replace maize in chick rations. It neither reduced the feed-conversion efficiency nor the rate of weight gain. Indeed, chickens eating pearl millet actually grew faster and healthier than those eating maize, sorghum, triticale, or wheat.

This was an important discovery because although maize is the Southeast's main poultry feed, it grows poorly there and the local poultry industry has to import maize from the Midwestern states. Some observers now conclude that as transportation costs increase, locally grown pearl millet could soon replace the imported maize as the poultry feed of choice. Several other areas of the country where maize is difficult to grow seem likely to switch over as well.

Cattle

Metabolism and feedlot trials in both the Midwest and the Southeast have shown that pearl millet is also good for feeding

Plainview, Texas, 1992. Hybrid pearl millet being harvested. In both the Southeast (mainly Georgia) and Midwest (Kansas and Nebraska) hybrid pearl millets have been created. At about 1 m tall, they are half the normal height and can be harvested by combine. Big improvements in production have been achieved; grain yields of 3,000 kg per hectare are not uncommon. Commercial varieties, like the one shown, have been released for farm use. (M. Marley)

cattle. The grain's oil content, which is more than twice that of maize or sorghum, gives it a relatively high energy density. Pearl millet has also proved potentially useful as a source of protein.* Compared with maize, it had higher concentrations of both crude protein (about 14 percent of the dry matter) and essential amino acids.

* Christensen et al., 1984.

TEMPERATE-ZONE TYPES

Traditionally, pearl millet has been grown within about 30° of the equator, but these days certain types are already growing each year in various parts of the United States—in Georgia, Kansas, and Missouri, for example—that are far from the equator. Moreover, although the plant is almost synonymous with drought and deserts, it is also growing well in mild and humid locations such as the sandy coastal plains of south Georgia and Alabama.

In these temperate areas of America, pearl millet is potentially invaluable as a summer annual grain crop. Maize is poorly adapted to this region where its own shallow roots (blocked by the acid subsoils) and the common summer droughts result in low yields. Hybrid pearl millet develops deep root systems in these acid soils, resulting in much more dependable yields. Pearl millet also resists midges and the lesser cornstalk borer, two insects that severely affect sorghum. Moreover, no aflatoxin problems have been observed with pearl millet.

In addition, pearl millet is giving the farmers in the Southeast undreamed of flexibility. Whereas maize must be planted within a two-week window in April, pearl millet can be planted at any time between April and July. This means that it can skirt the hazards of summer and still mature a crop before winter chills cut off all growth.

EARLY TYPES

A driving force behind U.S. pearl millet research is the chance that pearl millet might make double-cropping possible. This is now approaching reality. Rapidly maturing cultivars are soon to be released, and these are the types now seen as promising for the belt stretching from the Carolinas to Colorado. Planted in spring, just after the winter wheat has been harvested, they can ripen a crop before autumn, when the next winter-wheat crop needs to be planted. Key to this rotation is pearl millet's inherent ability to tolerate heat as well as drought. The plant survives and yields grain even during the sweltering summer and on the (often meager) moisture left unused in the soil by the preceding wheat crop. No other cereal can do that.

The global value of such precocious pearl millets could be substantial.

TROPICAL TYPES

Although pearl millet is the quintessential dryland cereal, it is also found in some of Africa's wet and humid tropical zones. Much pearl millet is grown, for example, in relatively high rainfall areas of Ghana. The types there are entirely different from those of West Africa's nearby dry zone. In general, they have seedheads (spikes) that are shorter and fatter; grains that are bigger, rounder, and whiter; and plants that mature much earlier. These differences are so conspicuous that the plants were previously classified as a separate species.[4]

Such types there have been little studied or appreciated by the world at large. Yet they appear to be promising in their own right and are good sources of genes for earliness and large grain size.[5]

The potential of pearl millet for the tropics can be seen in Ghana, where early millet is extremely important to rural people. The type grown there normally matures at the peak of the rainy season, a time when farmers have exhausted their food stocks from the previous harvest. At first, they gather pearl millet when the grains are in the dough stage and are soft and sweet. Usually, the freshly harvested heads are steamed, threshed, and dried. This process—the exact reverse of normal practice—probably makes it possible to recover the immature grains that would otherwise turn to mush when threshed.[6]

SUGARY TYPES

In India, as in Ghana (see above), pearl millet is sometimes roasted and consumed like sweet corn. Here, too, the grain is harvested in the milk or dough stage. This is a facet of pearl millet that has received little (if any) investigation. Yet it is reminiscent of the situation with maize a century or so ago. At that time the practice of eating maize grain in the soft, sweet, doughy stage was known only to a few Indian children and perhaps some adventurous farmers.[7] Today, "sweet corn" is a major food of North America, and a huge research effort has been expended on selecting strains whose grains convert sugar to starch only slowly so that they stay sweet. Canned sweet corn is in fact America's favorite preserved vegetable and has been outselling all the others since World War I.

Pearl millet, too, should have a big future as a sweet treat to be eaten more like a vegetable than a cereal.

[4] *Pennisetum gambiense* Stapf & Hubb. Today, however, they are considered to belong to race *globosum* of *Pennisetum glaucum* native to Ghana, Togo, Benin.

[5] Appa Rao et al., 1982.

[6] This is a fascinating tradition, well worthy of study and emulation. See section on parboiling, page 301.

[7] Sweet corn was not seen by the first colonists who reached North America, and when it was found in a valley in central New York in 1799 it was not appreciated at first. Some was planted along the coast but evoked no particular interest. Sweet corn began to be cultivated widely only after the Civil War (that is, in the 1860s).

POPPING TYPES

In India pearl millet is commonly popped. Dry grains sprinkled onto hot sand burst like popcorn. The pops are sometimes eaten with powdered sugar or brown sugar (*jaggery*).

The types that pop best have been given little or no special study. But popping is a promising method for bettering this crop (see Appendix C, page 297) and should be investigated further. Select types with round grains and impervious seed coats (so that the steam building up inside can reach the explosive levels necessary for good popping) will probably prove best.

LIGHT-COLORED TYPES

Although most of the pearl millets so far grown are tan or brown, white-grained types for the large-scale commercial production of food for people are now under development. These are attractive to look at and are sweet to the taste. Some have high protein contents. Also known are some yellow-grained pearl millets that are rich in carotene, the precursor of vitamin A. So far, however, they have been little appreciated.

EASY-PROCESSING TYPES

As noted earlier, pearl millet is among the more difficult grains to prepare. For one thing, the whole grain (caryopsis) contains a high proportion of germ. But more important, the germ is embedded inside the kernel and is difficult to remove. It is for this reason that traditional hand decortication often produces low yields of flour (not to mention its tendency to go rancid during storage).

The need for cultivars with improved dehulling properties is critical. Indeed, varieties with large, spherical, uniform, hard kernels that produce high milling yields already exist, but have not been documented systematically or brought into large-scale commercial production.

When pearl millets are processed into food products, there will be a need for larger supplies of more uniform grain with desirable milling properties and acceptable flavor, color, and keeping properties.

CUISINE-SPECIFIC TYPES

Most of the world's cereal breeding is done with foods such as bread, cakes, cookies, crackers, canelloni, or various breakfast concoctions in mind. But for pearl millet to sell in a big way in Africa it must be good for very different foods. In Africa (as well as in India), the major pearl millet foods are unfermented bread, fermented breads, thick and thin porridges, steam-cooked products, beverages, and snacks. Little or no information is currently available on which pearl millets have the best properties for each of these foods. This is a handicap. Undoubtedly, superior types exist and collections and investigations should be made in the houses of the users themselves. As we have said in the previous chapter, however, it is difficult to quantify, let alone breed for, the organoleptic properties of certain foodstuffs.

QUALITY-NUTRITION TYPES

Contrary to general opinion and oft-repeated statements in textbooks, pearl millet is one of the more nutritious of the common cereals. As has been noted, its grain has more fat than most, and its level of food energy (784 kilocalories per kg) is among the highest for whole-grain cereals. It also has more protein, and its level of the essential amino acid lysine is better than in most cereals.

However, some pearl millet grain may suffer (nutritionally speaking) because it is low in threonine and the sulfur-containing amino acids. Also, its lysine level could still be improved. Of course, the other major grains have the same defect, but in the last few decades high-lysine types have been found in maize, sorghum, and barley. It seems likely that a diligent search through the world's pearl millets with an amino-acid analyzer could disclose something similar.

HYBRIDS

As already mentioned, the development of maize hybrids in the 1930s led to a quadrupling of yields. A similar breakthrough, allowing the practical production of pearl millet hybrids, came in the late 1960s, when the first hybrids were created.[8] High-yielding, hybrids have been in use in India since 1966. Heterosis (hybrid vigor) in pearl millet can be substantial.[9] Indian scientists have succeeded in developing hybrids that can almost double the yield of local cultivars.[10]

[8] These were developed in the United States by Glenn Burton.

[9] Naturally, the types used to make the hybrid must be genetically diverse. The common finding that the hybrids show no increase in vigor apparently is owing to the fact that the types crossed were too closely related. Information from W. Hanna.

[10] In India, hybrid millets are used almost exclusively in irrigated farming. The yields can be spectacular, but they are not relevant to most of Africa's pearl millet production. Even in India, dryland farmers still use the nonhybrid forms.

Today, hybrid pearl millets are being planted in Kansas and Georgia. They are half the normal height—only a meter or so tall—and are capable of producing more than 3,000 kg grain per hectare. Their short stature and uniform growth make them amenable to harvest by combine. Commercial varieties are now being released to farmers.[11]

APOMICTIC TYPES

As is well known, hybrids have the limitation that farmers must buy new seed every year or so. Although in many countries this is now a routine part of farming and is seldom constraining, the farmer must be able to buy the seed and the suppliers must be able to produce enough and deliver it on time for the planting season. In rural Africa that can be a problem.

Forms of hybrids that maintain their production potential from generation to generation are being developed in pearl millet (see box, page 123). These forms, known as apomictic types, are on the verge of being perfected.

TOP-CROSS HYBRIDS

Crop varieties sometimes come to disastrous ends when circumstances change or a new disease arrives. In the case of hybrids, the disaster can be particularly severe because creating a replacement is a long and uncertain process that must start afresh with new genetic material. The whole operation might well take 10 years or more of diligent and dedicated effort. But plant breeders at the International Crops Research Institute for the Semi-Arid Tropics (ICRISAT) in India have developed a strategy to keep pearl millet hybrids going indefinitely, even when new diseases arise or conditions change.[12]

Normally, hybrids are developed using two inbred parents of known and uniform qualities. ICRISAT's strategy is to replace one parent with an open-pollinated variety of broad genetic background.

The resulting products, called "top-cross" hybrids, are now being tested. So far they have yielded as well as the best of the old hybrids and yet have shown greater resistance to disease (presumably because they have a wider range of genes).

[11] In Georgia, hybrid seed is produced and sold by a company that raises chickens. It provides the seed to farmers and contracts to buy their crop. The company's incentive is that pearl millet makes a better chicken feed than maize and can be grown locally. (As noted above, summer droughts and acid soils make maize uncompetitive in this corner of the country.)

[12] See Research Contacts for ICRISAT address.

This is all well and good, but it is in the prevention of future difficulties that top-cross hybrids really shine. Should one of them ever succumb to disease, plant breeders can introduce resistance through the open-pollinated parent in just a generation or two (in, say, not more than 2 years). It is possible, therefore, to keep a hybrid strong and secure by performing parallel breeding on the open-pollinated parent as a sort of ongoing genetic preventive maintenance.

The ICRISAT plant breeders are now taking the strategy a stage further and replacing even the sole remaining inbred parent with a hybrid of broad genetic background. This means that the resulting hybrid has even more genetic variability within it. This method helps, too, in reducing the cost of seed production.

WIDE CROSSES

Pearl millet (that is, *Pennisetum glaucum*) will hybridize with a few wild *Pennisetum* species, some of them very distantly related. Crosses with close relatives produce fertile hybrids, thus permitting extensive modifications to the genomes of both. Some hybridization work has already been done involving napier grass (*Pennisetum purpureum*). Pearl millet x napier grass hybrids have been released for perennial fodder supplies in India, the United States, and various other nations.

Two wild and weedy subspecies (*Pennisetum glaucum* subspecies *monodii* and *Pennisetum glaucum* subspecies *stenostachyum*) also readily cross with pearl millet. The useful characteristics they can confer include disease- and insect resistance, genes for fertility restoration of the A_1 cytoplasm, cytoplasmic diversity, high yield under adverse conditions, apomixis, early maturity, and many inflorescence and plant morphological characteristics.

Among other possibly useful wild species are *Pennisetum squamulatum, Pennisetum orientale, Pennisetum faccidum,* and *Pennisetum setaceum*.

Pearl millet has also been crossed with species of completely different genera, including buffel grass (*Cenchrus ciliaris*).[13]

In an approach that turns normal practice on its head, at least one researcher is using pearl millet to "improve" its wild relatives. The resulting tough, resilient, almost-wild *Pennisetum* hybrids appear useful for stabilizing desertifying environments, while giving those who live there a chance to get some food.[14]

[13] Read and Bashaw, 1974.
[14] Information from G.F. Chapman.

Genetic Jewels

Pearl millet is not now used as a genetic-research organism, but potentially it could be one of the best plants for illuminating details of both traditional and molecular genetics. A by-product from such fundamental science is likely to be new forms that increase the crop's value for meeting food needs.

THE PLANT WORLD'S DROSOPHILA

As a tool for investigating genetic interactions, pearl millet has the promise to rival drosophila, the fruit fly with which researchers have plumbed the details of animal genetics since the 1930s. Consider the following.

- The pearl millet plant is robust and demands so little space that it can grow in a 5-cm pot.
- It matures so quickly that four generations a year are possible. (Some genotypes flower just 35 days after planting; others can be induced into this by employing short daylengths and high temperatures.)
- It produces masses of progeny. A single inflorescence can produce 1,000 or more seeds, and a single plant (if unrestrained) can produce 25 or more inflorescences.
- Its flowers are small but are ideally set up for genetic manipulation. Unlike those on most plants, they are receptive to fertilization *before* shedding their own pollen; researchers can therefore readily cross-pollinate a given flower or merely leave it to self-pollinate.
- Its chromosomes are large and easy to count.
- The plants resulting from cross-pollinations usually grow with pronounced hybrid vigor, so that the genetic interactions are clear.
- There is abundant natural genetic diversity: pearl millet's gene pool encompasses about 140 species or subspecies belonging to the genus *Pennisetum*.
- Many genetic states can be obtained. The different *Pennisetum* species have chromosome numbers in multiples of $x=5$, 7, 8, and 9. For each of these, there are various ploidy levels, ranging from diploid to octoploid and beyond. In addition, both annual and perennial species occur. And there are types that are sexual or apomictic (see below).

FATHERLESS GRAINS

Like most plants, pearl millet produces seeds that have the characteristics of both parents. Certain of its relatives, however, produce seeds with only their mother's genes. For them, each new generation is identical to the last.

This situation is known as apomixis. It is not unusual in wild grasses, but to introduce it into crops has been considered too complicated, too expensive, or just too far-out. However, all that is now changing. Within the genus *Pennisetum*, apomictic types have been located in a number of species. If their trait for self-replication can be transferred to pearl millet, profound benefits would result.

For one thing, with apomictic pearl millet the farmer's fields would be safe from genetic drift. No longer would pollen blowing in from wild and weedy relatives downgrade the elite varieties.

For another, seed from different apomictic varieties could be mixed, and the farmer would retain the security of natural diversity as well as the productivity of man-made varieties.

For a third, apomictic pearl millet hybrids could be propagated by seeds for an unlimited number of generations without losing their genetic edge. Farmers would no longer have to buy fresh seed every year to enjoy the benefits of a hybrid.

The apomictic types of the wild *Pennisetum* species are not themselves promising as crops. They produce few seeds and have many weedy characteristics. But their gene for apomixis can be transferred to the pearl millet plant. Indeed, significant progress has already been made transferring this gene from the wild African grass *Pennisetum squamulatum* to cultivated pearl millet.* This development could catapult pearl millet into being a leader in high-tech agriculture.

Progress is being made in finding molecular markers associated with apomixis. This association will allow researchers in the future to isolate the gene(s) controlling apomixis and possibly use them to produce true-breeding hybrids in many crops. Indeed, in this way pearl millet's genes have the potential to revolutionize food production around the world.

* This work has been performed by Wayne Hanna at the Coastal Plain Experimental Station in Tifton, Georgia. The full address, as well as a contact for more information on apomixis, can be found in Appendix G, page 342.

Recently, researchers in Zimbabwe have crossed pearl millet with 49 different accessions of napier grass (*Pennisetum purpureum*). In doing so, they generated over 200 hybrids between the two near relatives. This was done not to improve the pearl millet, but to raise the yield and protein content of the napier grass, a widely used forage. The new hybrids yield 30–40 percent more dry matter but also 20–40 percent (9 percent versus 11–13 percent) more protein than the forms presently cultivated by local farmers. (ICRISAT)

SWEET-STALK TYPES

At least two grasses—sugarcane and sweet sorghum (see page 198)—produce stems filled with sugar. Apparently, nobody thought to look for this trait in pearl millet until the 1980s, when some Indian scientists stumbled on some during a germplasm-collecting expedition in the southern state of Tamil Nadu.[15] In the area around Coimbatore and Madurai, they found types that at maturity contained more than twice the normal amount of soluble sugars.

These sweet-stalk types had long narrow leaf blades, profuse nodal tillering (with asynchronous maturity), short, thin spikes, and very small grains. They could be easily identified by chewing them at the dough stage.

The sweet-stalk pearl millet is used as a fodder that is usually harvested in September, and a subsequent ratoon crop can be taken for grain and straw. The farmers consider them to be superior feedstuffs because livestock love the sweet stalks.

TYPES OF THE FUTURE

As can be seen from the above, pearl millet contains a wealth of genetic strengths and offers almost countless opportunities for innovation and advancement. Eventually, biotechnology could have a huge impact on such a diverse crop. It could, for example, be used routinely to transfer pieces of DNA from variety to variety or from the large numbers of wild *Pennisetum* relatives (or even from other genera). Probably, it is only a matter of time before techniques for this (by using vectors or electrophoration, for example) are developed.

Such transfers are most effective when the crop's protoplasts (wall-less cells) can be regenerated into whole plants. Although it is not yet possible to regenerate protoplasts in pearl millet, it is possible to regenerate suspension cultures (including those of pearl millet x napier grass hybrids) into whole plants.[16]

Perhaps the best way to codify the enormous diversity of this crop is to create a chromosome map (see box, page 34). This is likely to help make possible all sorts of advances in pearl millet. The task should be easier than with many crops. Pearl millet is a diploid with seven fairly large chromosomes and a large number of genes that are already known and definitively mapped.

[15] They include R. Appadurai of Tamil Nadu University, Coimbatore, and S. Appa Rao, M.H. Mengesha, and V. Subramanian, of ICRISAT. Their first test was to chew on the stalk. Later, they found that Brix readings can vary from 3 to 16 percent.
[16] All information from W.W. Hanna.

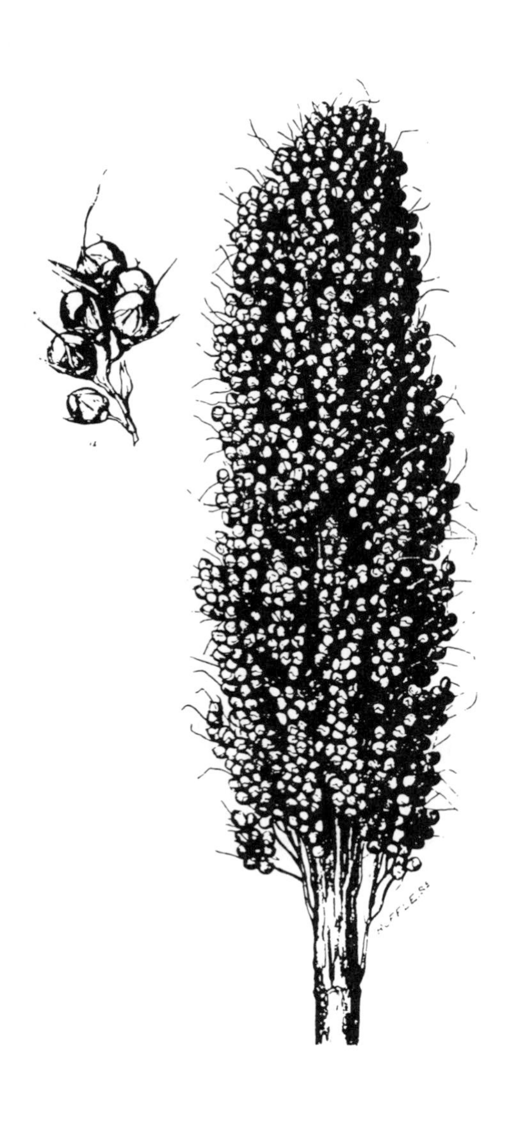

7

Sorghum

To include sorghum in a book on "lost" crops, on the face of it, seems like a gross mistake. After all, the plant is Africa's contribution to the world's top crops.[1] Indeed, it belongs to the elite handful of plants that collectively provide more than 85 percent of all human energy. Globally, it produces approximately 70 million metric tons of grain from about 50 million hectares of land. Today, it is the dietary staple of more than 500 million people in more than 30 countries. Only rice, wheat, maize, and potatoes surpass it in feeding the human race.

For all that, however, sorghum now receives merely a fraction of the attention it warrants and produces merely a fraction of what it could. Not only is it inadequately supported for the world's fifth major crop, it is under-supported considering its vast and untapped potential. Viewed in this light it is indeed "lost."

But this situation may not continue much longer. A few researchers already see that a new and enlightened era is just around the corner. Accorded research support at a level comparable to that devoted worldwide to wheat or rice or maize, sorghum could contribute a great deal more to food supplies than it does at present. And it would contribute most to those regions and peoples in greatest need. Indeed, if the twentieth century has been the century of wheat, rice, and maize, the twenty-first could become the century of sorghum.

First, sorghum is a physiological marvel. It can grow in both temperate and tropical zones. It is among the most photosynthetically efficient plants.[2] It has one of the highest dry matter accumulation rates. It is one of the quickest maturing food plants (certain types can mature in as little as 75 days and can provide three harvests a year).

[1] The amount produced is not known for certain because sorghum's production statistics (at least in some countries) are lumped together with millet's. Annual world production of the two together exceeds 100 million tons, of which 60 million is certainly sorghum. Based on the FAO figures for 1985, the number of hectares under sorghum are: Africa, 18 million; Asia, 19 million; North and Central America, 9 million; South America, 3 million. The main grain production (in millions of tons) was in the United States (28.70), India (10.30), China (an estimated 6.80), Mexico (6.60), Argentina (6.20), the Sudan (4.25), and Nigeria (3.50).

[2] Sorghum uses the C_4 "malate" cycle, the most efficient form of photosynthesis. This fundamental advantage of using sunlight efficiently is found in very few food crops—among the main ones, only sugarcane and maize.

It also has the highest production of food energy per unit of human or mechanical energy expended.[3]

Second, sorghum thrives on many marginal sites. Its remarkable physiology makes it one of the toughest of all cereals. It withstands high rainfall—even some waterlogging.[4] Recent research in Israel has shown that it also has some tolerance to salt—an increasingly useful feature for any crop these days.[5] But most importantly, it can endure hot and dry conditions. Indeed, it can produce on sites so burning and arid that no other major grain—with the exception of pearl millet—can be consistently grown.[6] Its massive and deep-penetrating roots are mainly responsible for this drought tolerance, but the plant has other drought-defying mechanisms as well. For instance, it apparently conserves moisture by reducing its transpiration when stressed (by rolling its leaves and possibly by closing the stomata to reduce evaporation) and it can turn down its metabolic processes and retreat into near dormancy until the return of the rains.

Third, sorghum is perhaps the world's most versatile crop. Some types are boiled like rice, some cracked like oats for porridge, some "malted" like barley for beer, some baked like wheat into flatbreads, and some popped like popcorn for snacks. A few types have sugary grains and are boiled in the green stage like sweet corn. The whole plant is often used as forage, hay, or silage. The stems of some types are used for building, fencing, weaving, broom-making, and firewood. The stems of other types yield sugar, syrup, and even liquid fuels for powering vehicles or cooking meals. The living plants are used for windbreaks, for cover crops, and for staking yams and other heavy climbers. The seeds are fed to poultry, cattle, and swine. On top of all that, sorghum promises to be a "living factory." Industrial alcohol, vegetable oil, adhesives, waxes, dyes, sizing for paper and cloth, and starches for lubricating oil-well drills are just some of the products that could be obtained.

Fourth, sorghum can be grown in innumerable ways. Most is produced under rain-fed conditions, some is irrigated, a little is grown by transplanting seedlings as is done with rice. Like sugarcane, it can also be ratooned (cut down and allowed to resprout from the roots) to

[3] Exceeding even maize silage, sugarcane, and maize grain. Heichel, 1976.

[4] At least some sorghums can survive standing in water for several weeks. Growth resumes when the water recedes.

[5] Information from D. Pasternak. Sorghum, however, is not as salt tolerant as several millets—selection and management will be needed to get good yields under saline conditions. See companion report, *Saline Agriculture*, for background on the importance of salt tolerance. (For a list of BOSTID publications, see page 377.)

[6] In one drought year the maize (corn) crop was so poor in Mitchell, South Dakota, that the annual "Corn Palace" had to be built out of sorghum. It was a humiliating comedown, but no maize could be found—only sorghum had survived.

provide crop after crop without replanting. It is ideal for subsistence farmers on the one hand and can be completely mechanized and produced on a vast commercial scale on the other.

Finally, sorghum is relatively undeveloped. It has a remarkable array of untapped variability in grain type, plant type, adaptability, and productive capacity.[7] Indeed, sorghum probably has more undeveloped and underutilized genetic potential than any other major food crop.

With all these qualities and potentials, it is small wonder that certain scientists regard sorghum as a crop with a great future. Undoubtedly, as the world moves towards the time when its supplies of food will be insufficient for its supplies of people, this plant will increasingly contribute to the happiness of the human race. This will happen sooner rather than later. Population is projected to almost double within most of our lifetimes. How to feed billions of newcomers on diminishing amounts of prime cropland will likely be the overwhelming global issue of the period just ahead. Obviously, vast amounts of the less fertile and more difficult lands must be forced to produce food. Moreover, if the much feared greenhouse effect warms up the world, sorghum could become the crop of choice over large parts of the areas that are today renowned as breadbaskets, rice lands, or corn belts.

In sum, it seems certain that no matter what happens sorghum will assume greater importance, especially to backstop the increasingly beleaguered food supplies of the tropics and subtropics. For a hot, dry, and overcrowded planet, this crop will be an ever-more-vital resource.

This is in fact already starting. Despite only modest international support, sorghum even now seems to be verging on a global breakout. In the United States, its yield improvements have outstripped those of all other major cereals.[8] In India, it is increasingly employed. And in Mexico, Central America, and the Caribbean—a most unexpected part of the world for this African plant—the most rapid growth of all is occurring.

Indeed, the rapidity with which Mexico has embraced sorghum is little short of spectacular. Before 1953, the crop was so little used in Mexico that, as far as international statistics were concerned, it didn't exist there. However, by 1970 it was being planted on nearly 1 million hectares, and by 1980 on nearly 1.5 million hectares. The reason is a pragmatic one: sorghum is not only cheaper to produce, it yields about twice as much grain as maize in Mexico (2,924 kg per hectare versus 1,508 kg per hectare in one recent test). Also, where rainfall is unreliable, sorghum is proving the more dependable of the two.

[7] There is such diversity in this crop that as many as 18 subspecies were once recognized by botanists.
[8] Leng, 1982.

Mexico uses most of its sorghum grain for animal feed, but it is increasingly relying on new, food-quality sorghums. These produce grains suitable for making tortillas, the round flat bread that is Latin America's staple food. In addition, sorghum is also being used to make breakfast cereals, snacks, starch, sugars, and other products that currently come from maize. It is even the basis for some (European-type) beers in Mexico, a country renowned for its brewing skills.

Although these developments demonstrate sorghum's capabilities and almost certainly portend a coming boom in production throughout much of the world, much remains to be done before this crop can truly fulfill its international potential. At present, it has several drawbacks, including the following:

- Lack of status. In global terms, sorghum is being held back by the mistaken prejudice that it is a "coarse" grain, "animal feed," and "food of the peasant classes."
- Low food value. In its overall nutrient composition—about 12 percent protein, 3 percent fat, and 70 percent carbohydrate—sorghum grain hardly differs from maize or wheat. However, sorghum has two problems as far as food *quality* is concerned. One is tannins, which occur in the seed coats of brown sorghum grains. When eaten, tannins depress the body's ability to absorb and use nutritional ingredients such as proteins. Unless the brown seeds are carefully processed, some tannins remain, and this reduces their nutritional effectiveness.

  The other problem is protein quality, which affects all sorghums, both brown and white. A large proportion of the protein is prolamine, an alcohol-soluble protein that has low digestibility in humans.[9]
- Difficulty in processing. Sorghum is harder to process into an edible form than wheat, rice, or maize.

Ultimately, none of these drawbacks is a serious barrier to sorghum's grander future, but each is a drag that—like a sea anchor in the tide of progress—is holding the crop from its destiny. Moreover, all of them can be overcome, as the following chapters demonstrate.

This plant's potential is so great that we have devoted the following four chapters to its various types. The next chapter highlights sorghum's promise for subsistence farmers—the millions in Africa and Asia (not to mention Latin America) to whom the plant means life itself. The subsequent chapter highlights commercial sorghums—the types that are increasingly grown by farmers who produce a surplus. The chapter that follows highlights specialty sorghums—unusually promising food

[9] The alcohol-soluble fraction makes up about 59 percent of the total protein in normal sorghum. The amount of this indigestible protein is lower in other cereals.

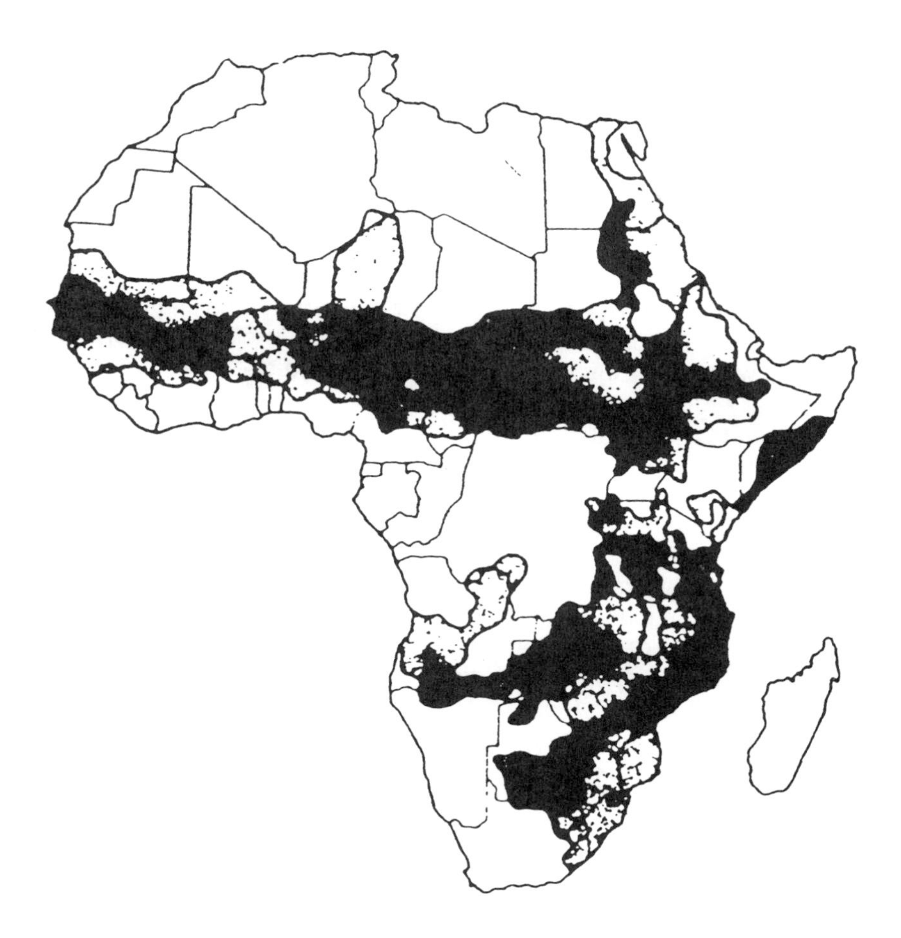

The extent to which Africa stands to benefit from sorghum research can be seen from this map. The crop is perhaps the continent's most widespread and important staple. Beyond the fact that yields can be raised far above the present average, sorghum's adaptation to a wide range of ecological conditions is an enormous asset.

Over the millennia, this ancient food was probably domesticated several times. At least four major types arose in different places. These are shown. One of the oldest, the durra (crook-necked) variety, was eaten in Egypt more than 4,000 years ago. Ethiopia is its center of diversity, and durra sorghum is still the staple food for most of the populace of the Horn of Africa. The region from eastern Nigeria through Chad and western Sudan is a center of diversity for the caudatum race. The region from western Nigeria to Senegal gave rise to the guinea race. The area from Tanzania to South Africa is the center for the kafir race. All of these separate sorghums have fed countless generations.

Africa's Gift to Mexico

The rise of sorghum in Mexico has been so spectacular that it has been called "the country's second Green Revolution." The crop has become the third largest in terms of area (after maize and beans) as well as in terms of value (after maize and cotton). Between 1958 and 1980, the number of hectares sown expanded by almost 1,300 percent and the amount of sorghum production increased 2,772 percent. More than 1.5 million hectares of sorghum were sown in 1980—more than double the amount of land planted to wheat, Mexico's first Green Revolution crop. Mexico has become the sixth largest sorghum-producing country in the world; only the United States and China used more of this originally African grain.

The fact that sorghum requires less water than maize or wheat is a significant advantage in Mexico, which has large areas of arid land. This has been true even in irrigated areas because the government has sometimes had to limit irrigation water owing to depleted reservoirs. Also, sorghum is now grown in some areas where irrigation has salinized the soil. It requires between two and four irrigations per year, compared to wheat's six or seven. Although average yields per hectare are not as great as those of wheat, they are substantially higher than those of maize.

At the beginning, most of Mexico's sorghum was grown for animal feed. Already, this grain forms a substantial part of the diet of all the chickens, pigs, cattle, sheep, and goats that are raised in the country. Although the animal feed industry also uses

types that are now little known in a global sense but that have outstanding merits for the future. Finally, there is a chapter on sorghum's promise as a source of energy as well as on other special qualities that can benefit farms and farmers.

These divisions are of course arbitrary. They are simply a convenient way to present the vast range of this plant's possibilities. There are many areas that overlap and much common ground between the different types, different purposes, and different users. In addition, major advances specifically in Africa's sorghum production are likely to come from methods and technologies that are beyond the scope of the following chapters: from controlling birds, locusts, and parasitic weeds to new approaches to milling, grain storage, and erosion control. These are discussed in appendixes A and B.

Morelos, Mexico. Farmer with his sorghum crop. (D.H. Meckenstock)

maize, barley, wheat bran, soybeans, and other products, sorghum supplies 74 percent of the raw material used in animal feed in Mexico.

Now, however, more and more food-grain sorghum is being grown (see box, page 166).

NUTRITION

Like other cereal grains, sorghum is composed of three main parts: seed coat (pericarp), germ (embryo), and endosperm (storage tissue). The relative proportions vary, but most sorghum kernels are made up of 6 percent seed coat, 10 percent germ, and 84 percent endosperm.

In its chemical composition, the kernel (in its whole-grain form) is about 70 percent carbohydrate, 12 percent protein, 3 percent fat, 2 percent fiber, and 1.5 percent ash. In other words, it hardly differs from whole-grain maize or wheat. When the seed coat and germ are separated to leave a stable flour (from the starchy endosperm), the chemical composition is about 83 percent carbohydrate, 12 percent protein, 0.6 percent fat, 1 percent fiber, and 0.4 percent ash.

The nutritional components are given in the tables and charts (next page), but some of the details are discussed below.

NUTRITIONAL PROMISE

Main Components		Essential Amino Acids	
Edible portion (g)	100	Cystine	1.3
Moisture (g)	9	Isoleucine	4.0
Food energy (Kc)	356	Leucine	13.5
Carbohydrate (g)	71	Lysine	2.1
Protein (g)	12.0	Methionine	1.3
Fat (g)	3.4	Phenylalanine	4.9
Fiber (g)	2.0	Threonine	3.3
Dietary Fiber (g)	8.3	Tryptophan	1.0
Ash (g)	2.0	Tyrosine	3.1
Vitamin A (RE)	21	Valine	5.0
Thiamin (mg)	0.35		
Riboflavin (mg)	0.14		
Niacin (mg)	2.8		
Vitamin B6 (mg)	0.5		
Biotin (μg)	7		
Pantothenic acid (mg)	1.0		
Vitamin C (mg)	0		
Calcium (mg)	21		
Chloride (mg)	57		
Copper (mg)	1.8		
Iodine (μg)	29		
Iron (mg)	5.7		
Magnesium (mg)	140		
Phosphorus (mg)	368		
Potassium (mg)	220		
Sodium (mg)	19		

In composition sorghum is similar to maize. Starch is the major component followed by protein, fat, and fiber. Compared with maize, however, sorghum generally contains 1 percent less fat and more waxes. Its complex carbohydrates have properties similar to those from maize.

The protein content is quite variable. The American literature reports several instances of levels ranging from 8.3 to 15.3 (these were measured on the milo sorghum that is grown throughout the Midwest). Most samples fall in the 9 percent protein category and are almost always 1 or 2 percent higher than in maize.

However, for human nutrition sorghum protein is "incomplete." It is deficient in critical amino acids, most importantly lysine. Today's standard sorghums provide about 45 percent of the recommended lysine requirement.

Although a primary food for millions of Africans, Asians, and Latin Americans, sorghum is low in protein digestibility. It must be properly processed to improve its digestibility. It is perhaps for this reason that much of Africa's sorghum is subjected to fermentation before it is eaten.

Carbohydrates

Carbohydrate is the grain's major component, with starch making up from 32 to 79 percent of its weight. The remaining carbohydrates are largely sugars, which can be quite high in certain rare varieties of sorghum grains.

The starches in most sorghums occur in both polygonal and spherical granules, ranging in diameter from about 5μ to 25μ (average 15μ). Chemically, the starch is normally made up of 70–80 percent branched amylopectin (a non-gelling type) and 20–30 percent amylose (a gel-forming type). However, some sorghum starches contain as much as 100 percent amylopectin; others, as much as 62 percent amylose.

In its properties, sorghum starch resembles maize starch, and the two can be used interchangeably in many industrial and feed applications. When boiled with water, the starch forms an opaque paste of medium viscosity. On cooling, this paste sets to a rigid, nonreversible gel. The gelatinization temperature ranges from 68° to 75°.

Protein

Sorghum's protein content is more variable than that in maize and can range from 7 to 15 percent.[10] In most common cultivars, as mentioned above, the kernel contains about 12 percent, which is 1–2 percentage points higher than maize.

The protein's amino-acid composition is much like that of maize protein. Lysine is the first limiting amino acid, followed by threonine.[11] Tryptophan and some other amino acids are a little higher than in maize.

The protein contains no gluten. A large proportion of it is prolamine, a cross-linked form that humans cannot easily digest. In fact, prolamine makes up about 59 percent of the total protein in normal sorghum. This is higher than in other major cereals, and it lowers the food value considerably.

In the long term, sorghums that have less prolamine may come available for routine use. A few of these more nutritious types have already been found: two in Ethiopia (see page 181) and one in the Sudan (page 183), for instance. Until such quality-protein sorghums are perfected, however, sorghum grain needs to be processed if its full protein value is to be realized.

[10] As much as 25 percent has been reported, but these appear to have been in seeds from stressed plants.

[11] Lysine provides about 45 percent of the recommended requirement. (5.44 g lysine per 100 g protein) FAO/WHO (1973).

Fat

Generally, sorghum contains about 1 percent less fat than maize. Free lipids make up 2–4 percent of the grain and bound lipids 0.1–0.5 percent. The oil's properties are similar to those of maize oil. In other words, the fatty acids are highly unsaturated. Oleic and linoleic acids account for 76 percent of the total.

Vitamins

Compared to maize, sorghum contains higher levels of the B vitamins pantothenic acid, niacin, folate, and biotin; similar levels of riboflavin and pyridoxine; and lower levels of vitamin A (carotene). Most B-vitamins are located in the germ.

Pellagra—a disease caused by too little niacin in the diet—is endemic among certain sorghum eaters (as it is among some maize eaters).

Minerals

The grain's ash content ranges from about 1 to 2 percent. As in most cereals, potassium and phosphorus are the major minerals. The calcium and zinc levels tend to be low. Sorghum has been reported to be a good source of more than 20 micronutrients.

Nutritional Concerns

Recently, the status of sorghum's future as a global food was thrown into disarray by nutritional experiments conducted on malnourished children in Peru. The conclusion was reached that sorghum was "unfit for human consumption."

Part of the problem was due to the fact that the samples used in Peru came from milled flour (comprising only the grain's endosperm) and they were merely boiled into porridge and fed directly. In Africa, by contrast, the whole grain is ground up (so that the protein- and vitamin-rich germ is also included) and often some form of fermentation is also employed.

At the heart of the issue of sorghum's nutritive effectiveness is the above-mentioned fact that almost 60 percent of the protein is in the highly cross-linked form called prolamine. Human digestive enzymes are unable to break up this indigestible protein. Even bodies desperately in need of more muscle, enzymes, blood, and brain continue passing prolamine that might otherwise provide the necessary amino acids.

However, sorghum has a second problem as far as food quality is concerned. Tannins, which occur in the seed coats of dark-colored sorghum grains, block the human body's ability to absorb and use proteins and other nutritional ingredients. Unless the grain is a low-tannin (yellow or white) type or unless brown seed coats are carefully removed, some tannins remain, and this reduces sorghum's nutritional effectiveness.

Yet a third problem is that when sorghum grain is germinated, a cyanogenic glucoside is formed. In the shoots, enzymes act on this to produce cyanide. This is a potential hazard only with germinated sorghum, and not with the grain itself.

SPECIES INFORMATION

Botanical Name *Sorghum bicolor* (L.) Moench

Synonyms *Sorghum vulgare* Pers., *S. drummondii*, *S. guineense*, *S. roxburghii*, *S. nervosum*, *S. dochna*, *S. caffrorum*, *S. nigricans*, *S. caudatum*, *S. durra*, *S. cernuum*, *S. subglabrescens*.

Common Names
China: kaoliang
Burma: shallu
East Africa: mtama, shallu, feterita
Egypt: durra
English: chicken corn, guinea corn
India: jola, jowar, jawa, cholam, durra, shallu, bisinga
South Africa: Kafir corn
Sudan: durra, feterita
United States: sorghum, milo, sorgo, sudangrass
West Africa: great millet, guinea corn, feterita
Middle East: milo

Description

Sorghum comes in many types. All, however, are canelike grasses between 50 cm and 6 m tall. Most are annuals; a few are perennials. Their stems are usually erect and may be dry or juicy. The juice may be either insipid or sweet. Most have a single stem, but some varieties tiller profusely, sometimes putting up more than a dozen stems. These extra stems may be produced early or late in the season. Plants that tiller after the harvest has occurred can be cut back, allowed to resprout, and grown without replanting (like sugarcane).

Soil permitting, the plant produces a deep tap root (see picture, opposite). However, a large number of multibranched lateral roots

For a plant with such a modest leaf area, sorghum's roots are huge. This underground "survival tool" seeks out moisture deep in the soil, equipping the crop for good growth in semiarid climates. The resulting ability to yield grain under dry conditions makes sorghum a crucial tool in the fight against world hunger. (A.B. Maunder, courtesy DeKalb Plant Genetics)

occupy the upper soil levels, particularly the top meter. They can spread laterally up to 1.5 m.

The leaves look much like those of maize. A single plant may have as few as 7 or as many as 24 leaves, according to cultivar. At first they are erect, but later curve downward. During drought they roll their edges together. Rows of "motor cells" in the leaves cause the rolling action and provide this unusual method of reducing desiccation.

The flower head is usually a compact panicle. Each carries two types of flowers: one type has no stalk (sessile) and has both male and female parts (perfect); the other is stalked (pedicellate) and is usually male (staminate).

Pollination is by wind, but self-pollination is the rule. The degree of cross-pollination depends on both the amount of wind and the panicle type, open heads being more liable to cross-pollination than compact ones.

Grains are smaller than those of maize but have a similar starchy endosperm. Most are partially covered by husks (glumes). The seed coat varies in color from pale yellow through purple-brown. Dark-colored types generally taste bitter because of the tannins in the seed coat. The endosperm is usually white and floury as in normal maize, but in some types the outer portion is hard and corneous, as in popcorn.

The crop is always grown from seed. Some seeds show dormancy and will not germinate for a month or so after harvesting. It is a little-known fact that the plant can also be propagated by stem cuttings: nodes along the stem have tissues (primordia) that can produce both roots and sprouts and thereby grow a new plant.

Sorghum is a diploid ($2n = 20$).

Distribution

This African crop is now known almost worldwide. Dhows, which have been crossing the Indian Ocean for some 3,000 years, probably first carried it away from Africa and took it to India more than 2,000 years ago. It was almost certainly put on board as seamen's provisions. The sorghums of India are related to those of the African coast between Somalia and Mozambique.

Sorghum probably traveled overland from India and reached China along the silk route about 2,000 years ago. It might also have gone by sea directly from Africa: Chinese seamen reached Africa's east coast more than about 1,000 years ago (probably in the eighth century AD), and they may well have carried some seeds home. Cross-pollination with a wild Chinese sorghum[12] seems the most likely reason why the

[12] *S. propinquum*, a diploid member of the *Halepensia* group.

Jowar

For perhaps 20 centuries, sorghum has been a staple of South Asia. Today, for example, it occupies at least 20 million hectares in India, more area than any other food crop except rice. In monetary terms "jowar," as it is locally called, is perhaps India's third most valuable food plant, exceeded only by rice and wheat.

Outsiders have often dubbed this African grain "the great millet of India." And no wonder. Jowar is an important food over much of the country, and especially in the dry areas of the central and southern states. Millions of Indians eat it. Some use it like rice, but most jowar is milled into flour. More or less white in color, this flour is used especially for making traditional unleavened breads (*chapatis*). Usually the whole-grain flour is employed, but some jowar is also polished to remove the germ and create a flour with a long shelf life. This can be blended with wheat flour (up to 25 percent) for preparing even Western-style raised breads.

Jowar grain is also malted (germinated), and in this form it finds its way into various processed products, including beer and baby foods. The grains of certain varieties pop like popcorn when heated. Indians eat the light and tasty product directly or as a flavoring in baked goods.

And sorghum feeds more than just India's people: its stalks are a major source of fodder. According to some reports, nothing can match its combination of high yield and nutritional quality. Varieties with juicy, sweet stalks have been developed. Cattle find those particularly delicious.

Perhaps 80 percent of India's cultivated sorghums are those (known as "durras") that are the dominant type in Ethiopia, North Africa, and along the Sahara's southern fringes. Many improved strains have been developed. They are grown mainly in the plains and rely on the summer rains, although some are grown under irrigation.

Jowar is notably important on the black-cotton soils, which are notoriously difficult to farm. It is one of the few crops that withstands the wildly fluctuating water tables that produce bottomless mud in the wet season and something resembling cracked concrete in the dry. An ability to extract moisture from deep in the heavy vertisol clay is among the crop's greatest qualities for India.

sorghum now found in China (the kaoliang group) has its own distinctive character.

Broomcorn was first grown in Italy in the 1600s and later spread elsewhere in southern Europe. This form of sorghum has produced most of the Western world's brooms and brushes ever since. Today, Mexico is a major producer.

Horticultural Varieties

This crop comes in such an array of widely different types that various botanists have previously recognized 31 species, 157 varieties, and 571 cultivated forms. However, these all cross readily and without barriers of sterility or differences in genetic balance, so it seems preferable to group them into a single species, *Sorghum bicolor*. Some botanical authorities also include certain wild sorghums, designating them as varieties within the species.

The ease with which cultivated sorghums cross with wild species (such as *S. arundinaceum*) may be a headache for the taxonomist, but it provides great scope for the plant breeder. Indeed, to synthesize new cultivars, a vast range of genetic characters can be brought together in bewildering numbers of combinations. As a result, many cultivars are recognized in Africa, India, the United States, and elsewhere, and new ones are being continually produced (see later chapters and notably page 191).

Environmental Requirements

Sorghum is adapted to a wider range of ecological conditions than perhaps any other food crop. It is essentially a plant of hot, dry regions but takes cool weather in stride and may also be grown where rainfall is high and even where temporary waterlogging can occur.

Daylength Although many cultivars are insensitive to photoperiod, sorghum is basically a short-day species. Most traditional varieties differentiate from vegetative to reproductive growth when daylengths shorten to 12 hours. This switch to flowering often happens just when the rains diminish, and the crop matures in the dry season that follows, a feature that greatly helps the farmer. Some of these traditional forms are extremely susceptible to photoperiod and reach impossible heights if not planted as daylengths shorten. On the other hand, the dwarf sorghums of the temperate zone are unaffected by daylength and can be planted year-round where climates permit.

Rainfall Although part of the crop is grown in rainy regions, sorghum is remarkably drought-resistant and is vitally important where the climate is just too dry for maize.

Altitude Sorghum is grown from sea level to above 3,000 m.

Low Temperature The plant is killed by frost. Optimum growth occurs at about 30°C.

High Temperature It is essentially a plant of the tropics or subtropics, roughly between 40° of the equator. However, in the United States it is being pushed ever farther into the cooler latitudes.

Soil Type Sorghum tolerates an amazing array of soils. It can grow well on heavy clays, especially the deep-cracking and black cotton soils of the tropics. It is equally productive on light and sandy soils. It can withstand a range of soil acidities (from pH 5.0–8.5) and tolerates salinity better than maize.

Sweet Sorghum

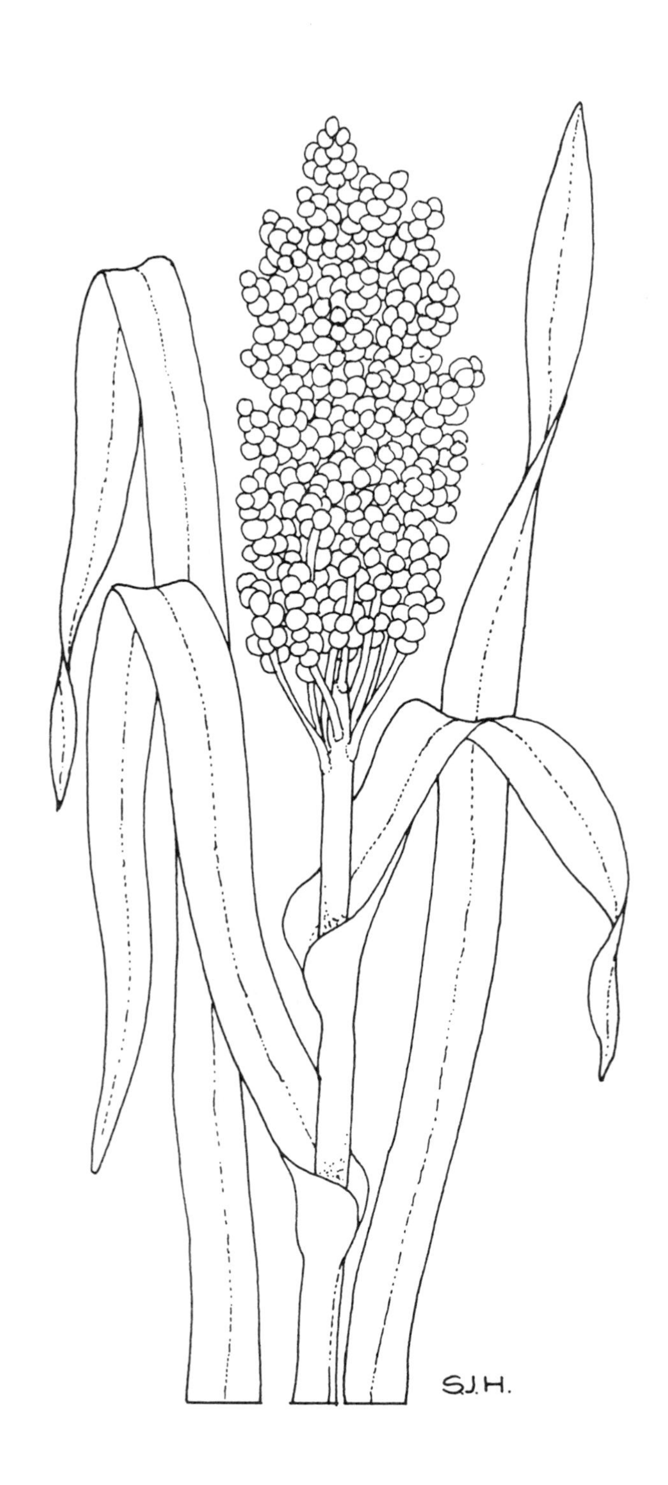
S.J.H.

8

Sorghum: Subsistence Types

Of all Africa's cereal grains, sorghum is the most important. It shares top billing with pearl millet in the drier zones and with maize in the wetter ones. In fact, Africa devotes more hectares to sorghum and millet than to all other food crops combined.

And sorghum is more important than the bald figures indicate. It is crucial to a substantial portion of the millions who coax from their meager and often declining lands barely enough to sustain life. Many—perhaps most—of those who grow it could hardly survive without this plant. For them, it provides the dietary energy and nutrients that make the difference between health and hunger.

Sorghum is vital in this way for the majority of the most poverty-stricken people in two huge belts that together look like the number 7 superimposed on the map of sub-Saharan Africa. One belt—spanning some 8 degrees of latitude (from approximately 7° to 15°N)—stretches like a giant sash across West Africa from Senegal to Chad. The other, equally huge, runs north to south covering the drier areas of eastern and southern Africa from the Sudan to South Africa (see map, page 131).

The recent past has not been kind to these two vast regions—especially the first. To many observers the picture is already bleak and getting bleaker. The sorghums that provide the subsistence for tens of millions yield on average less than 700 kg per hectare—sometimes much less. Yields have improved little or not at all in decades. Some observers question whether technology can ever make a difference.

The reasons are not unclear. Africa's farmers face a formidable web of interlocking constraints. There are constraints imposed by nature, which seems to take special delight in bedeviling Africa. There are constraints imposed by society and tradition. There are constraints imposed by poverty. And there are constraints imposed by politics, incompetent government, poor roads, and other infrastructural impediments. Subsistence farmers must somehow survive and produce their crops within all of them.

If the constraints were the same throughout Africa, they might be manageable; but they differ in degree and kind from farmer to farmer, village to village, valley to valley, and nation to nation. With all these localized and varying limitations, some people conclude that unified advances of the Green Revolution type that swept across India and Pakistan in the 1960s are inapplicable. Perhaps a different approach is needed.

Actually, that approach might come from Africa's own subsistence sorghums. During thousands of years, farmers have selected varieties to match their local conditions and food preferences. These traditional types are already remarkable for their diversity. In Sukumaland in Tanzania, for instance, a single researcher once counted 109 named cultivars—all of them in common use. In Samaru, Nigeria, more than 100 local types have been identified. And in the Lake Turkana area of Kenya there is such a variety of distinctly colored sorghums that just by looking at a grain, farmers claim that they can identify who grew it—a form of "natural bar-coding" that is said to ensure against theft.[1] For Africa as a whole, the number of distinct sorghums must range into the many thousands. Some have been reverently handed down from generation to generation.[2]

These traditional sorghums are not only varied, they can have remarkable qualities. Perhaps centuries of careful observation have gone into their selection. They incorporate features such as:

- Good seedling emergence and strong early root development (to compensate for the normal brevity of the early rains);
- Good tillering (to compensate for erratic early rains as well as shoot-fly attack);
- Long growing cycles (to make best use of infertile soils);
- Resistance to insects (particularly headbugs);
- Resistance to molds; and
- Tolerance of bird pests and striga, a parasitic plant that is an impossible pest in certain regions.[3]

In addition to the agronomic qualities mentioned above, subsistence sorghums have been carefully selected for features that affect the appearance, texture, taste, preparation, or shelf life of traditional foodstuffs. They are mostly grown by women, and are used primarily in the home to prepare local foods.

Traditionally, people consume the grain as a stiff porridge (*toh* or *ugali*), a thin porridge (*uji*), or in a range of fermented beverages.

[1] Information from D.J. Lowe.
[2] All this is made possible because sorghum is predominantly self-fertilizing and a given variety retains its distinctive qualities when it is planted year after year.
[3] Both of these troublesome organisms are described in Appendix A.

Scratching a living in Nigeria's dry northern region, a farmer plants seeds in soil turned to dust. Sorghum's adaptation to a wide range of such marginal conditions is an enormous asset in a crowded world. (Lynn R. Johnson)

Ethiopians form sorghum flour into dough balls that are boiled to form a staple food (*dawa*). In Nigeria, a similar type of dumpling as well as a flaked, dried sorghum-based food are staples. Many people cook the dehulled grain like rice, or grind it into flour like wheat and make biscuits, cakes, or unleavened breads. Some make couscous out of it. Sorghum is also important for brewing native beer or *pombe*.

As has been noted, Africa has two vast sorghum belts. Surprisingly, the conditions in each are so different that varieties perfected in one are seldom useful in the other.

The following conditions prevail in East and southern Africa:[4]

- Most of the crop is planted as a monoculture and laid out in rows.
- The rainy seasons tend to be short and (in most places) to come once a year.
- The plant varieties tend to have shorter stems, tight seedheads (panicles), and relatively high harvest indexes (the ratio of grain to other tissues).
- Birds are often such serious pests that they alone determine what variety is planted, how it is managed, and what level of inputs is applied (see Appendix A).
- The main striga species (notably in southern Africa) is the Asian type (*Striga asiatica*), so that plant breeders can use genes from striga-resistant Indian sorghums.
- Sorghums for brewing and for animal feed are increasingly important.
- Both modern varieties and hybrids have been used, at least on a modest scale, and some types introduced from India have proved extremely successful.[5]

In West Africa, on the other hand, the following conditions apply:

- Little of the sorghum is grown in monoculture; most is planted in mixtures with cowpea, pigeonpea, roselle, and other crops.
- The plants are seldom grown in rows, but are scattered randomly and are often far apart. In the drier parts of this zone the land is neither plowed nor otherwise prepared before planting, except that it is sometimes weeded or burned.
- The plants tend to be tall and lanky and have a low harvest index.[6]
- The plants flower toward the end of the rains, thereby helping the grains escape fungi and sucking bugs, which are prevalent while the rains persist but disappear during the dry months that follow.

[4] Information based on Carr, 1989.
[5] In Zimbabwe, for instance, this has led to the release of SV1 and SV2, both of which have considerable promise. In Zambia, some equally useful hybrids are in the pipeline.
[6] This is hardly a grave limitation because to most subsistence farmers the stalks are a vital fodder and no less valuable than the grains.

- The rainfall can be very erratic.
- Local sorghums are able to produce grains even when severely stressed by drought. (The types grown in higher rainfall areas produce dense, vitreous grains; those grown in dry areas produce floury grains.)
- The seeds are borne in open panicles—a feature generally inimical to high grain yields but one that helps avoid grain molds.
- The main striga is *Striga hermonthica,* a native species. Most of the striga-resistant sorghums from India or even from eastern Africa are susceptible to this parasitic pest.

NEXT STEPS

Actions to open the vast and promising future of subsistence sorghums include those discussed below.

Sharing Varieties[7]

As noted earlier, truly outstanding sorghums can be found throughout Africa. Many are exquisitely fitted to specific niches for subsistence farmers. Much good could be done merely by making these more widely available. Most are now unknown beyond the valley or village where they are treasured.

Local types are well proven, and moving them within ecological zones could be a powerful way to improve the long-term stability of farm production. Even moving them across ecological zones could become important because there may be increasing climatic change and uncertainty in the future. Farmers now plant cultivars suited to the existing rainfall pattern. However, if the pattern changes (as it did in West Africa in the 1970s) then all local cultivars may become inappropriate. Materials from another area may be the only way to stave off disaster.

Strengthening Farming Methods

To improve sorghum in subsistence production, research on farming methods seems likely to yield quicker benefits than research on breeding plants for higher yield.[8] Some improvements seem simple, obvious, and uncomplicated. For example:

[7] Information from S. Carr.

[8] A 4-year on-farm trial in the early 1980s demonstrated that none of the varieties carefully bred in research trials could outperform local types over all environments when the trials were conducted in the villages. In fact, despite the worldwide sorghum breeding done to date, less than 10 percent of Africa's sorghum area is being planted to nontraditional types from research stations.

Burkina Faso. Sorghum farmer inspects his maturing crop. (H.S. Duggal, courtesy ICRISAT)

- Watering. Studies conducted over the last 20 years in Burkina Faso, for instance, suggest that a little extra water applied when the grains are filling profoundly increases grain yield.
- Fertilization. In some areas, dramatic rises in sorghum grain yields can come from providing nutrients to the soil.[9]
- Legume rotations. Many lands where sorghum grows were infertile to begin with or are now worn out. Nitrogen-fixing leguminous plants could well be the key to rejuvenating most such sites.[10]
- Weed control.
- Water-harvesting and other water-conserving techniques.[11]
- Managing the fields to reduce devastating outbreaks of striga.

Tampering with tradition must be done with caution, however. Some seemingly obvious improvements can prove detrimental in the long run. For example, it is not for nothing that West African farmers grow sorghum plants wide apart. The crop is an excellent scavenger of nutrients and will grow successfully in soils in which maize fails completely, but it must then have room to develop large root systems. Typically, agricultural advisers recommend closer plantings, but where soil fertility is the limiting factor this can reduce the yield. (Of course, if fertility levels are increased, plant populations can be too.)

Another trap for the unwary is the preparation of the land. There is a strong interaction between the choice of variety and how the land has been prepared. In the moister areas, land is cultivated and ridged before planting; elsewhere, however, the seeds are broadcast onto unprepared ground. "Improved" varieties will usually outperform local material *only* where land is cultivated. Local varieties, on the other hand, show little response and cultivating the land before planting can be a waste of time.

Breeding Better Plants

Certain sorghums that yield almost as much as the best grain crops in the world are known (see next chapter). But for helping the subsistence farmer, an 8,000-kg-per-hectare crop is not a suitable target at the present time. Maximum yield is usually not the primary

[9] Unfortunately, however, most traditional sorghums have a low harvest index, and the effects of fertilizers can be disappointing compared with those on maize, for example. The responses vary depending on the poverty of the soil, but in most sorghum-growing areas of the drier zones the yield increase generally is less than half that for maize and is too little to attract many farmers at today's grain and fertilizer prices.

[10] This subject is covered in a companion report, *Tropical Legumes: Resources for the Future*. National Academy of Sciences. 1979. National Academy Press, Washington, D.C. For a list of BOSTID publications, see page 377.

[11] See companion report *More Water for Arid Lands: Promising Technologies and Research Opportunities*. National Academy of Sciences. 1974. National Academy Press, Washington, D.C. For a list of BOSTID publications, see page 377.

A "Cure" for Sorghum Borers

It is often hard to see how to improve on crops and methods that subsistence farmers have honed to their needs for hundreds or even thousands of years. However, modern ability to probe deeply into genetics, entomology, soil science, plant physiology, and other sciences can provide insights of great potential value. Here is a recent example.

Subsistence farmers value their sorghum stalks so highly that the grains are sometimes almost a secondary consideration. The stalks are vital for building houses, for fencing, and for firewood (see page 195). But there is a risk in employing them. Larvae of the sorghum stem borer (*Busseola fusca*) shelter inside. Thus, a farmer who keeps lots of stalks around is providing a haven for his worst enemy; eventually, the larvae will turn into adults that will come out in swarms to devastate the next crop.

A Nigerian researcher, A.A. Adesiyun,* has recently been looking into this long-standing problem. By monitoring the popula-

requirement. Reliability is more important. A yield that can be relied upon year after year is the primary goal of those whose life depends on their harvests. Thus, the immediate need is to improve the yield stability, together with whatever yield increase is compatible with that stability. Average yields of only 1,500 kg per hectare would double production in Africa (not to mention India).

Crop-breeding objectives for stabilizing yields for resource-poor farmers in Africa include:

- Raising pest and disease resistance (see below).
- Boosting tolerance to drought, humidity, and other changeable environmental stresses. (These tolerances, however, are pretty high already. In many locations it would be better to breed for higher yield at the existing tolerance levels.)
- Improving grain quality, especially those qualities that are important in storage and processing.

Some of these resistances and tolerances can be bred for outside the local area. "Hot spots" have been identified for many traits of economic importance. Midge, for example, is constantly severe at Sierra Talhada in northeast Brazil; *Busseola fusca* is severe at Samaru in northern Nigeria. An appropriate network of national or regional stations in similar areas could provide a powerful method for screening

tion inside the stalks he has come to appreciate the features that affect the pest and can thereby guide farmers on how to keep their stalks and have good harvests as well.

Naturally, farmers stack the stalks out of weather during the off-season. Adesiyun has found that this is good for the bugs: in the shade only 20 percent die, and all the rest eventually emerge, eager for more sorghum to bore into. However, Adesiyun then found that just stacking the stalks out in the open doubled the number of insects that succumbed. And this was nothing compared to warming the stalks over a fire or spreading them out thinly to bake in the sun for 3 days. This killed a whopping 95 percent of the larvae sheltering inside. The stalks could then be stored safely, even in the traditional stacks in the shade. Moreover, the "cured" stalks could be used around the house or the farm with little risk of infecting the fields with hordes of hungry hoppers.

* Institute for Agricultural Research, Ahmadu Bello University, Samaru, Zaria, Nigeria.

and mobilizing masses of useful local germplasm far more rapidly than at present.[12]

For subsistence use in Africa, it is usually important to breed multipurpose sorghums. Tall plants may be anathema to a cereal breeder, but to many small-scale farmers long stalks are resources vital for fencing, thatching, firewood, and other utilitarian purposes. Those farmers will not switch to a short-stalked type no matter how high-yielding.

Raising Pest Resistance

Among the traditional sorghums of the tropics are some with good resistance to foliar diseases and excellent tolerance to most of the indigenous insect pests. However, to maintain this happy position, research must be continued, especially on the use of systemic insecticides against borers and shoot-fly.[13] Unfortunately, the natural resistance is closely related to the amount of phenolic compounds (particu-

[12] Information from S. Carr.

[13] Extracts from the neem tree are promising in this regard. See the companion report, *Neem: A Tree for Solving Global Problems*. National Research Council. 1992. National Academy Press, Washington, D.C. For a list of BOSTID publications, see page 377.

The Dilemma of Daylength

It has been a tenet of modern crop breeding that eliminating sensitivity to daylength is a good thing—the resulting varieties can be grown at many latitudes and in different seasons. But West African subsistence farmers use daylength-sensitive sorghums in an ingenious and sophisticated manner that helps ensure a harvest even in the shortest and most erratic of seasons.

The actual week when the rains will start in the Sahel is unpredictable. The rains may be early, late, or sporadic. However, when the rains will cease is much more consistent. Unfortunately, though, once the rains have stopped, the ground rapidly dries out, leaving little chance for more growth. Thus, although the start of the planting season can vary, the crop must complete its cycle by the given time when the rains come to an end.

Traditional cultivars in West Africa have been selected to flower a little before the rains end, no matter whether the rains began early, late, or on time. The length of day triggers the flowering, not the age of the plant nor the status of the rains.

Local sorghums have evolved over centuries under those austere and fluctuating conditions. They fill out quality grains even under the stress of drought and the boom-and-bust cycles caused by sporadic showers. Introduced varieties and hybrids, by contrast, are "shocked" by the sudden onset and extreme stress of a Sahelian dry season. They seem to collapse physiologically and set floury grains that are useless as food.*

* Information from J.F. Scheuring.

larly the condensed tannins), and these compounds make it harder for people to digest the sorghum grain (see page 179).

Breeders can also help stabilize yields dramatically by breeding genotypes tolerant of striga. In fact, this is vital. Any "improved" materials lacking striga tolerance could be catastrophic to farmers in areas where this parasite is serious. A striga plant produces tens of thousands of seeds, each of which can remain viable for a decade or more. If susceptible sorghums are introduced, this terrible pest could quickly get out of hand and fill the soil with seeds that act like 10-year time bombs. Luckily, there now seem to be good possibilities for identifying and breeding striga-resistant types (see Appendix A).

Improving Bird Resistance

As noted elsewhere, birds prevent farmers from cultivating the most palatable sorghums in many parts of Africa. Today's bird-resistant types have seed coats containing tannins, which are both bitter and difficult to digest. If a more satisfactory solution can be found, it could be an outstanding contribution to Africa's future, and it would certainly help boost the production of sorghum. New possibilities have recently been discovered (see Appendix A).

Increasing Mold Resistance

In many parts of Africa, molds that destroy grain in the head (panicle) are holding sorghum back. If cultivars more resistant to such damage can be found, then earlier, fast-maturing types could be grown regardless of the humidity during the harvest period. Also, types with dense panicles (a better yielding and more efficient form) could be planted where now only quick-drying open-panicle types are practical. Some strains are inherently resistant to mold regardless of panicle type; these deserve much greater research attention.

Another, relatively easy, intervention is the treatment of seeds against smuts that affect the crop in the seedling stage.

Easing the Burden of Handling

The amount of hand labor needed to prepare the land, control the weeds, and scare away the birds is a serious limit to sorghum production in African subsistence farming. These are significant barriers to increased production. Thus, a major issue raised by any innovation is how much hand labor it demands. This is important to any farmer who has to work the fields by hand. In hoe agriculture one can literally "work oneself to death" by expending more energy than he or she gets out of the harvest.

End Use

As already noted, subsistence sorghums are able to meet the complex array of local requirements. The storage life, processing characteristics, and taste of *toh, ugali, uji, dawa,* and other traditional sorghum-based foods are paramount—more important than the absolute level of yield in the field.

Features that affect traditional foods are hard for scientists to quantify and breed for, especially when the research must be done in centralized research facilities. Subsistence-sorghum breeding is made

Threshing the harvest. (Zefa Picture Library, London)

even more difficult by the fact that Africa may have as many sorghum dishes as it has cooks.

Already, sorghums improved with exotic germplasm have been rejected because the *toh* they produced didn't keep its texture long enough. (The starch gel collapsed overnight.) The sorghum program in Niger, Burkina Faso, and Mali currently uses small diagnostic tests to evaluate advanced breeding materials for *toh* keeping quality. This approach, by which the plant breeders are directed as much by food technologists and home economists as by yield in the field, is a refreshing and much-needed innovation.

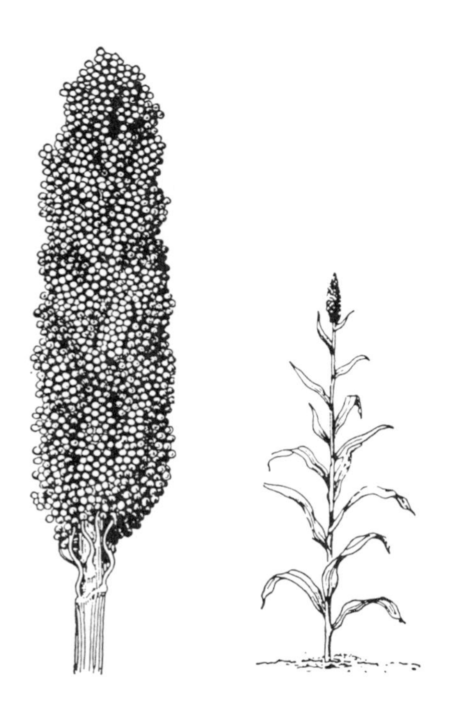

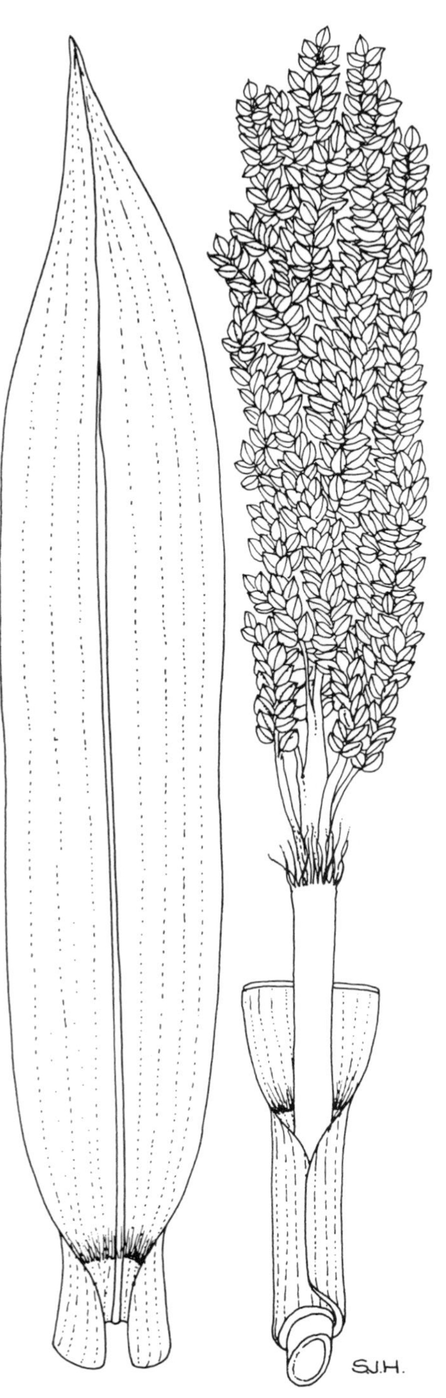

Sorghum caffrorum

9

Sorghum: Commercial Types

Today in Africa, sorghum is grown mostly for subsistence (see previous chapter). It feeds farmers and families who seldom, if ever, have any surplus to sell. But beyond Africa sorghum production is rising, mainly due to farmers who sell their grain so that others can eat. The United States, Mexico, Honduras, and Argentina are just some of the nations now taking advantage of this crop's powerful performance under pampered conditions. Indeed, it is paradoxical that while Mexico's maize is replacing Africa's sorghum in Africa, in Mexico itself the opposite is happening: sorghum is replacing maize in many areas (see box, page 133).

The commercial approach will eventually assist Africa as well. Growing sorghum the way commercial wheat and maize are grown can produce harvests of 3,000 rather than 700 kg per hectare. Indeed, the fact that sorghum has vast untapped commercial potential is important to the future of much of the world. Large areas in Central Asia, northern and central China, South America, and Australia have the potential for expanding the production of sorghum as a large-scale, high-tech competitor of the world's top three grains: wheat, rice, and maize.

Part of the problem in Africa is that so far sorghum has never been developed as a major food for urban areas. Lacking markets, it remains a crop of the small cultivator, consumed largely on the land where it is produced. But this need not—indeed should not—continue as the sole method of sorghum production. As with other crops, sorghum deserves the attention that governments give to any basic food commodity: stockpiling, purchase of surpluses, price supports, research, and policy support, for instance.

One particular restraint on sorghum has been the lack of processed foods—flour, meal, breads, or other materials—for use by those who are not farmers and are not prepared to devote hours of every day making flour from raw grain. The development of a sorghum-based food-processing industry would do much to offset Africa's shift in demand toward imported rice and wheat.

Sorghum in America

When introduced to the United States in the middle of the last century, sorghum was first cultivated on the Atlantic coast. By 1900, it had spread as far west as California. Today, Texas, Kansas, Nebraska, and Missouri are the leading producers. The crop's value now averages about $1.1 billion annually. Much is exported. In 1990, the United States shipped 7,239,000 tons of grain sorghum—almost half of all it produced. Japan was the largest buyer, followed by Mexico.

In the United States itself, grain sorghum is most commonly used as livestock feed. It is fed to cattle (both beef and dairy), poultry, pigs, lambs, horses, catfish, and shrimp. The grain has many industrial uses as well. It is used in foundry-mold sands, charcoal briquets, and oil-well-drilling mud. In addition, sorghum flour is used in the manufacture of plywood and gypsum to build houses as well as in the refining process of potash and aluminum. Some of the ethanol used to fuel American cars is made from grain sorghum.

Although it can be found from the Carolinas to California, sorghum is grown primarily in the Great Plains in the center of the United States. (One dot equals 2,000 hectares.)

Sorghum harvest on the High Plains of Texas. (A.B. Maunder)

There are good reasons for thinking that this may come about. And soon. For example, recent research has shown that sorghum grain can be parboiled to create a fast-cooking, convenient food just as has been done with rice. Also, various projects are under way to produce sorghum flour for sale in stores. In fact, in Botswana sorghum meal is already commercially available.[1] Nigeria, too, is pioneering the processing of locally grown sorghum to replace imported grains (see box).

By and large, the actions needed to boost commercial farming differ dramatically from those needed by subsistence farming. Whereas subsistence farmers may be tied (for reasons of precedent, poverty, environment, or fear of the unknown) to local varieties, commercial farmers are not. They can use newly created sorghum varieties, including hybrids and the best of research-facility results. Their grain is to be sold, probably to markets where the products of perhaps thousands of other farmers are pooled. In this case, the standard varieties demanded by the mass market take precedence, and the cash earned by selling them may pay for fertilizers and other inputs that are beyond the meager means of subsistence growers.

The evidence is persuasive that—just as in the cases of wheat, maize, and rice—sorghum responds dramatically to modern technology. For instance, although subsistence-sorghum yields have remained static at or below 700 kg per hectare, those of commercial sorghums have jumped like those of the Green Revolution crops in Asia. In the 1970s for instance, yields from India's rainfed sorghum increased 50 percent (from 484 to 734 kg per hectare) and Argentina's rose 55 percent. Irrigated yields are considerably higher: in India, about 1,800 kg per hectare is common. Hybrid sorghums can achieve even more: 4,500–6,500 kg per hectare are now not unusual yields in the United States, Europe, China, and on commercial farms in Zimbabwe.[2]

In a few cases, sorghum's yield ceiling has been raised to dazzling heights. For example, yields of 13,000 kg per hectare are being reported under special conditions in Mexico.[3] In Argentina and the United States 12,000 kg per hectare have been measured. Farmers in China are said to average 10,000 kg per hectare in certain areas.

Given such advances, sorghum's total global production may eventually match that of maize. And perhaps more important, much of the production will be at sites where maize can barely survive. This will greatly increase the food available in the world.

[1] For information on both parboiling and flour production, see Appendix B.

[2] Mean grain yields in the United States—around 1,200 kg per hectare before the release of hybrids—are now 4,200 kg per hectare.

[3] Vega, 1984.

The rest of this chapter highlights certain forms of sorghum that could help this plant reach its ultimate performance outside the confines and constraints of subsistence farming.

TYPES FOR ALL SEASONS

Sorghum's ultimate promise is perhaps best glimpsed in a research program in Texas and Puerto Rico. The Sorghum Conversion Project, as it is called, is a concentrated research effort that has catalyzed much of the present improvement in sorghum. It changes tall, late, or nonflowering varieties that produce well only in the tropics into short, early-maturing forms that can be used in many parts of the world, including the temperate zones. Its materials are already opening new horizons in sorghum production. Indeed, it is these materials that have led to the big jump in sorghum production in the United States, Mexico, Central America, parts of South America, and elsewhere. All in all, the result of this project could be one of the most significant advances in food production of this era.

In essence, the conversion program has vastly enhanced the source material available to sorghum breeders. It provides seeds of hundreds of types that are not only productive and adaptable, but also contain genetic resistance to insects and diseases and have desirable food qualities. Of the 1,300 lines in the program, more than 400 have been "converted" as of 1991.[4] These select lines are being used to develop gene pools from which breeders can draw genotypes that best fit their local needs and environments.

This development is described in more detail in the box (see page 172).

HYBRIDS

In the 1930s, America's maize yields were static. With the advent of hybrids, however, yields doubled and redoubled in just two decades. Maize quickly became not only a food but a "living factory," yielding feeds, sweeteners, starch, oil, and myriad industrial raw materials. It rose to such importance that today the U.S. economy would collapse without it.

Sorghum hybrids have much the same inherent potential, as their brief history shows. The first was produced only in 1957, but the effect was electric. Within 4 years, almost all American sorghum growers had switched, and the mean yield nationwide more than doubled from

[4] Information from F.R. Miller.

Texas. Field of hybrid sorghum. America's grain-sorghum production more than quadrupled between the early 1950s and the late 1960s, due primarily to higher productivity resulting from hybrids. (A.B. Maunder, courtesy DeKalb Plant Genetics)

1,280 kg per hectare to 2,750 kg per hectare. Within 10 years, as the hybrids improved, the yield had more than tripled to reach 3,810 kg per hectare. In a little over 20 years it had almost quadrupled to reach 4,190 kg per hectare. Seldom has there been such a rapid increase in grain yields in a cereal crop.

The hybrids were developed by crossing sorghums from southern Africa (the so-called kafir type) with others from Central Africa (caudatum types). The benefits come both from the hybrid vigor (which results when widely divergent strains of an organism are crossbred) and from the fact that the plant's heightened potentials and profits encouraged farmers to apply fertilizers and pesticides.

Hybrids have produced quantum jumps in production in India and Latin America as well, but so far, except in the Sudan, Zimbabwe, and South Africa, they are uncommon in Africa itself. In most of East Africa, for instance, only 5–10 percent of the crop is in the form of hybrids or other improved varieties, and in West Africa the percentage is even lower. This is not unexpected. Occasional U.S. hybrids, such as NK 300, prove productive over a wide range of conditions in Africa, but most do not. Also, most U.S. hybrids were developed for stock-feed and their grains make poor-quality foods. In addition, they lack the necessary resistance to striga, a parasitic plant unknown in most sorghum-growing areas of the United States.

These days, however, hybrids that produce food-quality grain are coming available. Moreover, it would appear that the problems of poor adaptability and striga resistance will be overcome. On the face of it, then, hybrid sorghums produced for sale rather than for subsistence should play a big role in Africa's future agriculture.

Of course, hybrids are not without drawbacks. They perform best under good production conditions and good quality control. They are suited only to sites where seeds and other materials can be readily delivered. (Farmers must purchase fresh seed for each planting.) Further, it has been found in Nigeria that during the rainy season the male-sterile plants used in making the hybrid seed are vulnerable to ergot.[5]

Some observers believe that problems such as these make hybrid sorghum appropriate for only a small part of Africa. This may be true, but as the following sections show, there are reasons for thinking that large-scale, efficient, productive, and very profitable sorghum production can indeed become a major part of Africa's agriculture mix.

[5] This fungal disease infects empty florets. It can be overcome by producing seed under irrigation during the dry season but, at least in West Africa, the areas where this is practical are limited.

Honduran Hybrids

The fact that sorghum hybrids can eventually benefit Africa and other regions is suggested by recent experiences in Central America, where a special kind of hybrid has been developed for peasant farmers.

Farmers in Honduras have in recent decades planted more than 60,000 hectares of sorghum, but harvested less than 1,000 kg of grain per hectare—the lowest yield in Central America. This may not be surprising considering that more than 90 percent is grown on marginal land and the varieties are nondescript landraces (locally called *maicillos criollos*).

These "mixed-breed" strains of unknown ancestry are low yielding, monstrously tall (3–5 m), and late maturing. On face value they should be replaced. However, the farmers resist. As with peasants everywhere, yield is not their top priority. The diverse "mongrels" are preferred because they are dependable. Also, they mature later than maize so that farmers growing the two crops together have time to harvest both conveniently.

But now a big change is beginning. Now researchers have crossbred *maicillos criollos* with elite germplasm from overseas.[6] This has produced new, souped-up forms of the traditional types, called *maicillos mejorados* (improved indigenous varieties) or *maicillos enanos* (dwarf indigenous varieties). They are still basically the dependable, convenient types of old, but the new genes have reduced their height, improved their disease resistance, and increased their yields.

These slightly renovated rustic relicts, still retaining the qualities that farmers value, have broken through the yield plateau that for years strangled greater sorghum production. Improved *maicillos* (the word means "little maize" and reflects the fact that sorghum and maize are not too distantly related) yield 24–58 percent more than their ancestors, even when little or no fertilizer is applied.

A second phase is now beginning. It involves hybrids made by crossing two local landraces. Although hybrid sorghums have been known for four decades, they have previously been made by crossing only elite parents. Honduras is unique in using the local "mongrels" as parents. For purposes of producing the necessary hybrid seed to sell to farmers, researchers there have created dwarf lines that can be mechanically harvested using combines. In trials throughout Honduras the resulting *maicillos* hybrids have outyielded the traditional landraces by 100 percent. Some have produced 6,000 kg per hectare under dryland conditions. The plants are taller than their dwarfed parents

[6] The elite materials were from Texas A&M University and ICRISAT and mainly comprised types developed in the Sorghum Conversion Project. The *maicillos criollos* were collected throughout Honduras, Guatemala, and El Salvador.

(because of complementary height genes), and they can still be used in the traditional maize/sorghum intercropping system.[7]

Unlike other new technologies that tend to benefit the affluent and progressive farmers most,[8] hybrid *maicillos* are targeted for the poor and less venturesome. They provide an alternative that may increase yields on perhaps 235,000 hectares throughout Central America. Cost to the farmer? Negligible, according to the researchers involved. The seed needed to plant a hectare (when cropped with maize) costs no more than a chicken or two, or about a third of the cost of a bag of fertilizer.

"Vybrids"

The criticism most commonly aimed at any hybrid crop proposed for poor farmers is that its seed is worthless for replanting. The fact that farmers must purchase new seed each year is often seen as a disastrous financial burden. Much of the criticism has been overemphasized.[9] However, in many developing countries logistical logjams and supply bottlenecks do make it difficult to produce hybrid seed and get it to the farmers on time and in good condition.[10]

With sorghums, however, there is a distant possibility of having the best of both worlds—to grow hybrids that also produce seed that can be planted. These so-called "viable hybrids" or "vybrids" are not yet available, but a few sorghum researchers are hot on their trail.

Vybrids are made possible by the fact that certain rare sorghums are apomictic—they produce offspring without the male and female gametes fusing. In other words, their seed arises from a nonfertilized nucleus, and for this reason each plant produces progeny genetically identical to itself. This special clonal propagation through seed retains the benefits of hybrid performance while not requiring a highly developed industry to produce and distribute seed each year.

The theoretical possibility of producing viable hybrids in crops was discussed as early as the 1930s. Nearly 60 years later, the various

[7] Information from D.H. Meckenstock.

[8] Although this is a widespread belief, it seems to be true mainly in the initial phase. After a new technology is established, all farmers—even the poorest—eventually benefit. In fact, despite much rhetoric to the contrary, the poor farmers of Asia benefited more from the Green Revolution than the rich ones—they got life itself when the widely predicted famines never materialized.

[9] For example, no hybrid can survive in the marketplace unless its improved performance and the farmer's increased income far outweigh the cost and bother of purchasing seed. Also, experience in India and in Africa is showing that farmers are fully prepared to pay as long as the cost is justified by the hybrid's performance. In addition, it takes very little sorghum seed to plant a hectare. Compared to maize, the cost should be much less.

[10] This, too, is often overemphasized. Hybrid maize is a success in Kenya and Ghana, for example. However, getting good seed to the right place and on time is likely to remain a real constraint in most African nations, at least in the near term.

Sorghum Beer

In Africa, as in many parts of the world, brewing uses vast amounts of grain. However, in Africa the raw materials are sorghum, maize, pearl millet, and finger millet, not barley, rice, or wheat. Also, the basic process is unique. African brewing includes a lactic-acid fermentation, known as souring. And the resulting beverage is something like a fermenting gruel and has the consistency of malted milk.

Normally called "sorghum beer" or "opaque beer," this drink already constitutes a considerable part of the diet in many areas, and it will likely become an ever bigger commodity. With so many people moving into the cities, it is even now shifting from an exclusively family enterprise to an industrialized one. In South Africa, for instance, sorghum-beer brewing is already a highly specialized industry. Annual production is about one billion liters.

Malting is the first step in brewing this or any type of beer. The grain is soaked and left to germinate. This activates amylases and other enzymes that hydrolyze the grain's starch and proteins to sugars and amino acids. After several days, when germination is complete, the sprouted grains are dried, ground to a coarse powder, mixed with cold water, and added to a preparation of ground-up grain that previously has been steeped in boiling water.* The enzymes continue working, this time turning the new source of starch into sugar. The souring process also takes place as bacteria act on part of the sugars to form lactic acid. The product—a thin gruel called "sweet wort"—may be drunk after less than a day. Its alcoholic content is negligible, but it contains some B vitamins and it is often given to children.

If the brewing is continued, various yeasts multiply, and within a day or so fermentation begins. This produces alcohol, B vitamins, new proteins, and more lactic acid. The resulting brew is normally drunk after 4 or 5 days. Suspended particles of starch, yeast, grain, and malt give it the characteristic milky body. High acidity (resulting from the lactic acid) prevents the growth of pathogenic microorganisms.†

Brewing raises the nutritional value of sorghum. It adds vitamins, neutralizes most of the tannins, hydrolyzes the starch to more digestible forms, and increases the availability of minerals and vitamins. South African studies indicate that iron is 12 times more available in sorghum beer than in a boiled sorghum gruel; riboflavin may be almost twice and thiamine almost a third more available; niacin's availability remains unchanged. In principle, 2 liters of sorghum beer could supply a person's daily requirement for thiamine and riboflavin and 40 percent of the requirement for niacin. However, many of these B vitamins are locked up in the yeast cells and cannot be digested unless the beer is first boiled. Unfortunately, this is seldom done.

Special varieties of sorghum are maintained for their brewing qualities. In many places, the dark brown grains are prized. Their most important characteristic is their high level of amylase activity. They have considerable potential as substitutes for barley, even for brewing lager-type (European-style) beers. Nigeria already uses them this way, at least on a semi-commercial scale.

Recently in South Africa, the Council for Scientific and Industrial Research (CSIR) developed three shelf-stable brewed-sorghum products: a pasteurized bottled beer, an aseptically packed still beer, and a wort concentrate that can be diluted and fermented to produce beer. These are safe to transport, and can be distributed to remote areas or even exported.‡

In South Africa, sorghum beer is the basis of a giant company that was formerly part of a government monopoly but has now been spun off to African entrepreneurs with amazing success (see box, page 305).

The beer is more than a mere drink. As one writer has stated: "The whole social system of the people is inextricably linked up with this popular beverage: the first essential in all festivities, the one incentive to labor, the first thought in dispensing hospitality, the favorite tribute of subjects to their chief and almost the only votive offering dedicated to the spirits. Beer is a common means of exchange or payment for services rendered, and in times of plenty it is not only freely consumed, but often is the principal or sole food of many men for days on end. It is evident in all ritual and ceremonial occasions binding together different groups or individuals and affecting a reconciliation when things go wrong. With most tribes, harvest thanksgiving takes the form of beer, preceded by an offering of beer to the ancestors of the chief."

* Grain for this purpose is usually sorghum or maize, but other grains and even banana are also widely employed. The boiling water gelatinizes the starch, rendering it readily hydrolyzable by the malt amylase enzymes.

† These beers are safer to drink than water because, at consumption, their pH is between 3 and 4, an acidity level at which no common pathogenic microorganisms grow. However, sorghum beer is not immune to spoilage; acetic-acid-producing bacteria (as in wine) can quickly turn it to vinegar.

‡ Information from John R.N. Taylor, Brewing and Beverage Program, CSIR.

attempts at producing them have resulted in some progress. One notable success has been in breeding buffel grass (*Cenchrus ciliaris*)—a native African species, distantly related to sorghum, that is used as a forage throughout the tropics. Another has been with forage grasses of the genus *Dichanthium* (*Bothriochloa*).

Work on sorghum apomixis has now reached the stage where apomicts and vybrids from crosses between them have been formed in research facilities. The scientists are confident that the vybrids can now be developed for farm use.

Vybrids will benefit more than farmers. For sorghum breeders of all stripes, vybrids offer exciting potential. Sexual types can be used in the normal way to develop hybrids with superior characteristics and then induced into apomictic forms that will retain the new qualities, generation after generation, from then on.

STRIGA-RESISTANT TYPES

One of the tragedies facing Africa is that a parasitic plant is cutting it off from the wealth of sorghums that have been, or are being, developed in a score of countries overseas. Indeed, striga is probably the greatest constraint to the production of foreign sorghums in Africa itself.

Recently, however, researchers have discovered a striga-resistant gene in sorghum. This could be a big breakthrough. For Africa, it will help open the door to the truly remarkable types developed in the Americas and China, for instance.

This topic is treated in Appendix A. It is made suddenly more relevant because a new test has been developed that can determine, within a few days, whether a certain sorghum (or other species) is resistant to striga. Tests in laboratories and greenhouses have been most encouraging. Should these results also prove practical in the field, it could open the way for overcoming the depredations of this vegetative parasite that victimizes desperately needed food plants. For the first time the crops will have the means to defend themselves.

DWARFS

The last 40 years have seen dramatic increases in the yields of wheat, rice, maize, and some other cereals. This has come not from boosting the plants' overall growth (as most people may think), but from rearranging their architectures so that the plants are shorter. With less energy going into stalk, more is left for growing grain. In

Sorghum researcher, Gebisa Ejeta, examines experimental plots of striga-tolerant cultivar SRN-39 (left) and local cultivars in Niger. The pernicious parasite occurs only among the striga-susceptible local type (right). Striga-tolerance has been known in India and East Africa, but SRN-39 is one of the first examples of a sorghum that can withstand the West African striga species. For more information on striga and the problems it causes, see Appendix A, page 276. (D. Rosenow)

Sorghum's Miraculous Conversion

At first glance, sorghum seems almost impossibly diverse. Seed banks hold more than 25,000 samples, all distinct and all able to produce fertile intercrosses. How to extract from the myriad combinations the particular ones most useful worldwide is a monumental problem that might seem beyond the realm of reason. However, remarkable progress is already being made, thanks to a project that exemplifies how many of the cereals in this book might be advanced in the future.

In the 1950s, U.S. Department of Agriculture scientist Joseph Stevens developed a "blueprint" for systematically enhancing the genetic base of the world's sorghum crop. Along with several colleagues in the United States and India, he began assembling, evaluating, characterizing, and classifying a base collection of sorghum samples. This collaborative effort was carried out in India and continued into the early 1960s. The Indian government, as well as dozens of African and Asian countries, contributed their germplasm and support. Eventually, about 11,000 different sorghums were on hand.

As a first step in sorting useful genetic materials out of the vast sorghum collection, a unique "shuttle-breeding" procedure was devised. The breeders produced and grew a first generation of random crossbreeds in the tropics (mainly at Mayagüez, Puerto Rico) where the days are short. They collected seeds from a wide range of the most desirable looking progeny and took them to a temperate zone (Texas) where days are long during the growing season. There, the seeds were grown out and a new generation of seeds were gathered again from the most promising specimens. This dual-latitude screening ensured that the resulting seeds (and their subsequent generations) could grow and produce grain under both tropical and temperate conditions.

The next step was to partially refine these genetically diverse populations. Again, a wide array of different specimens were grown and the most desirable selected, this time emphasizing short stature and early maturity. The final result was a cornucopia of various sorghums—all broadly adaptable to various daylengths, all short in stature, and all early maturing. Out of the myriad tall, slow, and sensitive types, suitable only for small farms in the tropics, have come universally useful types for use throughout the world, on any scale.

Although the resulting plants were selected for basic qualities, they were deliberately kept diverse. Now, that welter of gene types is being fine-tuned to meet the specialized demands of dozens of different localities. Specific characteristics now being "custom-designed" include:

- Resistance to disease (downy mildew, striga, anthracnose, and smuts)
- Resistance to insects (aphids, midges, worms, shootfly, and others)
- Resistance to stressful conditions (drought, heat, soil acidity, and salinity)
- Strong stalks (to stop the plants breaking or falling over in wet soil)
- Nonsenescence (to keep plants green and functional, even under stress)
- Twin-seed (making *both* florets in the grain-producing spikelet fertile)
- Easy threshing
- Erect leaves (to increase the amount of sunlight intercepted)
- Higher yield (more grains of good size in each seedhead)
- Greater root development (to help the plants withstand stresses)
- Faster grain filling (to reduce danger from drought and insects)
- Resistance to weathering (seeds that do not soften)
- Light colors (to make the most widely acceptable food products)
- Increased protein content (more than 10 percent)
- Superior amino-acid balance (high lysine, in particular)
- Improved flavor
- Greater digestibility
- Expanded diversity for food products (notably specialty types for convenience foods)

Materials from the sorghum conversion program are already helping transform this formerly obscure and often scorned grain into a major contributor to world food supplies. Indeed, their seeds have become cornerstones for much of the present rise in sorghum production worldwide.

All in all, the Sorghum Conversion Program has become one of the most successful plant-breeding programs ever; a model of achievement for crop scientists everywhere and with every crop. It provides populations that are reservoirs of genes, rather than a single, highly inbred variety.

technical terms, this is called raising the "harvest index." Thus, 50 years ago wheat had a harvest index of 32 percent; now it can be as much as 48 percent in some cultivars. In other words, almost half of the weight of the plant (above ground) is now grain.

Moreover, reducing the height makes the plants less likely to get top-heavy and blow over in a summer storm. In addition, the squat, strong plants are more able to benefit from fertilizer, which otherwise would make them spindly and top-heavy. And dwarfing not only boosts yields: wherever mechanical harvesting is practiced, short stature means that the seedheads can be efficiently captured by combine harvesters so that larger areas can be planted.

So far, only a few of the world's sorghums have had their architecture refashioned in this way. Nonetheless, an increasing number of short-stalked sorghums that mature at an even height and can be harvested by combine are becoming available. Most have been created in North America. Indeed, all of America's commercial grain types are now dwarfs.

Initially, sorghums in the United States were tall and had a harvest index of 21 or 22 percent (about the same as in the spindly subsistence types now grown in West Africa), but careful selection, followed by intensive breeding, has reduced the internode length. Now the harvest index for many improved types used in the United States, Mexico, and Argentina is 48–52 percent, as high as that of wheat.

Dwarf sorghums have also been created at research stations in Zambia. These local dwarfs, as well as those from overseas, could eventually usher in a new era for the continent.[11]

CONVENIENCE FOODS

As has been noted, commercial sorghum's major problem in Africa is that markets for flour and foods are undeveloped. If this were overcome, a large and healthy trade between a country's own sorghum farmers and its cities could operate to everyone's benefit. Today, ever-increasing numbers of city folk are being weaned onto wheat-flour bread and white rice, and any resulting economic benefits go mostly to farmers and traders a world away. The tragedy is that many of the city dwellers, especially newcomers, are accustomed to sorghum foods and would continue to purchase them if they could.

It is not inconceivable that Africa could produce vast amounts of sorghum flour and sorghum-based processed foods for sale in the cities

[11] Information from S. Carr. Dwarf types have also been introduced in West Africa but, so far at least, have performed poorly.

and towns (see Appendix B). This could result in opportunities for much innovation.

More than 30 years ago, for example, South African researchers developed a precooked sorghum product. They slurried raw sorghum flour with water and passed it through a hot roller that both cooked and dried it. The product proved very palatable and would keep for at least 3 months without deteriorating. Whole milk or skim milk could be used in place of water, producing a tasty flour rich in protein, calcium, and phosphorus. Processing costs reportedly were low.[12]

This is just one of many approaches by which sorghum might be produced for urbanized peoples. Many recipes using milled sorghum grits or flour have already been developed and tested by several universities.[13] And the recent development of parboiled products from sorghum could open up even more markets that could benefit millions of Africa's farmers (see Appendix B).

[12] Coetzee and Perold, 1958.

[13] These include the Home Economics Department at the University of Nairobi and the Department of Soil and Crop Sciences at Texas A&M University (see Research Contacts).

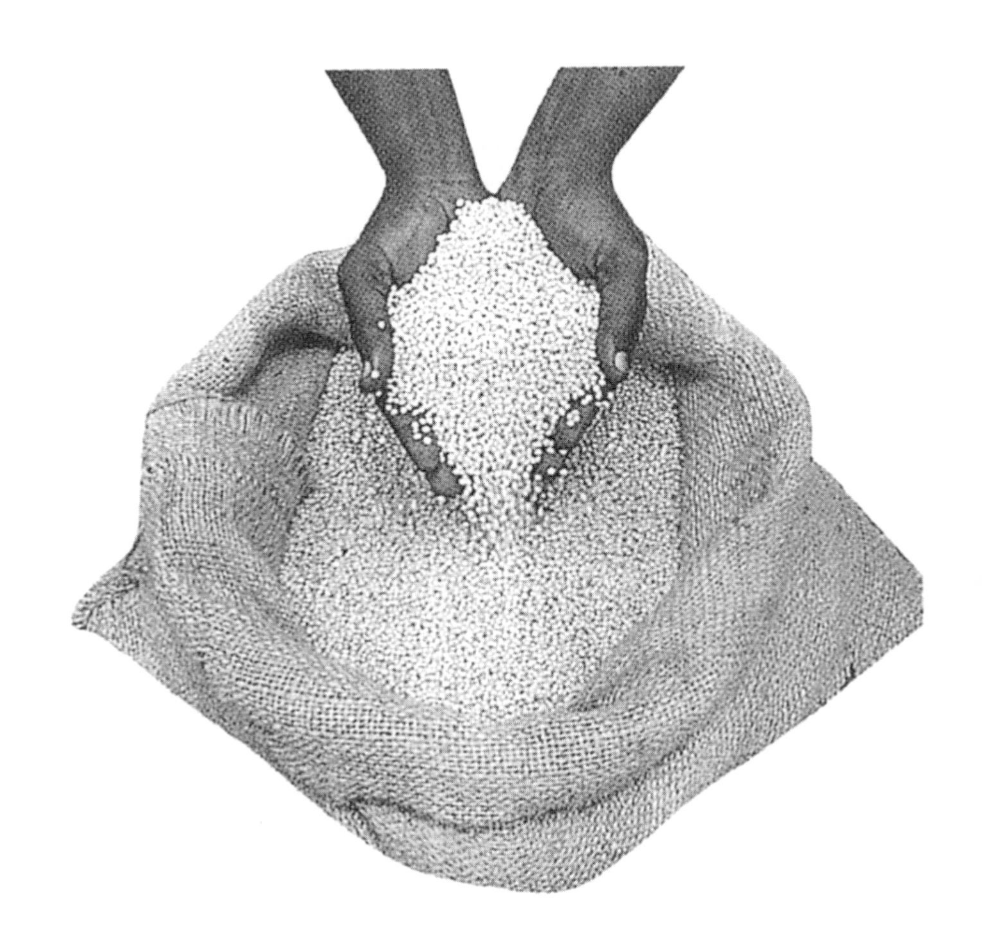

10 Sorghum: Specialty Types

Sorghum's range of genetic diversity is truly amazing. Some types look so abnormal that until recently they were classified as separate species. However, all of them cross readily with one another, all have a chromosome complement of $2n=20$, and all are recognized today as variants of the same plant, *Sorghum bicolor*.[1]

Many of the unusual types are promising resources in their own right. Some have properties and uses quite unexpected of a cereal. A few hold out the possibility of producing far better grains than those of today's major sorghums. Others could provide entirely new types of sorghum foods. Yet others can yield feed, forage, fertilizer, fiber, fuel, sugar, and raw materials for factories of many kinds. In this array of plant types, the vast potential of this remarkable species can be seen. Examples of promising, but little-known, food types are discussed below.

POPPING SORGHUMS

In parts of Africa and Asia, sorghums that pop like popcorn can be found. These have seldom received much scientific or entrepreneurial recognition. There is probably, however, a huge latent market for them. They make tasty foods, and they may have worldwide promise. Popping boosts the flavor of sorghum, and it is energy efficient and nutritionally desirable. (Compared with boiling, for instance, popping is so rapid that it takes little fuel and it denatures or hydrolyzes the proteins and vitamins only slightly.)

Popped sorghum is already a favorite in central India, and it is starting to find favor in several other countries as well. In India, people

[1] Synonyms include *Sorghum vulgare* (for the entire species complex) and *Sorghum caffrorum*, *Sorghum caudatum*, *Sorghum conspicuum*, *Sorghum arundinaceum*, *Sorghum dochna*, and *Sorghum durra* (for what are now considered subspecies, or "races"). There are hundreds of common names. Those in widespread use include: guinea corn, jowar (India), kaoliang (China), kafir corn, milo (United States), sorgho, and maicillo (Central America).

sprinkle a handful of dry grain onto a bed of hot sand or a hot sheet of metal. The popped kernels are brushed off as they form. Most are consumed by school children as a snack. They may be balled with crude sugar (jaggery). They may also be pounded into a nutty-flavored flour, which is typically mixed with milk and sugar, buttermilk, salt, or chilies.

A world collection of sorghums is maintained at ICRISAT. Of 3,682 accessions tested, 36 have shown good popping qualities. Most originated in India. These could be the starting point for breeding popping sorghums on a scientific basis. Indeed, they could create a new and very tasty food that could quickly establish itself in most of the 30 or more nations that grow sorghum as a staple—not to mention in at least that many more nations that now look on sorghum as "barely fit for cattle."[2]

As with popcorn, the best popping types usually have small grains with a dense, "glassy" (corneous) endosperm that traps steam until the pressure builds to explosive levels.

VEGETABLE SORGHUMS

In certain countries, sorghum is eaten like sweet corn. The whole seedhead (panicle) is harvested while the grain is still soft (dough stage). It is roasted over open coals, and the soft, sweet seeds make a very pleasant food. These strains are found notably in Maharashtra, India. Like sweet corn, they have sugary endosperms containing 30 percent glycogen as well as grains that shrivel when dry. They are a treat for anyone.

This unique method turns sorghum into a vegetable crop—more like broccoli than like barley. It has so far received little or no serious study from scientists, but it could be a powerful way to capitalize on the plant's ability to produce food in sites where most crops fail. The types that perform this way should be collected, compared, and cultivated in trials. The traditional processes by which they are used should be analyzed, as should the nutritional value. Seedheads in the dough stage may have a better-than-expected food value.

VITAMIN-A SORGHUM

In some developing countries a lack of vitamin A in the daily diet blinds many children. However, certain sorghums with yellow grains

[2] Even popcorn was neglected until quite recently. Although it has long been a popular treat in the United States, only in the last 10 years—since microwave ovens made it convenient for the home and office—have modern breeding techniques been applied in force. Sales have since skyrocketed. The increasing popularity of microwave ovens could also boost the use of popping sorghums.

may solve the problem, at least among sorghum-eating societies. The color comes from xanthophyll and from the carotene pigments that are vitamin-A precursors. People eating them have a better-than-normal production of vitamin A.

Yellow sorghums are especially well known in Nigeria but probably can be found elsewhere, too. The carotene levels are typically only a fraction of those normally found in yellow maize. However, because of poverty or locality, sorghum eaters often have no chance to vary their diets. Yellow varieties may be the most practical way to protect their eyesight.

TANNIN-FREE SORGHUMS

Some sorghum types contain invidious ingredients that "lock up" protein and starch so that a person's body cannot fully get at them. Traditionally, these ingredients have been called "tannins," although strictly speaking, this is not an exact term.[3]

Many sorghums, especially those now grown in East Africa, are high in tannins. To a large extent they have been deliberately selected because birds hardly touch them (see Appendix A). These birds include the quelea—a small, rather nondescript weaverbird that has replaced the locust as the most serious pest of small-grain crops in parts of Africa. This voracious seed-eater may well be the most abundant bird species on earth, and its importance as a pest has increased in recent years despite all the control operations that have been mounted against it.[4]

Today, people can eat the dark-seeded sorghums only if the tannins are first removed. There are two approaches for getting around this. One is to use the seeds in processes that neutralize tannins—making beer or fermenting the grain with wood ash are examples.[5] The second relies on the fact that the tannins are located primarily in the grain's outer layer. Milling this off makes the rest of the grain edible. This is not easy to do, however, and the seemingly endless task of pounding seeds with heavy poles causes untold hours of daily drudgery throughout most of rural Africa. Indeed, it is one of the fundamental barriers to the wider use of this crop (see Appendix B).

Overcoming the tannin problem would open new possibilities for

[3] Recent research has shown that the antinutritional ingredients are more than just the pigments known as tannins.

[4] Quelea remains a threat to crops in Zimbabwe, for instance, even though more than 521.5 million of them were killed in the country between 1972 and 1987 (an average of 32.6 million a year).

[5] Information on the wood-ash treatment is from G. Graham, who noticed that campesinos in Peru had developed it as a way to make sorghum more palatable.

sorghum as a world food grain. Research in the 1980s has demonstrated that the genes controlling tannin production can be reduced through crossbreeding. Tannins can be eliminated or at least reduced to negligible quantities. White-seeded, tannin-free types are known and are particularly promising for the future.

BIRD-RESISTANT SORGHUMS

Removing tannins makes sorghum a far better food for humans, but in parts of Africa, unfortunately, it would seem also to be good for the birds. However, some white-seeded types that are both tannin free and shunned by birds are already available.

Two sorghums that are bird resistant and free of tannin were identified in 1989.[6] These two genotypes (Ark 1097 and a Brazilian hybrid) were assayed and found to contain absolutely no tannin throughout the whole time their seeds were developing. In addition, both showed good bird resistance in trials in Indiana, USA. In Puerto Rico, where bird pressure is greater, each was damaged, but only in one of two replications; in the other, it remained untouched. All in all, these white-seeded, tannin-free genotypes appear to be slightly less bird resistant than the standard, strongly resistant, high-tannin types. Nonetheless, the level of resistance was enough that these sorghums can be very useful in areas where bird damage is normally severe.

The nutritional quality of these two is not yet fully determined, but all indications are that both are fully comparable to the low-tannin (bird-susceptible) sorghums. In a feeding trial, for example, laboratory rats grew much faster and showed more efficient feed utilization than the (high-tannin, bird-resistant) control. Remarkably, they were even better than the low-tannin types. Indeed, there were no apparent nutritional problems associated with consuming the grain.

Trials of these sorghums are under way in Kenya.

QUICK-COOKING SORGHUMS

The starches in the grains of most sorghums have gelatinization temperatures around 70°C. They must reach that temperature to become cooked and edible. However, research has shown that some sorghums have starches whose gelatinization temperature is only about 55°C. This can reduce the cooking time required. These sorghums have waxy kernels (endosperm) rather than hard vitreous ones. Thus,

[6] Information in this section from L. Butler.

they cannot always be used in the normal manner. Nonetheless, there is a good possibility that they will make nontraditional quick-cooking products that will appeal to many.

These unusual types are found especially in East Asia. The starch in their grains is entirely amylopectin, rather than amylose and other normal forms.

AROMATIC SORGHUMS

Some sorghums in Sri Lanka and northeastern India are said to have the aroma of basmati, the fragrant rice preferred by millions of Asians. Although bland-tasting rice has dominated international markets, the basmati type has always been tropical Asia's favorite, and it is now increasingly sold worldwide (even in the United States) as a high-priced specialty. The discovery of sorghum counterparts opens up similar opportunities. They, too, might become specialty foods of high value. Also, they might help boost the acceptance of sorghum—normally the blandest of grains—even where it is a staple.

All in all, flavorful types like these present good opportunities for improving markets and increasing consumption, not to mention boosting the returns to farmers.

QUALITY-PROTEIN SORGHUMS

Deep in the misty green valleys of Ethiopia's highlands is hiding a unique sorghum that, in both nutrition and palatability, far surpasses the thousands of types found elsewhere.

Ethiopians call these types "milk in my mouth" (*wetet begunche*) and "honey squirts out of it" (*marchuke*). To anyone who has tasted normal, bland, sorghum flour, the names alone indicate something special. Both varieties produce somewhat lower yields than normal but everyone likes to eat them. The taste of roasted *marchuke,* for instance, has been likened to that of roasted chestnuts. People gather the grains, roast them over a fire, and pop them down like peanuts. Both are often used to enhance the flavor of local dishes made from regular sorghums. The taste comes from the reducing sugars that caramelize as they are roasted.

Until 1973 these two varieties were restricted to a tiny upland area of north-central Ethiopia. The growers hid them in the middle of their sorghum fields (mainly so the landlords wouldn't find out and raise the rents based on the extra income from these elite types). In 1973, however, researchers analyzing different sorghums for their food value stumbled onto them.[7] Of 9,000 varieties tested, these two were unique. They contained 30 percent more protein, but more important, their protein had about twice the normal level of lysine, an amino acid critical to nutritional quality.

This finding is significant because the more than 500 million people for whom sorghum is the main source of sustenance are relying on a food that is not great, nutritionally speaking. Its protein content is modest (averaging about 9 percent), and its protein quality is among the lowest of any cereal—mainly owing to its dismal lysine level.

In the years since 1973, neither of the two quality-protein sorghums has fulfilled its promise. There are several reasons for this. Both types produce floury grains with small and soft endosperms, a feature that makes them more susceptible to birds, fungi, and insects. More important, however, soft grains are not favored for traditional purposes. Upon pounding or milling in a machine, they form a paste rather than a flour. Also, there is not much endosperm there to make a flour from in the first place.

This fundamental problem with grain type is a big barrier: either a laborious breeding program is needed to transform the grains into the hard-endosperm form[8] or people must use the soft form in foods differing from their normal grain-sorghum fare.

A promising immediate use of these remarkable varieties is as feed. Animals are less fussy than humans, and lysine-rich feeds, which are particularly necessary for pigs, are critically short in many places. Fish meal and soybean meal (the main lysine sources for livestock) are often unavailable or too expensive, especially in remote Third-World areas. High-lysine sorghum with its inbuilt robustness and drought tolerance could well become a vital feedstuff for northern China; large, dry areas of the Soviet Union; much of the Middle East; the semiarid zones of India and Pakistan; substantial portions of Mexico; and other places that are dry, salty, and lacking in lysine-rich feeds.

Moreover, the single gene responsible for the high lysine may be invaluable for boosting the quality of conventional sorghums. Researchers at several research facilities are trying to transfer this gene. They hope to enhance the nutritional value of normal sorghums without affecting the grain structure or other important traits.

SORGHOS

Sorghum and sugarcane are fairly closely related, and certain sorghums (often termed "sorghos") have stems that are just as rich in

[7] The researchers were John Axtell and Rameshwar Singh of Purdue University.
[8] This has been done with a high-lysine maize (see the companion report, *Quality-Protein Maize*).

The Super Sorghum of the Sudan

Although it is perhaps the most important grain in Africa, sorghum still has tremendous untapped potential. Many remarkable types are yet to be discovered by science, as the following example shows.

When word leaked out in 1984 that a disastrous famine was impending in Dafur and Kordofan, the horror that swept the world energized many people into action. No one took a more original approach than the organizers of "Band Aid," a project in which rock and roll stars staged a free concert for worldwide television. The donations from dozens of countries then went to help those stricken provinces of the Sudan. Part ended up in a far-sighted study of sorghum.

With Band Aid funding, David Harper, Omar Salih, and Abdelazim Nour visited 150 villages in the drought-devastated area, checking on the people's welfare and gathering samples of the local crops—especially those that had best survived the drought. A sorghum variety called "Karamaka" proved to be truly remarkable.

For one thing, Karamaka had a protein that was unusually nutritious. It had more than the normal amount of protein but, more importantly, its protein had about twice the nutritional value of other sorghum proteins. Its lysine content (3.4 percent) was 62 percent above normal, and the other essential amino acids were not diminished to any significant extent. As a result, Karamaka protein had a chemical score of 62 rather than the 30-40 figure of regular sorghum protein. Its nutritional value was therefore almost two-thirds that of milk protein, the usual standard of protein perfection.

For another, Karamaka grain possessed an unusual combination of carbohydrates, containing less starch and much more sugar than normal. Indeed, the total sugars in the grain amounted to 35 percent. The individual sugars were composed of both sucrose and reducing sugars, but the sucrose level alone was approximately twice normal.

The ultimate star of the Band Aid concerts may be this drought-tolerant crop, whose palatability and protein might lead sorghum into a new era of significance for feeding the world at large. Karamaka not only foiled the famine, it proved a nutritional gem, on a par with the best quality cereals.*

* More information is available from D. Harper (see Research Contacts).

sugar as sugarcane's. These sweet sorghums are surprisingly poorly known compared with sugarcane and sugar beet. Nonetheless, they have a big potential in a world increasingly in need of renewable sources of energy (see next chapter). Also, as food crops they deserve more attention.

Unlike sugarcane, sweet sorghum grows in a wide geographic range. It can be considered "the sugarcane of the drier and temperate zones." It has a production capacity equal or superior to sugarcane's, at least when considered on a monthly basis.

Two types have been developed by breeders:

- Syrup sorghums, which contain enough fructose to prevent crystallization; and
- Sugar sorghums, which contain mostly sucrose and crystallize readily.

RICELIKE SORGHUMS

The *shallu* type of sorghum (the margaritifera subrace of the guinea race) has small, white, vitreous seeds, which are boiled like rice.[9] As of today, little or nothing is known about this interesting form of sorghum, but it could have a good future and deserves exploratory research.

TRANSPLANT SORGHUMS[10]

In certain regions of semiarid West Africa, various special sorghums are transplanted like rice. These are used particularly by peoples living in the bend of the Niger, including parts of Cameroon, Chad, Niger, and Nigeria.

Little is known about these. However, transplant sorghums are produced in the dry season—growing and maturing entirely on subsoil moisture. They are ephemerals that must get through their life cycle before the soil dries back to powder or pavement. They must mature quickly to survive. Some can produce a crop in 90 days—merely half the time the rainfed types require in that area.

One fascinating example has been identified at Gao in northern Mali. It is cultivated by ex-nomad Tuareg, and yields more than 1,000 kg per hectare on residual moisture from the runoff water remaining after light rains.[11] Two others are masakwa and moskwaris.

[9] Information from J. Harlan.
[10] Information largely from R.K. Vogler.
[11] Information from J. DeVries.

These dry-season sorghums have special traits including:

- Large, hard, high-quality grains, locally considered special delicacies;
- Heat tolerance at the seedling stage;
- Drought resistance or tolerance; and
- Ability to flourish on residual moisture in heavy clay soil.

Transplant sorghums grow only on clay pans with a high water table. They are often cultivated on vertisols, which are among the world's most refractory and frustrating soils to deal with. Wet, these soils become soft, sticky, and plastic; dry, they become iron hard and deeply cracked. At least once a year they go from one extreme to the other. Few plants can withstand the trauma. For all that, however, vertisols have high fertility. Any crop that can perform in such recalcitrant sites could be a boon to several parts of the tropics that are now languishing for lack of a crop suited to vertisols. Transplant sorghums therefore deserve international attention.

The yields from transplant sorghums depend on the amount of moisture stored in the soil, but are relatively high by the standards of the very difficult sites where they are grown. (Their high yields probably result from the fertility of the swamp clays.)

These transplant types apparently are uniquely adapted to the unusual conditions of inundated clays and perhaps are unsuited to dry or infertile soils.

FREE-THRESHING SORGHUMS

Despite general opinion, some sorghums thresh easily. The heads hold onto the seeds during the harvest as well as during drying and transport; however, the farmer can separate the seeds from the heads with hardly more effort than is used to thresh wheat or rice. For example, the sorghum variety called "Rio" has an "easy thresh" characteristic. Another variety line being used currently in U.S. breeding programs is SC599. It is both free threshing and tolerant of drought in the post-flowering stage.[12]

The term "free threshing" is also applied to the involute glumes of some West African guinea sorghums. Their seeds are completely exposed and they easily thresh completely free of the glumes.

[12] Information from F.R. Miller. The glumes (chaffy bracts) in these free-threshing types cover about 30 percent of the seed.

Sorghum Comes to America

Sorghum has been in the United States for a long time. The grain types commonly called "guinea corn" and "chicken corn" were introduced from West Africa at least two centuries ago. Both were probably packed as provisions on slave ships and reached the New World only inadvertently. Americans first grew these grains along the Atlantic coast but later took the crop westward where it found a better home in the drier regions. Later-arriving grain types include some that were deliberately introduced by seedsmen and scientists towards the end of the 1800s. By 1900, sorghum grain was well established in the southern Great Plains and in California; indeed, it had become an important resource in areas too hot and too droughty for maize (see page 160).

The sorghum known as "broomcorn" was supposedly first cultivated in the United States by Benjamin Franklin. He is said to have started the industry in 1797 with seeds he picked off an imported broom. The stiff bristles that rise from the plant's flower head have produced many of America's brooms and brushes ever since (see page 209). By the 1930s, for example, American farmers were cultivating 160,000 hectares of broomcorn.

The so-called "sweet" sorghum, with its sugar-filled stems, reached these shores in about the mid-1800s. It landed first in the Southern states—supposedly introduced as a cheap treat for slaves. Within 50 years, however, it had spread so widely and become so popular that sorghum was known as "the sugar of the South." Each locality in the Southern farm belt had a mill to crush sorghum stalks. The resulting syrup, a little thinner than molasses, became *the* sweetener of the region: poured over pancakes, added to cakes, and everywhere employed in candies and preserves. Today, this golden liquid is not so well known, but many rural communities still hold annual sorghum festivals and crude old mills squeeze out an estimated 120 million liters of syrup each year.

Sudangrass was introduced in 1909. This form of "grass sorghum" is now used for animal feed throughout the nation's warmer regions (see page 211).

Opposite: A scene in the backhills of North Carolina. Sugar sorghum has all but disappeared from commercial cultivation in the United States, but it is still grown on a small scale, mostly for home use. The thin greenish liquid squeezed out of the stalks is boiled down into sorghum molasses, thinner in consistency than sugarcane molasses, but lighter in color—an almost transparent golden shade. (P. Mask)

Sorghum in China

In China, sorghum is amazingly popular. In the northern parts, especially, millions of villagers consider *kaoliang* a part of everyday living. Many employ every part of the plant—from top to bottom.

Grains. For millions of Chinese, sorghum is a daily staple. The grains are eaten at perhaps every meal. Certain types of waxy grains are baked into cakes. Other types are fermented and distilled into strong spirits. To connoisseurs, China's best liquors are those made from sorghum—the famous (or infamous) *mao-tai* and *samshu,* for example. Certain grains, particularly the darker-colored varieties, are vital for feeding horses, donkeys, and other livestock.

Seedheads. In some varieties, the empty heads are converted into brooms and brushes.

Stalks. Sweet-stemmed sorghums are a major source of sugar to millions of Chinese. Some are also harvested green and cut up like sugarcane batons. (Children are particularly fond of chewing on them.) The stalks of more woody varieties are bound together, cemented with clay, and used for partitions and walls and fences. The supple green stems are split and woven into baskets and fine matting. The strong dry stems are widely used in making handicrafts and many types of small household utensils, including plate-holders and pot covers. Sorghum stalk is, more-over, a favorite for making children's toys and many types of containers. (Sorghum cages are used to keep pet birds and insects, for example.) In some places, woody sorghum stems are the basic fuel for cooking.

Leaves. In parts of China the leaves are frequently removed before the grain harvest and used for fodder. They are vital for raising cattle, goats, horses, and rabbits.

Roots. The roots are grubbed out and dried for fuel.

All this is not just an ancient traditional practice. In modern China, hybrid sorghum has played a vital role in increasing food supplies. These days, sorghum is a high-yield crop—both for grains and for stems. In sum, the experiences in China demon-strate just how universally valuable this African grain can become.

CHINESE SORGHUMS

All sorghums are indigenous to Africa, but the plant reached Asia so long ago that thousands of cultivars developed there. Indeed, the Far East devotes a huge area to this crop. It is especially surprising to find this tropical crop in chilly climes as far north as Manchuria. Throughout northern China, however, farmers rely on sorghum not only to keep themselves fed when wheat fails but also for many of their household needs (see box opposite). Even when wheat is available, the people often eat a cheap and rather coarse sorghum bread. Special steamed breads are made from sorghum in some areas. Sorghum also goes into noodles, porridges, and boiled (ricelike) dishes. A significant proportion is used to produce strong liquor. Sorghum is also eaten, although to a lesser extent, in Japan.

China contains a cornucopia of types that are unknown elsewhere. *The Flora of Chinese Sorghum Varieties*, for example, lists more than 1,000 local varieties: 980 for food, 50 for industrial use, and 14 for sugar. All of these should be rapidly gathered and tested elsewhere in the world. They undoubtedly offer many genetic benefits. Eventually, they and their genes may become critical to human survival in many areas outside China.

Reuniting the genes of these Far Eastern types with those of Africa after a 2,000-year separation could be an extremely powerful genetic intervention leading to a whole new line of "Chinaf" hybrids.

COLD-TOLERANT SORGHUM

When CIMMYT first tried growing sorghum in the Valley of Mexico, the crop would not set seed. The problem was low temperatures at night. The researchers then got some high-elevation sorghums from Ethiopia, made crosses, and now have types adapted for that upland valley with its chilly nights. Cold tolerance is available in the germplasm but has not yet been fully exploited.

HEAT-SHOCK SORGHUM

Sorghum thrives under searing conditions. Air temperatures of 45°C leave it unfazed. Even at that temperature, young plants have been known to grow 20 percent in height in a single day. But sorghum has its limits. When soil temperatures climb above 50°C, its seedlings struggle to survive. Such temperatures are not uncommon at the soil surface in semiarid areas, and the consequences for sorghum farmers are often dire, sometimes even disastrous.

Now, researchers at ICRISAT have found that certain sorghums withstand heat better than others. No one has paid attention to this quality before, and almost all of today's sorghums produce seedlings susceptible to burning hot soils.

By sowing seed in hot fields and seeing which survived, lines with heat-tolerant seedlings have been identified. But such tests are expensive, time-consuming, and subject to hosts of uncertainties. Now, researchers at the Welsh Plant Breeding Station[13] are devising mass-screening techniques that can be performed in a laboratory and with much more precision.

One Welsh technique, already adopted by ICRISAT, monitors the amount of protein synthesized by the germinating seeds. In hot surroundings, the most heat-tolerant types produce the most protein. However, this test is expensive and cumbersome to run on thousands of samples, so now the Welsh researchers are developing a second-generation test based on "heat-shock proteins" (HSPs).

All living things make HSPs when exposed to temperatures above their normal range. They do it quickly—often within 15 minutes. Once made, the proteins—which are similar in plants, animals, and bacteria—seem to confer an ability to prosper in the heat. Their exact function is still uncertain, but they may protect the organism's proteins, messenger RNA, or membranes from damage. One HSP—often called HSP70 because it has a relative molecular mass of 70,000—may ensure that heat-damaged proteins regain their proper shape so that they can continue working as enzymes, muscles, and antibodies.

The researchers now have found that briefly exposing a sorghum seedling to temperatures between 40°C and 45°C induces it to produce a characteristic set of HSPs. From then on, the plant can tolerate temperatures of 50°C or even more without suffering damage.

Although all sorghum seedlings make HSPs, those that tolerate heat best make HSPs much sooner after germinating. Speed is the secret of their success.

This response is being studied in the hope of finding an easily recognizable feature that can identify heat tolerance without torturing the seeds. If successful, this will open the way to mass screening so that farmers in the hottest areas will no longer face the heartbreak of seeing their fields wilting in the blazing sun before the plants have even grown more than knee-high.

Another approach is to find the regions of the chromosomes which are important for survival of heat stress. DNA probes are being used as markers by the researchers in Wales to follow regions of the chromosomes linked to the thermotolerance trait from parents to subsequent generations.

[13] Led by Cathy Howarth and Chris Pollock.

TROPICAL SORGHUMS

A few sorghums grow in the humid lowland tropics. Although they are not well studied, the guineense and other related groups (roxburghii and conspicuum, for example) could be useful as genetic sources for improvement of genotypes for humid tropical regions.

WILD SORGHUMS

At least two undomesticated forms show extremely robust growth under the harshest of conditions.

One, the verticiliflorum form (previously known as *Sorghum verticiliflorum*) is a wild grass, distributed from the Sudan to South Africa. It is often found in damp areas (along stream banks and irrigation ditches, for example) or as a weed in cultivated fields. On the other hand, it is also a dominant climax species in many of the area's dry, tall-grass savannas. It is thought to be a progenitor of the modern bicolor, caudatum, and kafir races of sorghum but has seldom been considered a genetic resource in its own right. Nonetheless, in research now under way, this plant is proving extremely useful in forage-breeding programs. No doubt it contains disease-fighting abilities and pest resistances that could be deployed to help sorghum.

The other (previously known as *Sorghum arundinaceum*) is a wild and weedy rainforest species that flourishes in Africa's wet tropics, where today's domesticated sorghums are poorly adapted. Although very little information is available, it appears to be more photosynthetically efficient at low light intensities than cultivated sorghum.[14] As of now it is not cultivated, but it may have a future as a domesticated crop for humid and forested regions. It is a robust species, very common along roadsides, vacant lots in cities, and other "wastelands."

WIDE CROSSES

Sorghum can be crossed with grasses genetically distant enough to be classified in different genera or even in different subfamilies. It is certainly highly speculative to think that these crosses might have any economic merit, but exploratory research efforts seem well worth undertaking. A few possibilities are discussed here.

Crosses between sorghum and certain types of *Chrysopogon*, *Vetiveria*, and *Parasorghum* are possible. Crosses with *Pseudosorghum*

[14] Downes, 1971.

Will Sorghum Go High-Tech?

Since the 1960s, when tissue culture was developed for replicating plants such as potato and tobacco on a mass scale, researchers have attempted to apply this technique to grasses. For a decade or two it was considered an impossibility, but recent discoveries have changed that, and a few grasses can now be propagated this way. In 1989, for example, Indian researchers L. George and S. Eapen of the Bhaba Atomic Research Centre in Bombay reported replicating certain cultivars of sorghum using tissue culture. This development could open a new world of understanding and advancement for the world's fifth major food crop.*

The Indian scientists studied seven sorghum cultivars (CO21, CO22, CO23, CO24, TNS24, TNS25, and TNS30). Cells from the stems refused to form callus (the first step in the tissue-culture process), but cells from the base of the leaves formed callus in every case. Also, cells from the seeds of one cultivar (CO23) formed callus in about one-third of the samples.

When the researchers added hormones to induce the undifferentiated callus tissues to produce plantlets, all the callus samples formed roots. However, only three of the cultivars (CO23, TNS24, and TNS25) formed shoots, and then only in 10–15 percent of the samples.

This discovery, while limited, is one upon which further refinements and higher efficiencies can be built. With tissue culture, powerful techniques such as restriction fragmentation length polymorphisms (RFLPs, see page 34), the production of pathogen-free plants, and challenge breeding can be applied to understanding and improving this crop, which is so vital to Africa and the world.

Techniques like these could open possibilities even for far-out developments such as introducing into sorghum the gluten genes from wheat, adding virus-resistance genes, making somaclonal selections, and sorting through the crop's massive genetic diversity in ways that are far more efficient than any imaginable even just a few years ago.

* These results are reported in *Current Science* (India) 58(6):308–310. The researchers used the standard Murashige and Skoog medium to grow the leaf cells and added 2,4-D as a hormone to stimulate growth. They used a combination of kinetin and tri-iodobenzoic acid to induce root growth.

and selected members of the Bothriochloeae and the Sorgheae also seem possible. Crosses between subtribes might be possible if certain members of *Chrysopogon* and *Capillipedium* were used.[15]

American researchers are currently performing experimental crosses between sorghum and johnsongrass (*Sorghum halepense*), a perennial forage that has already introgressed with sorghum to become a pernicious weed in the United States. It is hoped the grain qualities of sorghum can be united with the rhizomatous habit of johnsongrass to create a powerful new perennial cereal.

Recently, crosses between sorghum itself and its sudangrass subspecies (*Sorghum bicolor* subspecies *sudanense*) have produced hybrid grasses with outstanding vigor. Their productivity and performance have boosted even more the acreage and overall yield of forage sorghum, a main part of the livestock-grazing industries of America and Argentina. They also promise to help in reclaiming salt-affected lands (see next chapter).

It has long been known that sorghum can be crossed with sugarcane. Chinese researchers now report developing a hybrid between the two that contains more sugar and produces more stalk and grain than either parent.[16] Research along these lines might turn up fascinating new resources of undreamed-of usefulness.

[15] These speculations were put forward decades ago by Robert P. Celarier, who was thinking in terms of clarifying taxonomic relationships in the subtribe Sorgheae. However, the economic potential of these man-made crosses might be substantial.

[16] S. Wittwer, Y. Yu, H. Sun, and L. Wang. 1987. *Feeding a Billion*. Michigan State University Press. Such a cross might prove a method for boosting sorghum's grain yield. In a sorghum flower, only one spikelet of each pair is fertile. In sugarcane and its relatives, both spikelets of a pair are fertile. Moreover, this trait can be transferred to sorghum, at least at the tetraploid level. See Gupta et al., 1978.

11

Sorghum: Fuel and Utility Types

Few people heretofore have paid much attention to the idea of growing sorghum to burn. Cereal scientists, quite naturally, have regarded the plant exclusively as a food. But these days, feeding the fire can be as hard as feeding the people. Certain sorghums might help, and they warrant research.

Moreover, fuel is fundamental to many other parts of modern living. Indeed, most of the human race is so hooked on flammable liquids for running factories and powering trains, trucks, cars, and buses—not to mention providing electricity—that life would be impossible, or at least intolerable, without them.

For all that, however, the prime liquid fuel, crude petroleum oil, is in jeopardy. Perhaps the greatest challenge of the coming century will be the development of sustainable alternatives. Surprisingly, sorghum might be one of them. Indeed, sorghum could well bring many countries a giant step toward the renewable-energy future everyone is hoping will eventuate to keep life livable in the post-petroleum era.

This chapter highlights sorghum's potential to produce both solid fuels and liquid fuels, to yield industrial products, and to help maintain the overall sustainability of agricultural production.

FIREWOOD

Although food is fundamental, fuel is almost as basic to the modern diet. Without it food cannot be cooked, and today's main grains, pulses, roots, and tubers, as well as many vegetables, must be cooked to be edible.

These days, millions cook over open fires. Indeed, for more than a third of the world's people, the real energy crisis is a frantic scramble for firewood. In the poorest countries, up to 90 percent of the population depend on wood to cook their meals. In parts of Africa and Southeast Asia, an average user may burn well over a ton a year.[1]

[1] See companion report *Firewood Crops: Shrub and Tree Species for Energy Production*. For a list of BOSTID publications, see page 377.

Although the search for food soaks up a major part of the daily lives of billions, the search for fuel to cook it with is becoming equally time-consuming. Firewood is more and more difficult to find. In an increasing number of places, gathering fuel now takes more time than growing food. There is a saying in Africa that it costs more to heat the pot than to fill it.

Although in recent years much effort has been expended on developing firewood crops, few advisers or administrators have ever thought of developing sorghum for the fire. It is a fact, however, that certain types have woody stems that put out surprising amounts of heat. They could well become part of the mix of the firewood crops of the future.

Although these solid-stemmed sorghums have received almost no study as fuel resources, one type has been tested in a preliminary way. It comes from Egypt, where its stalks are more valued than its grains. Egyptians use them as fuel. Called Giza 114, it has solid lignified stalks that burn at an especially high temperature for the stem of a grass.

Little is known about Giza sorghum but, based on results from preliminary trials, it could have a glowing future. It has shown promise in Peru, for example, where it was produced to fuel cookstoves and brick kilns. It is now being tested in Haiti, where it also seems to have good potential as fuel.[2]

It is not inconceivable that sorghums like this could become a standard part of farming in fuel-short nations. Their annual biomass yield is likely to equal or better that from trees. The yield of sorghum stalks has been measured in China as 75 tons per hectare, probably representing more than 10 tons per hectare of dry biomass. This would be a respectable annual production for even the fastest growing trees. The overall yield in fuel-calories per hectare may also be comparable, although even the densest sorghum stem will not equal the caloric output of a wood sample of equal volume. Perhaps, too, a modest harvest of grain can also be achieved.

Compared with trees, sorghums have the advantage in that they produce fuel within months—even weeks. Several crops a year may be possible in appropriate locations. This may help relieve not only the frenzied foraging for firewood that goes on today, but also the destruction of woodlands and forests that seems to end only when desert or degraded soils remain. People who can find fuel in fields close at hand will not hike to far-off forests and haul bulky wood all the way back. Their need is not for large-diameter tree trunks but for small stems that can be easily cut, carried, and fed into the space beneath a pot perched on rocks. For such a purpose, solid-stalked sorghums could become vital resources of the future.

[2] Information from M. Price.

Sorghum stalks make a poor fuel, but for millions they are all that is available. To this extent, they provide a vital contribution toward relieving hunger because all the starchy staples must be cooked to become edible. (Nathan Benn)

LIQUID FUELS

For the economic stability and expansion of nations, liquid petroleum fuels—kerosene, gasoline, and diesel, for instance—have become essential. As noted, these liquids not only power factories, trains, trucks, and buses, they also generate electricity and produce thousands of items from machines to medicines. Moreover, maintaining mobility is critical to the public welfare: police, fire fighters, ambulances, mass transit, and construction fleets all depend on liquids that will explode in the cylinders of internal combustion engines.

For these and other reasons, the growing dilemma over future petroleum supplies makes it imperative to investigate renewable fuels, especially those suited for use in existing engine types. Of all the nonpetroleum possibilities, ethanol is the only one now significantly used in motor transport.[3]

At present, ethanol is made from either sugarcane or maize. In the future, however, sorghum is likely to also be a prime supplier. The stalks of certain sorghums are just as packed with sugar as are sugarcane's. Their juice contains 13–20 percent total fermentable sugars. They can yield about 6 percent alcohol.

Sweet-stalk types are sparingly distributed across sorghum-growing areas of Africa and India, where people chew the green and tender stems like sugarcane or make syrups, molasses, sugar, or confections from them. They were once a major source of sweeteners in the southern United States. Now, however, they have a rising potential as sources of fuel.

All in all, sweet sorghums are important for future ethanol production because they have:

- High biomass yield;
- High percentage of fermentable sugars;
- High percentage of combustible materials (for fueling the processing);
- Comparatively short growth period;
- Tolerance to drought stress; and
- Relatively low fertilizer requirement.

Moreover, sweet sorghums may produce some grain for food or feed. Indeed, as sorghum is one of the most efficient plants, and as it produces fermentable sugars as well as grain, it seems almost ideal for

[3] See companion report, *Alcohol Fuels: Options for Developing Countries*. National Research Council, Washington, D.C. 1983. Research on another renewable-energy alternative, vegetable oils, is described in E. Griffin Shay. Diesel fuel from vegetable oils: status and opportunities. *Biomass and Bioenergy*. Vol. 4, No. 4, 1993. pp. 227–242.

producing both energy and food. Technologies used in the sugarcane industry can be applied virtually without modification.

Sweet sorghum has a number of potential advantages over sugarcane. For example, it is adapted to many growing conditions, unlike sugarcane, which is restricted to tropical climates. It requires less water and fertilizer. It can be planted more easily (from seeds not stems). And it also has a potential for low unit costs because it can be fully mechanized and the fields need not be burned (unlike sugarcane fields).

Sorghum's advantage over maize (in which the grain is converted to alcohol) is that it produces sugar rather than starch. As a result, sorghum juice can be directly fermented without the expense or delay of an initial hydrolysis.

Recently, researchers in at least three countries have begun to appreciate the potential of sorghum as a fuel as the following examples show.

India

In southern India, the potential of sorghum varieties that yield both grain and sugar-filled stems is being explored.[4] Engineers at the Nimbkar Agricultural Research Institute (NARI) have found that these dual-purpose varieties solve three problems: they yield food, the fuel to cook it with, and the fodder to feed the farm animals that help produce it. From the top of the plant comes grain for food; from the stalk comes sugar (and hence alcohol) for fuel; and from the pulp remaining after the sugar is extracted comes animal fodder.

In the past, multipurpose sorghums were dismissed or at least overlooked, probably in the expectation that the individual yields of the various products would be low. But the NARI researchers are showing that this may not be the case. Indeed, they claim that 1 hectare of their sorghums can annually yield 2–4 tons of grain, 2,000–4,000 liters of alcohol, and enough crushed stalk to feed from three to five cattle year-round.[5]

The idea of "growing" fuel alcohol is of course not new. However, most other programs have faltered because the cost of the fuel needed to distill the alcohol rendered them economically unattractive. NARI engineers circumvented this by designing a solar-powered still, incorporating a solar collector and a distillation column that can run at 50–70°C—temperatures that the solar collector can easily provide.[6] Also,

[4] Rajvanshi et al., 1989.

[5] For the fermentation, NARI uses strains of *Saccharomyces cerevisiae*. The average fermentation efficiency was 90 percent and the fermentation process took 48–72 hours to complete.

[6] The pilot model consists of flat-plate solar collectors (38 m^2 in area) coupled to a hot water storage tank (2,150 l capacity).

they have developed pressurized and unpressurized lanterns as well as a wickless stove that will run on aqueous alcohol taken directly from the still.

NARI suggests that this combination of multipurpose sorghum and appropriate technology could, in theory, meet all the automotive fuel requirements in India by the year 2000, completely replace the kerosene now used in Maharashtra, and supply 80 percent of the fodder for all the cattle in Maharashtra. Although such levels will never be approached in practice and it seems axiomatic that grain yields will tumble when sugar is also produced, the NARI concept is a powerful one that could be a big breakthrough that boosts sorghum into an energy resource worldwide. And perhaps, after all, it is not too far-fetched to envisage sorghum producing both high contents of sugar in the stem and high yields of grain.[7]

United States

A large sorghum-for-alcohol project was carried out across the United States between 1978 and 1984. As part of this project, the University of Nebraska developed a demonstration farm based entirely on renewable fuels. Sweet sorghum was the principal crop for alcohol production. Hybrids that grew rapidly and produced large amounts of sugar were created.[8]

A major constraint of sweet sorghum in the temperate zone is the harvest period. Wherever the potential of a freeze exists, the harvest period is greatly reduced because the crop must be gathered before any freezing weather. Sugar in the damaged stalks begins to ferment.

Brazil

Of all the nations in the world, Brazil is the ethanol-fuel pioneer. It already has fuel alcohol in large-scale nationwide use. So far, however, this has come almost entirely from sugarcane. Now, Brazil's scientists are exploring the use of sweet sorghum. The two crops, it has been found, supplement one another: sorghum can provide alcohol during the season in which sugarcane is unavailable.[9] Therefore, using the two together increases the period of production, decreases the unit cost, and increases the total amount of alcohol that a distillery can

[7] Researchers in Texas have also discovered that high yields of sugar are not incompatible with high yields for grain. Information from F.R. Miller.

[8] Information from M.D. Clegg.

[9] Sugarcane in Brazil is normally ready to harvest between June and November; the harvesting period for sweet sorghum is from February to May. Information from R.E. Schaffert.

Will Brazil's Cars Run on Sorghum?

Brazil leads the world in the use of fuel alcohol. In 1993, about 4.3 million vehicles—one-third of the country's total fleet and about 40 percent of its car population—operate on ethanol. Almost all that alcohol now comes from sugarcane, but in the future it may come from sorghum as well.

Brazilian researchers have shown that sweet sorghum can yield from 22 to 45 tons of raw biomass per hectare in 110 days. Fermentable solids (80 percent sugars and 20 percent starch) in the stalks amount to 2.5-5 tons per hectare. To optimize the output, enzymes are added so that the starch in the stems is also converted to alcohol. Research has shown that in this way 1 ton of sweet-sorghum stalks has the potential to yield 74 liters of 200-proof alcohol.

Such discoveries have implications for countries everywhere. In that distant but inevitable day when the world's petroleum runs out, maybe people will turn to sorghum to keep civilization humming. Brazil is showing us yet another way this remarkable plant will be important in our future.

produce each year. The same equipment is used to process both sugarcane stalks and sweet-sorghum stalks.

The Brazilian scientists are also extending their studies to incorporate sorghum into an integrated system in which the by-products are used as food, feed, fertilizer, and fiber. Further, they are adapting this technology to a microscale to allow the economical production of fuel in a decentralized industry. This reduces transportation costs and may perhaps allow the farmers to generate their own energy.

SORGHUM IN SUPPORTING ROLES

Around the world, sorghum is mostly grown for food or feed and (as just mentioned) a little is being grown for fuel. However, there are several interesting uses in which sorghum is grown not for its own sake but for the benefit of other crops. Below are three examples.

Soil Reclamation

Saline Soils It has recently been found that crosses between sorghum and sudangrass (a special race of sorghum), have the capacity to repair saline soils made crusty by sodium compounds. David L. Carter, director of soil and water management research at the U.S. Department of Agriculture station in Kimberly, Idaho, predicts that "they are going to produce some good forage on these marginal lands and at the same time will reclaim some of these soils for crops for human consumption."

Acids released by the sordan roots dissolve calcium carbonate or lime, and in so doing they release calcium. The calcium then displaces sodium in the soil. The newly released sodium reacts with carbon dioxide to form sodium bicarbonate, a soluble salt that is less injurious to plants and mostly washes away in the rain.

After growing sordan on sodic lands for about 2 years, farmers can often re-use the soil for conventional crops.[10]

Reclaiming Toxic Soils U.S. Department of Agriculture scientists in Lincoln, Nebraska, have found that sorghum has an exceptional ability to absorb pollutants out of soil. According to their research, sorghum strips excess nitrogen out of soils with such efficiency that it may solve waste disposal problems for cities and livestock operations (such as feedlots) that generate nitrogen-laden wastes. "We've been able to capitalize on sorghum's natural ability to act as a scavenger," says Kenneth J. Moore. "Sorghum thrives in toxic soils that kill less resilient plants and its penetrating roots can capture the nitrogen in a vast volume of soil."

Moore, an agronomist, and his colleague Jeffrey F. Pedersen, a plant geneticist, are now developing a system in which nitrogen is not only removed but is returned to use safely and economically. They plant sorghum in highly contaminated soils, cut the crop several times through the growing season, and feed the foliage to livestock. The key to the process is sorghum's robust growth and extensive root system.

Such an environmental tool could be very valuable these days. In Nebraska, for instance, municipal and livestock wastes are commonly disposed of by applying them to fallow cropland. An excessive buildup of nitrogen is one of the resulting hazards. "By planting forage sorghum in well-managed cropping system, producers can safely recycle that nitrogen," says Moore.

Two years ago, Moore and Pedersen began their project at a sewage sludge disposal site by planting several types of sorghum: grain types, forage types, tropical types, sweet sorghums, and sorghum-sudangrass hybrids. Soils there contained 400 kg per hectare of nitrogen. The tropical sorghums and hybrids absorbed the most nitrogen from the soil, removing an average of 200 kg and yielding more than 20 metric tons of dry matter per hectare in one season.

"We hoped for more, but the first year's growing season proved to be short and cool," says Moore. "Under normal conditions, some

[10] The companion report *Saline Agriculture* has more information on sordan grass and salt-tolerant agriculture. For a list of BOSTID reports, see page 377.

New Life for Salty Soil

Over the last few decades, irrigation has saved the world's food supply from catastrophe. But irrigation has a fundamental flaw: in the drylands where it is most used, evaporation leaves the site with a surplus of soda and salt. In their worst forms such "sodic" soils become self-sealing: their internal structure collapses so that water just sits uselessly on the surface. Sorghum, it turns out, can help.

Sorghum roots ooze large amounts of sugars. Ordinarily, soil microbes gobble these up, but sodic soils tend to be anaerobic and lack the right organisms. Instead, chemical processes break down the sugars in a way that releases carbon dioxide. A weak natural acid carbon dioxide reacts with the soluble alkalis (sodium carbonate and sodium bicarbonate) to form acetic acid and a little formic acid. These stronger acids, in turn, react with the insoluble alkalis such as calcium carbonate. Sorghum's overall effect is therefore to reduce the alkalinity and convert minerals into more soluble forms. When those wash away, the soil's natural porosity is reopened.

This process occurs with amazing efficiency. Researchers at the U.S. Department of Agriculture have reclaimed marginal sodium-affected soils using sorghum (mainly the forage types called sordan and sudangrass) after just one season. In fields so toxic that crops would not grow, they get respectable stands of barley and alfalfa after just one season of sorghum. Beans, a highly salt-sensitive plant, can be grown after two or three seasons of sorghum. Within one season it not uncommon for the alkalinity to drop a full pH unit and the calcium solubility to increase tenfold.

At first, however, the plants come up scraggly, stunted, and yellow. This has been traced to iron deficiency, to which sorghums are very sensitive. But when the "acidification mechanism" kicks in, the iron concentration in the plant shoots up, they turn green and grow rapidly.

The process is much more than a way to reclaim soils. The researchers are also getting some of the highest dry-matter production recorded in feed-sorghum, especially during the hottest of the summer months. Dry weight up to 67 tons per hectare.*

* Information in this section is from Charles W. Robbins, U.S. Department of Agriculture, 3793 North 3600 East, Kimberly, Idaho 83341, USA.

tropical sorghums absorb as much as 300 kg of nitrogen and yield 25 tons of dry matter per hectare."

Sorghum is so efficient a scavenger that nitrogen levels in the foliage can actually build up to levels harmful to livestock. To address this possibility of nitrate toxicity, the researchers rated their sorghums for nitrate content. Most were at or near toxic levels, but the ensiling process (a lactic-acid fermentation; see Appendix C) removes any threat to the animals.

With further refinement, this process could prove to be a method for continuously stripping nitrogen (and perhaps other pollutants, both useful and hazardous) out of the wastes from cities and industries. "Sorghum-sudangrass hybrids are very popular now in Nebraska and other Central Plains and Midwest states," says Pedersen. "They could be put immediately to work consuming organic wastes."

WIND EROSION

Researchers the world over are working hard to keep sorghum alive, but James D. Bilbro, Jr., is more interested in sorghum dead. He wants to foil the winter winds that pick up soil from Texas farmland and whirl it away across the American landscape. Dead sorghum, it seems, is an answer.

Bilbro, a U.S. Department of Agriculture agronomist in Big Spring, Texas, is exploring ways to protect farmland during a long, cold, blustery winter when the crops have been harvested and the land is bare. Today, farmers in his part of the country normally put in a special crop to cover the land and keep the soil pinned down. The plants survive under the snow, and to get the land back for planting the main crops again, the farmers must eventually kill them with herbicides.

Bilbro asks: Why spend money on herbicides and risk the environment when nature could do the work? In late summer or fall he plants warm-weather crops and finds that they serve very well. Although dead by December, they cover at least 60 percent of the ground, thereby eliminating wind erosion.

Of the 16 crops Bilbro has tested, forage sorghum is the most promising. He thinks that farmers will soon start using it to protect soil because it will save them money, help the environment, and (because the sorghum plants live such a short time before the frost arrives) leave more moisture behind for the subsequent crops.

The technique is being developed in the Texas High Plains, but it may prove useful wherever wind erosion is a problem in the cold-weather zones.

This may seem like a minor use for a major food crop, but the potential is actually vast. Wind damaged 1.74 million hectares of cropland and rangeland in the 10-state Great Plains area during the last wind-erosion season (November 1991 to May 1992). And more than 6 million hectares were reported to be vulnerable to losing their topsoil to the wind. And that was just in the United States.[11]

Weed Control

In previous times, farmers used many plants in crop rotations to control weeds. With the advent of modern herbicides, this practice was dropped in favor of continuous cultivation of the most profitable cash crop. Science is now documenting what these farmers knew—and perhaps too often have forgotten. One example from the United States involves sorghum.

Despite the fact that U.S. farmers apply nearly 200 million kg of herbicides every year, they lose $10 billion worth of crops to weeds. But one Nebraska farmer, Gary Young, doesn't buy any herbicides and his 100 hectares of crops are doing just fine. About 10 years ago, Young noticed that his fields produced fewer than normal weeds the year after he grew sorghum. Since then he has relied on sorghum, not chemicals.

Now there is increasing proof that sorghum is a weed killer that works. Frank Einhellig, a biologist at the University of South Dakota, and James Rasmussen, an ecologist at Mount Marty College of Yankton, South Dakota, recently completed 3 years of field trials on Young's farm. On test plots covering 6 hectares, they had Young plant strips of sorghum, maize, and soybeans, and they measured the number of weeds that came up in the following year's crop. The strips that had been planted with sorghum produced only one-third as many weed seedlings at crop-planting time. Even in midsummer—without herbicides or cultivation—the total weed biomass was still 40 percent less than that on the plots that had been planted to maize and soybeans the previous year.

The surprise is that sorghum suppressed broad-leaved weeds without affecting grasses. It is a selective "herbicide" and thus has special importance for cereal farmers. (It is also well known that broad-leafed crops following sorghum are likely to give poor yields.)

The active ingredients are thought to be phenolic acids and cyanogenic glycosides given off by sorghum's roots. Phenolic acids affect plant-cell membranes and thus reduce a plant's ability to absorb water. They also disturb cell division and hormonal activity, and seem to inhibit seed germination as well as the seedling's early growth and

[11] For more information, contact James D. Bilbro, Jr., USDA-ARS Conservation and Production Systems Research Unit, P.O. Box 909, Big Spring, Texas 79721-090, USA.

Sorghum Saves the Season

As this book shows, sorghum is a remarkable crop, but even we were surprised to learn of the following recent experience.

In the area around Lubbock, Texas, cotton has long been king. The rains there fall in the spring (as well as fall) and the cotton thrives in the hot, dry months that follow. But in the spring of 1992, the rains and record low temperatures came during the planting season. Throughout the region, more than 800,000 hectares were lost because of the unusual conditions. The cool and damp released the soil diseases and pests that had built up over the years and the cotton seedlings quickly succumbed.

The Federal government declared the crop a total loss and authorized disaster payments for the farmers. The farmers, however, faced an unexpected problem: their land was bare and could blow away in the summer winds or wash away in later rains. They needed a ground cover. In desperation they decided to sow nearly 600,000 hectares to sorghum.

Even in this seemingly simple task there was a difficulty. The cotton fields had been treated with a weed killer that is both persistent and designed to kill grasses. Sorghum obviously could not survive. Then someone suggested that an old-fashioned farm implement called a "lister-planter" might work. Fifty years before, farmers used these double-moldboard plows but had since given them up as too old fashioned and too energy consuming.

Now, however, in the 1992 emergency, the countryside was scoured for any of the old plows that were still lying about. Some were found quietly rusting away behind various barns.

Instead of planting sorghum seed in the normal way on the ridges left by the lister, the farmers planted it in the furrows. There, the roots had better access to the soil moisture, but more importantly the toxic topsoil had been scraped aside.

Nothing more was done. The sandy land had already been treated with nitrogen for the cotton crop and—although most observers believed that the rains had probably already leached the fertilizer below root depth—everyone hoped that the combination of furrow-planting and sorghum's deep roots would ensure at least a solid stand to cover the land. A few went beyond that and hoped for a modest harvest of sorghum grain.

The crop was harvested in the fall of 1992. Even with the late planting, minimum preparation, and no inputs, it was a record for that parched area. The figure—4,500 kg per hectare—actually matched the national average for sorghum. Elevators overflowed

Typical scene near Lubbock, Texas in 1992. Part of the unexpected near-record harvest. (A.B. Maunder)

with the unexpected bounty and grain had to be mounded in huge piles in the city streets. Some of the piles were half a kilometer long. Coming on top of their disaster payments, the farmers made more money than ever!

Is it any wonder, therefore, that the cotton farmers of Texas now look on sorghum with new respect? Years before they had used it as a rotation crop, and now they would like to use it that way again. Planting sorghum one year in four, they think, should break the buildup of cotton pests and diseases in the soil and help avoid future failures of the cotton crop. It might also improve soil tilth, decrease erosion, and diversify the local agriculture.

development. Cyanogenic glycosides are known to break down into secondary substances that include cyanide. "Cyanide," Einhellig notes, "is a pretty strong inhibitor of any growth system."

In his latest technique, Gary Young plants sorghum in the fall and allows it to freeze during the winter. The dead sorghum almost completely suppressed weeds, particularly broad-leaved weeds, throughout the year. Snap beans and other crops planted in the residue the following season required almost no weed control.

Now, many of Young's neighbors also plant sorghum and are finding reasonable weed control without herbicides.

In Africa, these effects may be especially important. Today, weeding is perhaps the greatest of all drudgeries in African farming. Most is done by hand—some of it on hands and knees. Returning to the old ways might just solve the problem.

With the new findings in mind, it is possible that the ongoing switch from sorghum to maize may be exacerbating Africa's weed problems. In the future, though, sorghum may become the maize farmer's best friend. Rotations of the two may benefit both.

Crop Support

West African farmers use sorghum for supporting yam plants. They employ a special kind that has stalks like ramrods. The yam plants are extremely heavy so the fact that sorghum can hold them up is graphic evidence of its strength.[12]

Actually, it is even more remarkable than it appears at first. The sorghums support the crushing weight of yams even 8 months after their grain has matured and they have died. Farmers bend the sorghum stalks over to create an intertwined "trellis" about 1.2 m high. The yams are grown on this woven wall of *dead* stalks from the previous season's sorghum crop.

Few plants could withstand such treatment. The tentlike canopy of clambering yam plants entraps heat and moisture and fosters molds, mildews, and rots of many kinds. These sorghums, therefore must be very fungus-resistant, even when dead.

Little attention has ever been given to yam-staking sorghums. Latin America's traditional use of maize plants to hold up climbing beans has been extolled, but Africa's even more remarkable counterpart is little known.

These strong-stalk sorghums might be excellent for use with many climbing annuals, including, for example:

- Macroptilium—an extremely promising tropical forage legume

[12] The yam vines can be 3 m tall and weigh perhaps 50 kg.

whose yields rise dramatically if it can be kept off the ground, where it becomes affected by mildew.

- Winged bean—a climbing bean that could become a major crop of the tropics if cheap ways to support it can be found.
- The viny types of lima beans, common beans, common peas, and runner beans that tend to be the highest yielding varieties but are seldom grown because of the expense of staking them or the lack of poles.
- Beans, squash, or other climbing plants traditionally grown on maize. Switching to sorghum might extend this useful practice to locations too dry for maize.

SORGHUM IN INDUSTRIAL PRODUCTS

Strictly speaking, this book is about plants that produce food, but we cannot resist rounding out the sorghum story with a glimpse at this plant's actual and potential utility as a source of everyday items for industry and for people in their homes.

Fiber Resources

In the rural regions of Africa and Asia, people have devised many uses for sorghum stems. These include:

- Roof thatching;
- Sleeping mats and baskets (made from the peeled stems); and
- Strings in traditional musical instruments (in Nigeria, for example, the peeled bark is used this way[13]).

In China, a particularly strong type has been developed for its pliable, dense stalks. Usually known as galiang sorghum, it is used for constructing fences, walls, and many household items, including grain bins bigger than the beds of pick-up trucks.

Brooms

Broomcorn belongs to this special galiang group of sorghums. It is a special sorghum that is grown not for food, forage, or fuel but for the bristles that rise from its flower head (inflorescence). These stiff, very strong, strawlike projections can be up to 60 cm long. For several centuries, people have used them to make brooms and brushes.

[13] Information from S. Agboire.

Broomcorn was apparently developed in the Mediterranean region during the Middle Ages. (The original sorghums are thought to have come from Africa or India.) It was growing in Italy before the year 1596, and soon thereafter it was being cultivated in Spain, France, Austria, and southern Germany.

Before this sorghum's arrival, Europe's houses, warehouses, front steps, streets, and other places that accumulate dust, dirt, leaves, and horse manure were swept with loose bundles of straw. These not only fell apart quickly, they lacked the strength and springiness to properly flick dust and dirt out of cracks and crevices. Broomcorn, therefore, may well have been one of the most beneficial advances in European public health.

In the United States broomcorn became, if anything, even more important than in Europe. Benjamin Franklin is credited with introducing this strange sorghum. He apparently brought the seed from England in 1725 (when he was only 19) and grew the first broomcorn in North America. It took hold, however. In 1781, Thomas Jefferson listed broomcorn among six important agricultural crops of Virginia. It has been the basis for billions of long-lasting brushes and brooms ever since.

In the competition with man-made fibers and the vacuum cleaner—both of which should in theory have swept it aside—broomcorn is holding its own in the United States. Today, products made of this sorghum are used in millions of American households, warehouses, stores, factories, steel mills, smelters, cotton mills, and barns. They range from whisk brooms to yard brooms for rough sweeping and special purposes.

Considerable development of broomcorn subsequently took place in the United States, but apparently few (if any) other countries have given the crop much attention. This is certainly surprising and should be investigated. Dozens of countries—from Rwanda to Russia—still sweep with bundles of straw. For them, too, this sorghum with the wiry flowers might be a boon.

The broomcorn plant is unlike other sorghums. The stem is dry and hard. The kernels are small and are often enclosed in long ellipsoid husklike coverings (glumes).

The plant has been typecast as a source of brooms and brushes, but it could very well have other equally important uses. For instance, broomcorn stalks are used for paper in France. Reportedly, excellent yields of fiber are obtained by planting the crop very densely. The pulp is used to manufacture kraft paper, newsprint, and fiberboard.

Danish scientists have also made a good paneling using the chips from internodes. Similar products are beginning to be explored in Zimbabwe as well. However, insufficient work has been done to really know the possibilities.

Chinese researchers are using tall sorghums for making plywood. The process apparently works well and gives a product stronger than wood.[14]

Dyes

Moroccan leather is said to get its color from red dye extracted from special sorghums. These red-seeded varieties were raised in sub-Saharan Africa and in the old days were sent across the Sahara to Fez or elsewhere by caravan. Natural dyes (especially red ones) are increasingly in demand these days, so perhaps these types could be commercially produced once more (see box, next page).

Resins

There is a black-grain sorghum from Africa called "shawya" that shows promise in producing industrial resins.[15]

ANIMAL FEED

The United States probably leads the world in developing sorghum as a feedstuff. The plant is now a vital animal feed throughout the nation's warmer regions (see page 160).

Although it has been in the United States since the earliest days (see page 186), grain sorghum first became a major American crop in the 1930s, when dwarf cultivars were bred. These lent themselves to large-scale operations and combine harvesting, and the acreage began increasing. The grains were used exclusively for feeding livestock and became so valuable for this purpose that by shortly after World War II, sorghum had become the most important cash crop in Texas and was a valuable resource in several other states as well.

Then in the late 1950s male sterility was discovered in sorghum (see page 163). This made hybrids possible. Sorghums that had originated in South Africa, Ethiopia, and the Sudan were bred together to create hybrids, and yields jumped as much as 40 percent. This led, in turn, to vastly more plantings and even more American animals were soon living off sorghum grain.[16]

Today, the country produces about 19 million tons of sorghum grain each year, and millions of American cattle, pigs, chickens, and turkeys

[14] Information from F.R. Miller.
[15] Information from L.W. Rooney.
[16] In 1957, about 15 percent of U.S. sorghum had been the hybrid form; within 2–3 years, the figure exceeded 90 percent.

Red Sorghum Rising

In parts of West Africa people grow a form of sorghum that is inedible (and may even be poisonous). The plant provides a windbreak around huts and along the edges of fields, but more importantly it provides masses of leaf sheaths. These rusty-colored, parchment-like wrappings, which surround the leaf stems, provide pigments that are traditionally used to color leather goods. Millions of suitcases, shoes, hats, baskets, book covers, and other products get their brilliant red hues this way. The scarlet flame of the famous "Moroccan leather" and of the fez have their origins in this particular sorghum plant (race caudatum).

Traditionally, bundles of leaf sheaths were extracted in a difficult and laborious cottage-industry process. Now, however, this time-consuming and uncertain technique is being updated. In Burkina Faso, Mouhoussine Nacro, head of the Organic Chemistry Laboratory at Ouagadougou University, has been developing a new and more versatile version since 1989. Indeed, he is opening up the potential for producing sorghum dyes on a massive scale.

Nacro's dye-extraction process uses simple techniques but modern materials. Basically, he and his colleagues crush the sorghum sheaths, add a solvent, separate the liquid emulsion, and centrifuge the result. This produces the pure pigment as a burgundy-red powder that is ready for use and can be safely stored.

The pigment, Professor Nacro has discovered, is a mixture of anthocyanins. The main component, apigenin, is the same natural coloring used by food industries in many parts of the world. Moreover, it is increasingly sought these days because synthetic food dyes are suspected of causing harm.

Red-sorghum leaf sheaths contain over 20 percent of the apigenin and are said to be the only known source of such large concentrations. They contain more than four times the amount in the skin of the red grape, currently the most common source.

Burkina Faso's new process can easily be reproduced on an industrial scale, and commercial production of dyes could result in a new and valuable use for sorghum—one that has widespread application throughout the developing world, but especially in West Africa.

are fattened on it. Production is centered in the Great Plains, and extends over a vast area from the Gulf of Mexico to the Dakotas (see map, page 160).

But the crop is a more important feedstuff even than that. Only about two-thirds of America's sorghum plants are harvested for grains, and most of the rest also goes for animal feed. They, however, are turned into forage or silage or are left in the fields for grazing. This use of foliage rather than grain developed after sudangrass was introduced in about 1909. This grass sorghum has since been hybridized with grain sorghums to yield the "sorghum-sudan" hybrids. These crossbreeds are now widely used in the dry regions of the Plains states as well as in the Southeast, where other forages are sometimes hit hard by midsummer droughts and pests.

Although sorghum has advanced rapidly during the last 50 years, the fact that Americans developed it mainly as a livestock feed is in some ways unfortunate: the varieties typically had brown or red seed coats and are only peripherally relevant to food production. Moreover, in the public mind the crop became stigmatized as "animal food." Only now is there a nationwide glimmering of appreciation for sorghum as something people can eat. Today, American farmers are growing more and more of these food-grain sorghums, abandoning the brown and red types and switching to those with yellow or white seeds.[17]

Broom Corn

[17] Even people who work with the crop think the name "sorghum" has too many bad connotations in the American public's mind. Researcher Bruce Maunder has suggested the name "sungrain," on the basis that the white, cream, and yellow grains are "sunlike" and the grain is directly exposed to the sun's rays from pollination to harvest.

12

TEF

Tef (*Eragrostis tef*) is a significant crop in only one country in the world—Ethiopia. There, however, its production exceeds that of most other cereals. Each year, Ethiopian farmers plant almost 1.4 million hectares of tef,[1] and they produce 0.9 million tons of grain, or about a quarter of the country's total cereals.[2]

The grain is especially popular in the western provinces, where people prefer it to all other cereals and eat it once or twice (occasionally three times) every day. In that area, tef contributes about two-thirds of the protein to a typical diet.

Most tef is made into *injera,* a flat, spongy, and slightly sour bread that looks like a giant bubbly pancake the size of a serving tray. People tear off pieces and use them to scoop up spicy stews that constitute the main meals. For the middle and upper classes it is the preferred staple; for the poor it is a luxury they generally cannot afford.

Unlike many of the species in this book, tef is not in decline. Indeed, farmers have steadily increased their plantings in recent years. The area cultivated rose from less than 40 percent of Ethiopia's total cereal area in 1960 to more than 50 percent in 1980.

Tef is so overwhelmingly important in Ethiopia that its absence elsewhere is a mystery. The plant can certainly be grown in many countries. Some has long been produced for food in Yemen, Kenya (near Marsabit), Malawi, and India, for example. Also, the plant is widely grown as a forage for grazing animals in South Africa and Australia.

Now, however, the use of tef as a cereal for humans is transcending the boundaries of Ethiopia. Commercial production has begun in both the United States and South Africa, and international markets are opening up. This is because Ethiopian restaurants have recently

[1] The common name is often spelled "teff" or "t'ef" in English. We recommend "tef": it is simple, unconfusing, and perhaps a good marketing name that might help the crop's future expansion and acceptance worldwide.

[2] According to statistics of the mid-1980s, tef produced 23 percent of Ethiopia's cereal grain. The others were sorghum (26 percent), maize (21.7 percent), barley (17 percent), and wheat (12.4 percent).

Tef production in the Ethiopian Highlands. Tef is a reliable cereal for an unreliable climate. Its straw (left) is as important to the farmer as is its grain (pile at right). (International Livestock Centre for Africa)

become popular in both Europe and North America. Many cities (including Washington, New York, Chicago, San Francisco, London, Rome, and Frankfurt, not to mention Tel Aviv) now have restaurants that rely on *injera,* as well as the convivial communal dining it fosters. And only tef can make authentic *injera.*[3]

The new appreciation of tef is also extending into the research community. These days scientists in Ethiopia and a few other countries are beginning to seriously study the plant and its products.

This is all to the good. Tef has much more promise than has been previously thought. It provides a quality food. It grows well under difficult conditions, many of them poorly suited to other cereals. Even in its current state it gives fairly good yields—about the same as wheat under traditional farming in Ethiopia. And it usually produces grain in bad seasons as well as good—an invaluable attribute for poor farmers and of special benefit to locations beset by changeable conditions.

However, along with its advantages tef has serious drawbacks, mainly stemming from its tiny seeds, high demands for labor, lack of development, and difficult cultural practices. All in all, at this stage at least, it is neither easy to grow nor easy to handle.

PROSPECTS

To chart tef's future—both its course and final destination among world cereals—cannot now be done with confidence. This will become clearer as the current research efforts begin producing more results. Nonetheless, there are good reasons for optimism that tef's technical limitations can be overcome and that it can rise to be a specialty crop in a number of nations. It could happen quickly. Indeed, *injera* is such a fascinating food (half pancake, half pasta) that it has the potential to eventually become well-known worldwide.[4]

Africa

In Ethiopia, the plant's stable yield under varying conditions, as well as the grain's good storage properties, palatability, and premium prices, will likely make tef ever more attractive.[5] However, although

[3] *Injera* can be made from other grains, but when made from tef it keeps its soft and spongy texture for 3 days; when made from wheat, sorghum, or barley it hardens after only a day. Buckwheat is perhaps the closest substitute.

[4] Something similar is now happening with the tortilla, the round flat bread of Mexico and Central America, which is being sold in supermarkets throughout the United States and is also showing up ever more frequently in other parts of the world.

[5] On the open market in Ethiopia, its grain always commands a price substantially above that of other cereals.

Injera

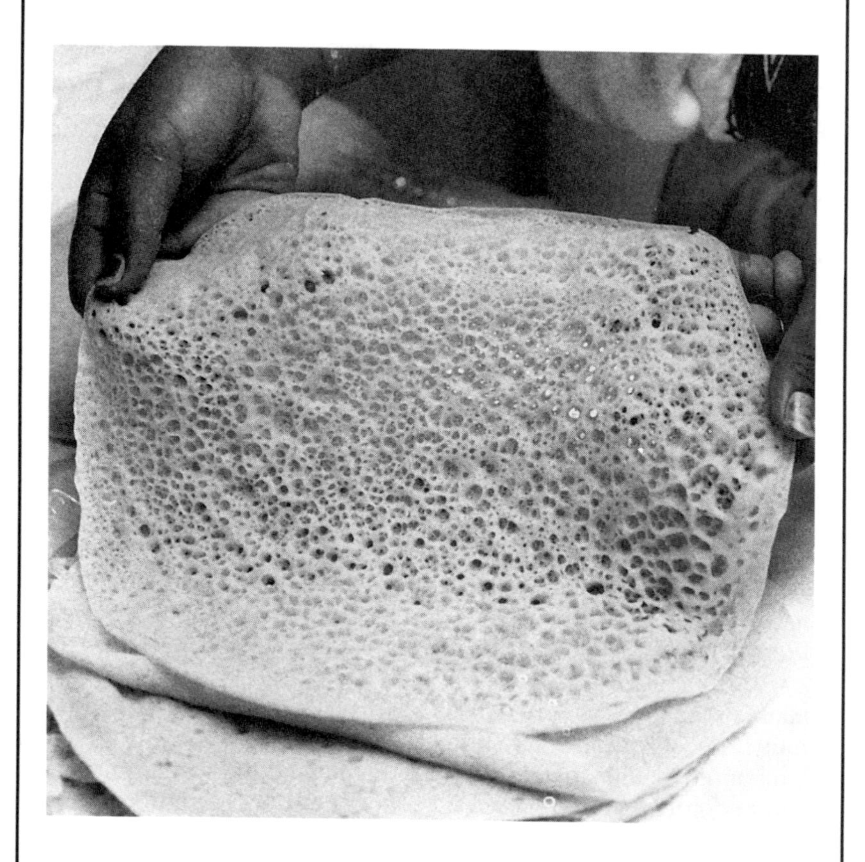

Perhaps the most intriguing of all the world's staples, *injera* is a bread like no other. Moist, chewy, and almost elastic, it has a unique look and feel. A very correct British gentleman visiting Ethiopia in the mid-1800s tried to explain the experience of eating *injera*: "fancy yourself chewing a piece of sour sponge," he said, "and you will have a good idea of what is considered the best bread in Abyssinia." But these days people are not so closed-minded. Indeed, the search for new tastes and new culinary sensations is becoming a force that is opening up the food industries of affluent nations. *Injera* is now winning converts all over the world. It is served in fine restaurants in Europe, North America, and Israel and is receiving an enthusiastic welcome.

prospects for raising its production seem good, substantial increases will probably occur only after its labor requirements are reduced.

Tef may also come to benefit other African countries, notably some that today face food-production problems. The plant's resistance to diseases, pests, and heavy soils give it special appeal.

Several of tef's relatives are valued forages in the world's arid zones,[6] and tef itself might also have a future as a fodder. Indeed, in southern Africa it is already used extensively, having originally fed the horses and oxen of the Boer War almost a century ago. Tef hay is of such quality that South African farmers prefer it over all others for feeding, their dairy cattle, sheep, and horses (see page 230).

Moreover, this grass is exciting South Africans as a "quick fix" for holding down bare soil and thereby baffling erosion while more permanent ground covers establish themselves.

Humid Areas Prospects probably low. For Africa's humid areas, tef's prospects are unknown because trials have not been conducted (or at least not reported). However, the crop comes from a relatively dry environment and probably has little or no potential in a hot and steamy one.[7]

Dry Areas Good prospects. Tef is a reliable cereal for unreliable climates, especially those with dry seasons of unpredictable occurrence and length.

Upland Areas Good prospects. Most of Ethiopia's tef is produced at moderate elevations, but it has long been common on the high plateau and is being slowly introduced to higher and higher locations. Its future contribution to the rural economy of these and other African highlands appears to be substantial.

Other Regions

Tef holds promise for many countries beyond Africa. Mexico, Bolivia, Peru, Ecuador, India, Pakistan, Nepal, and Australia might well adopt it. In addition, this plant's rapid maturity and inherent cold tolerance may open new areas of grain cultivation for high latitudes where growing seasons are short—Canada, Alaska, the Soviet Union,

[6] These are usually called "love grasses" as reflected in the botanic name, which is derived from Eros (god of love) and grostis (grass). Weeping lovegrass (*Eragrostis curvula*), from southern Africa, is widely planted in the southwestern United States, for example.

[7] It is true that it is grown even in Ethiopia's Ilubabor province where the rainfall is very high, but mostly on steep slopes that quickly shed the runoff.

and northern China, for instance. It might also become important to Israel, which has a rising Ethiopian population.[8]

Some observers see tef as a promising new grain for the United States as well. They point out that it is nutritious enough to be a "health" food and tasty enough to be a gourmet food.[9] A company in Idaho already produces it on a commercial scale and supplies markets nationwide (see box). Tef is also being produced on farms in Oklahoma, where it is harvested by machine and sold under contracts from food companies eager to buy it.[10] These experiences, limited as they are, are probably laying the groundwork for a mass-produced specialty grain that will remain a part of the American food system.

USES

Tef grain comes in a range of colors from milky white to almost black, but its most popular colors are white, red, and brown. By and large, the darker the color, the richer the flavor. Although blander in taste, the white seeds command the highest prices. However, the red and brown seeds come from plants that are hardier, faster maturing, and easier to grow. In addition, tef aficionados prefer their more robust flavor.

Tef contains no gluten—at least none of the type found in wheat. For this reason, Americans with severe allergies to wheat gluten are among those buying tef these days. Despite the seeming lack of this "rising" protein, *injera* is a puffy product, somewhere between a flat bread and a raised one.

In Ethiopia, tef flour goes into more than just *injera*. Some is made into a gruel (*muk*), some is baked into cakes and a sweet dry unleavened bread (*kita*), and some is used to prepare homemade beverages. In the United States, it is recommended as a good thickener for soups, stews, and gravies, and, at least according to one promotional pamphlet, "its mild, slightly molasses-like sweetness makes tef easy to include in porridge, pancakes, muffins and biscuits, cookies, cakes, stir fry dishes, casseroles, soups, stews, and puddings."[11]

As fodder, the tef plant is cheap to raise and quick to produce. Its straw is soft and fast drying. It is both nutritious and extremely palatable to livestock. Its leaf:stem ratio (average 73:27) is high, its

[8] Israel in recent years has been importing tef from the United States, South Africa, and Ethiopia.

[9] "I am using it constantly in my cooking—it makes most wonderful waffles and pancakes, for example," notes botanist Fred Meyer of Washington, D.C.

[10] Information from C.L. Evans.

[11] A recent book, *Whole Grain Gourmet*, by R. Wood (William Morrow, 1991), includes many up-scale tef recipes.

NUTRITIONAL PROMISE

Main Components		Essential Amino Acids	
Moisture (g)	11	Cystine	1.9
Food energy (Kc)	336	Isoleucine	3.2
Protein (g)	9.6	Leucine	6.0
Carbohydrate (g)	73	Lysine	2.3
Fat (g)	2.0	Methionine	2.1
Fiber (g)	3.0	Phenylalanine	4.0
Ash (g)	2.9	Threonine	2.8
Vitamin A (RE)	8	Tryptophan	1.2
Thiamin (mg)	0.30	Tyrosine	1.7
Riboflavin (mg)	0.18	Valine	4.1
Niacin (mg)	2.5		
Vitamin C (mg)	88		
Calcium (mg)	159		
Chloride (mg)	13		
Chromium (μg)	250		
Copper (mg)	0.7		
Iron (mg)	5.8		
Magnesium (mg)	170		
Manganese (mg)	6.4		
Phosphorus (mg)	378		
Potassium (mg)	401		
Sodium (mg)	47		
Zinc (mg)	2		

Tef has as much, or even more, food value than the major grains: wheat, barley, and maize, for instance. However, this is probably because it is always eaten in the whole-grain form: the germ and bran are consumed along with the endosperm.

Tef grains are reported to contain 9–11 percent protein, an amount slightly higher than in normal sorghum, maize, or oats. However, samples tested in the United States have consistently shown even higher protein levels: 14–15 percent.

The protein's digestibility is probably high because the main protein fractions—albumin, glutelin, and globulin—are the most digestible types. The albumin fraction is particularly rich in lysine. Judging by the response from Americans allergic to wheat, tef is essentially free of gluten, the protein that causes bread to rise. Nonetheless, tef used in *injera* does "rise" (see page 219).

COMPARATIVE QUALITY

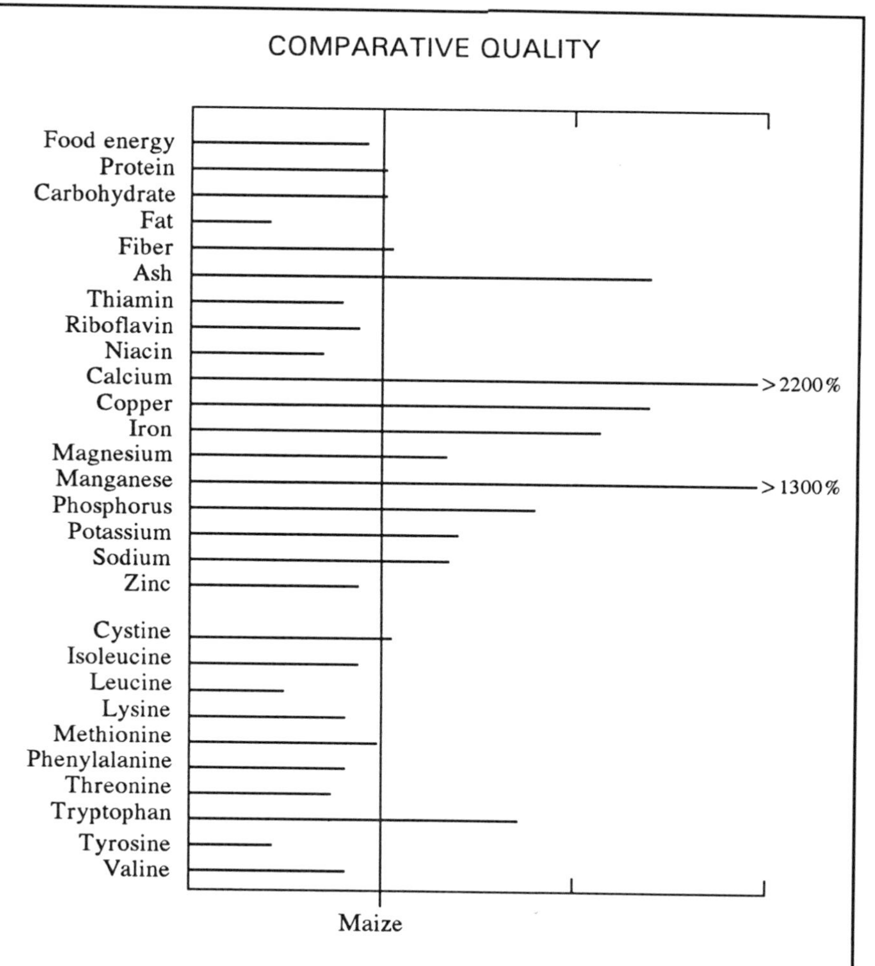

The level of minerals is also good. The average ash content is 3 percent. Tef is reported rich in iron, calcium, potassium, and phosphorus. The iron and calcium contents (11–33 mg and 100–150 mg, respectively) are higher than those of wheat, barley, or sorghum. In Ethiopia, an absence of anemia seems to correlate with the levels of tef consumption and is presumed to be due to the grain's high content of iron.

However, some samples of tef have failed to show the extraordinary levels of iron. Part of the iron may well come from dust and dirt that clings almost uncannily to these tiny grains. Washed seeds have shown a level of iron of about 6 mg, much less than the reported figures but still a remarkable amount.

digestibility (65 percent) relatively high, and its protein content (1.9–5.2 percent) low but nonetheless valuable. Ethiopian farmers rely on it to strengthen their oxen at the end of winter, a time when fresh grass is unavailable but the plowing season is coming on.[12]

In Ethiopia, tef straw is the preferred binding material for walls, bricks, and household containers made of clay.

NUTRITION

Tef seeds appear similar to wheat in food value; however, they are actually more nutritious. There are two reasons for this: (1) the seeds are so tiny that they have a greater proportion of bran and germ (the outer portions where nutrients are concentrated); and (2) because the seeds are so small, tef is almost always produced as a whole-grain flour.[13]

For a grain, tef is rich in energy (353–367 kcal per 100 g). Its fat content averages about 2.6 percent.

In most samples, the protein content is as good as, or better than, that of other cereals. It ranges from 8 to 15 percent, averaging 11 percent. The protein, as in most cereals, is limited by its lysine level. Otherwise, however, it has an excellent balance of essential amino acids.[14] Indeed, two nutritionists, having surveyed all the common foods of Ethiopia, commented: "[W]e want to draw attention to the high values for methionine and cystine found in tef . . . . The protein from a mixture of tef and a pulse will give a near optimal amino acid mixture with regard to both lysine and to the sulfur-containing amino acids."[15]

The vitamin content seems to be about average for a cereal, but making *injera* involves a short fermentation process, and the yeasts generate additional vitamins. The value of the grain is thus enhanced.

The mineral content is also good (average ash content 3 percent). The iron and calcium contents (0.011–0.033 percent and 0.1–0.15 percent) are especially notable. In Ethiopia, an absence of anemia seems to correlate with the areas of tef consumption, presumably due to the grain's good iron content.

[12] Tef fodder is therefore a vital component of Ethiopia's whole farming system, a point often overlooked by those who consider only the grain. Information from G. Jones.

[13] Refined flour can be made, however. With appropriate screening it can be sifted away from the bran and the germ. Information from W. Carlson.

[14] In Ethiopia, it is said that a daily intake of one *injera* pancake supplies enough of these amino acids to sustain life without another protein source; two are sufficient to ensure good health.

[15] It is notable that Ethiopians commonly mix fenugreek (*abish*), lentils, peas (*ater*), or faba bean (*bakela*) with *injera* batter, a practice that satisfies this nutritional criterion.

AGRONOMY

Ethiopian farmers grow tef either as a staple or as a standby. As a staple, they plant it like other cereals, but they normally sow it late and harvest it well into the dry season. As a standby, they wait until their main crop—maize, sorghum, or maybe wheat—shows signs of failing. Then they sow a fast-maturing tef as a backup source of sustenance in case of disaster.

Even where other cereals offer reasonable reliability and substantially higher yields, Ethiopian farmers still include a field or two of tef. Not only does it bring them high prices, its late sowing date allows them to grow and harvest both crops.[16]

In Yemen, tef is known as a lazy man's crop: the farmers merely toss seed onto moist soil following flash floods and then return after about 45 days to collect the grain.[17]

No matter how it is grown, tef requires little care once it is established. Its rapid growth stifles most weeds; few diseases and pests attack it; and it is said to produce well without added nutrients. However, in most places tef will respond to fertilizers.[18]

HARVESTING AND HANDLING

Tef threshes well with standard methods and equipment. Very early-maturing types are ready to harvest in 45–60 days; early types in 60–120 days; and late types in 120–160 days.

Yields range from 300 to 3,000 kg per hectare, or even more. Although the national average in Ethiopia is 910 kg per hectare, yields of 2,000–2,200 kg per hectare are considered routinely attainable if good agronomic practices are carefully followed. Yields of 2,000 kg per hectare have been achieved on South African farms also, although storms have sometimes leveled the fields, resulting in large losses.[19]

The grain is easy to store and will survive for many years in traditional storehouses without damage by insects. This makes it a valuable safeguard against famine.

[16] Information from Sue Edwards.

[17] Information from H. Moss.

[18] In South Africa it has been found that forage (and no doubt seed) yields improve dramatically when up to 80 kg of nitrogen are added per hectare. However, the current varieties put more growth into straw than seed, a feature not necessarily disliked by farmers. Information from N.F.G. Rethman.

[19] Information from N.F.G. Rethman.

Injera, the most important bread in Ethiopian cooking, is normally made from tef. It is a thick spongy pancake in which other food can be sandwiched or rolled. The pancakes are only a few mm thick but can be up to half a meter across. (Panos picture)

LIMITATIONS

The seeds are so small that this alone makes the crop hard to deal with. The fields are tedious to prepare, and it is difficult to get an even stand. Also, wind or rain can bury the minute seedling before it can establish itself. Threshing, winnowing, and grinding such tiny seeds by hand is very laborious. Handling and transporting them is also a problem because they tend to fall through any crack.

NEXT STEPS

Tef seems poised on the brink of becoming a resource for everyday foods, gluten-free specialty items, animal feeds, and erosion control. Ethiopian farmers, therefore, have much to teach nations the world over. The problem is that at this point few people have recognized tef's qualities. Activities are needed to spark interest and raise overall awareness of tef's status, potential, problems, and requirements. These could begin, for example, with conferences, monographs, newsletters, and publicity materials.

Although people in tef's homeland know more about the crop than anyone else, it is unrealistic to expect that Ethiopia can spearhead such activities, at least at present. An international global effort is called for. Luckily, tef is not a weed. Trials can be conducted in different parts of the world with little hazard. Although many countries could participate, the United States, South Africa, and Australia especially could help pioneer the selection of types for trials and eventual use worldwide.

Tef is also a challenge to the world's cereal scientists, agronomists, and food chemists. It is an interesting new cereal that few people know of at present. It seems to offer many possible benefits, but what its limits and potentials are in practice is still very uncertain.

Germplasm Collection and Evaluation The germplasm in Ethiopia is potentially of worldwide importance. Since Ethiopia is the center of origin and the center of diversity for this crop, preserving its diversity is a prerequisite for all tef improvement. Actually, several thousand samples have already been collected. Although more undoubtedly remain, perhaps the most urgent task is to characterize the tef lines already available.

Plant Breeding Until very recently, crossing tef was tedious. It was constrained to a few minutes at about dawn, and required supremely skillful personnel. Now, however, techniques have been developed that make the process quite straightforward and routine.

Tef Pioneers

Until recently, Ethiopia's official commitment to tef research has been small compared with its investment in wheat, maize, and sorghum. However, several organizations have devoted their own efforts to boost the crop.

Both the Debre Zeit Agricultural Research Centre of Alemaya University of Agriculture and the Institute of Agricultural Research at Holleta Research Station near Addis Ababa have produced high-yield strains. Some of these get so heavy with grain that the stalk collapses.* Research is now under way to develop varieties with short, stiff straw to create high-yielding tefs that can benefit from heavy fertilizer use and irrigation without collapsing.

The Institute for Agricultural Research has also done research on tef with encouraging results at Debre Zeit. It has developed a variety, DZ 01–946, which has given yields of 1.78 tons per hectare.

There has also been increasing international interest. In England, London University's Wye College is doing systematic breeding. In Israel, the Volcani Centre is carrying out tef research trials. And in the United States, Wayne and Elizabeth Carlson of Caldwell, Idaho, have been developing cultivars and processing techniques for farmers both domestic and foreign (see box, opposite).

* This was a problem with early high-yielding wheats until short-stemmed varieties were bred in Mexico—a combination that led to the Green Revolution varieties that for 20 years have fed the added millions in Asia who would otherwise have starved.

A program of tef improvement by plant breeding—combining the desirable qualities of several parents in a planned way—might well bring big advances. Objectives include early maturity, short and stiff straw, disease resistance, and higher harvest index.[20] One variety created in Ethiopia has yielded 3,560 kg per hectare.

Other targets for improving the crop, especially for large-scale commercial production, include larger grain size, less shattering of seeds, and quicker drying seeds.[21]

Agronomy In Ethiopia, large yield improvements can be achieved by applying techniques that are already known: careful land prepara-

[20] A variety called "munité" seems especially valuable in this regard because it is very short (40 cm), early maturing, and has a high harvest index.

[21] The ability to harvest drier seeds (than possible at present) would likely reduce harvest losses and increase tef's acceptance by farmers worldwide.

Tef in the United States

Wayne and Elizabeth Carlson are among the handful of non-Ethiopians who have begun growing tef for food. The crop is thriving on their farm near Caldwell, Idaho. In the harsh, dry valley on the Idaho-Oregon border, their fields are now producing Ethiopia's favorite food grain.

Wayne became aware of tef while working as a biologist in Ethiopia. On returning to the United States, he planted some. Within 5 years the Carlsons had progressed from growing a few varieties in their backyard to harvesting 200 acres of four selected strains, as well as threshing, milling, and packaging thousands of kilos of tef seed each year.

The Carlsons' tef flour now goes to natural-food markets nationwide as well as to the numerous Ethiopian restaurants that have been springing up in major cities to serve Americans as well as an estimated 50,000 Ethiopian immigrants and students. Their long-range goal is to make tef a new option among America's cereal crops.

Tef's homeland has not been overlooked. Each year the Carlsons return a portion of the grains they have bred to Ethiopia for trials and for farmers. Last year, they donated 16,000 kg of seed to a relief agency for planting in Ethiopia.

Wayne Carlson says that the Western world should pay more attention to tef. For centuries the plant's adaptability and nutritional value have helped Ethiopian highlanders maintain their independence in the harsh surroundings in which they live, he notes.

Tef in Transvaal

In 1886, the Royal Botanic Gardens at Kew, England, obtained tef seed from Abyssinia and distributed it to various botanic gardens and other institutions in India and the colonies. In its first issue (1887), Kew's *Bulletin of Miscellaneous Information* advocated introducing the crop "to certain hill stations in India, to elevated portions of our colonial empire, and indeed to all places where maize and wheat cannot be successfully cultivated."

These efforts stimulated tef trials in various parts of Africa, Asia, and Australia. As a result, many reports on the plant's performance were received.

Perhaps the most effective introduction was to the Transvaal (which was not then under direct British control). Growers there found that "it makes very rapid growth, maturing in seven or eight weeks from the time of sowing, and if cut before the seed develops, a second crop can be obtained from the same stand; it makes an excellent catch-crop for hay, two successive cuttings being obtainable during the summer on unirrigated land. The plants seed heavily, our yield of seed from a small plot has been at the rate of about three-fourths ton per acre [1.875 tons per hectare]; the seedlings are not readily scorched by the intense heat of summer. On account of the soft, thin straw, it dries and cures very quickly."

But despite the good results, tef took off only by a fluke. As is usually the case with new farm crops, it did not sell well when first offered. The story goes that a farmer, having more tef hay than he required, sent the surplus to the Johannesburg market. It sold poorly—none of the buyers knowing the stuff—and it finally went for animal bedding. It is softer than the ordinary bedding (normally cut from sedges and *Arundinella eckloni*), and a buyer

tion, use of selected seeds, fertilization, sowing and weeding at the optimum time, and disease and pest control, for example. Yields can also be increased by mechanization. Sowing methods require special attention.

Ornamentals There is now an explosion of interest in ornamental grasses in Europe, the United States, and Japan. With its upright, compact habit, its often brilliantly colored leaves (many color combinations are possible), and open feathery panicles, tef is exceptionally

selected one lot for a racing stable. Rumor has it that the stable owner found his racers eating their bedding in preference to their feed! To his surprise they also began to put on condition. Then he bought up all the tef on the market and called for more. Others soon got wind of this and the price rose. Tef was accepted and became a fodder of notable importance to the Transvaal in the early twentieth century. (For instance, during the Boer War it probably fed the horses on both sides.)

"Tef has raised scores of small Transvaal farmers from poverty to comparative comfort, and has been largely instrumental in putting the dairy industry of the Witwatersrand on its feet," wrote Joseph Burtt Davy in the *Kew Bulletin* of 1913. "The opinion has been expressed by our farmers that 'if the Division of Botany of the Department of Agriculture had done nothing else, the introduction and establishment of tef as a farm-crop would have more than paid South Africa the whole cost of the Division for the ten years of its existence.' "

In the Transvaal, as well as in other parts of South Africa, tef is often sown with its relative, weeping lovegrass (*Eragrostis curvula*). This perennial has been developed in South Africa into an almost incredible array of types for land protection and reclamation purposes. It is providing outstanding erosion control on toxic, dry, degraded, and infertile slag heaps and other problem sites where nothing previously would grow. As an erosion-fighting plant, weeping lovegrass is better than tef because it is a perennial whose natural staying power keeps the land covered as the seasons go by. But while tef may not be good at such a "long-distance event," it is very good as a "sprinter." Thus tef is used to produce a fast cover that protects the site while its slower cousin is finding its legs.

attractive. The development of selected strains might create a small but profitable market niche as an ornamental.[22]

Forages In South Africa various productive races have been selected for hay production. These deserve to be exploited elsewhere. Also, it seems likely that a wealth of new types, adapted to many

[22] This possibility is already being explored in the United States. Information from W. Carlson.

different conditions, can be created from Ethiopia's broad germplasm base.

Erosion Control It seems likely that demand will increase worldwide for non-weedy annual grasses that can serve as temporary ground covers. South Africans are now using tef as a "nurse crop" that quickly covers the ground and fosters the establishment of perennial grasses sown along with it. This should be tested elsewhere, too. In South Africa it is already used in mixtures to protect road cuts, open-cast mine workings, stream banks, and other erodible sites.[23]

Black Cotton Soils Tef has evolved on the Ethiopian highlands on vertisol (black cotton) soils that frequently get waterlogged. Few other cereals can be grown there. In fact, tef is able to withstand wet conditions perhaps better than any cereal other than rice. It even grows in partly waterlogged plots, as well as on acidic soils.[24]

Vertisols are a problem in many parts of the tropics. They are cracking clays that regularly heave and sag and split.[25] Few crop plants can withstand such soil abuse. Tef might be a savior for such sites. India, in particular, has vast areas of these "impossible" soils.

SPECIES INFORMATION

Botanical Name *Eragrostis tef* (Zucc.) Trotter

Synonyms *Poa abyssinica* Jacq.; *Eragrostis abyssinica* (Jacq.) Link

Common Names

Afrikaans: tef, gewone bruin tef (ou bruin)
Arabic: tahf
English: tef, teff, Williams lovegrass
Ethiopia: tafi (Oromo/Afar/Sodo), tafe-e (Had); t'ef, teff, taf (Amarinya, Tigrinya languages)
French: mil éthiopien
Malawi: chimanganga, ndzungula (Ch), chidzanjala (Lo)

[23] It is often sown with its relative *Eragrostis curvula*. This perennial has been developed in South Africa into an almost incredible array of types for land protection and reclamation purposes. They are providing outstanding erosion control on toxic, dry, degraded, and infertile slag heaps and other problem sites where nothing previously would grow. (Information from J.J.P. van Wyk, see research contacts.)
[24] Information from H. Kreiensiek.
[25] Early in the growing season, these soils become waterlogged and go anaerobic; later they crack and dry out, breaking off the roots of plants that have survived the trauma. They also get so gooey during the rains that machinery and people cannot move across them.

Description

Tef is an annual tufted grass, 30–120 cm high, with slender culms and long, narrow, smooth leaves. It is shallow-rooted. Its inflorescence is a loose or compact panicle. The extremely small grains are 1–1.5 mm long, and there are 2,500–3,000 seeds to the gram.

The plant employs the C4 photosynthetic pathway, using light efficiently while having low moisture demands. It is a tetraploid with a chromosome number of $2n = 40$.

Distribution

Tef was grown in Ethiopia before recorded history and its domestication and early use is lost in antiquity. Its most likely ancestor is *Eragrostis pilosa,* a wild species that looks very similar and has the same chromosome number. Samples claimed to be tef have been found in the tombs of the Egyptian pharaohs. The plant is still harvested in the wild—and wild tef is eaten, sometimes on a considerable scale, in mixtures with other wild grains (see wild grains chapter, page 251).

Cultivated Varieties

There are many different types of tef. The narrow panicled "muri" (rat-tailed) types and the dwarf, semi-prostrate and short-lived "dabi" types, for example. Both of these differ strikingly from the tall, loose-panicled varieties that are most commonly grown.

As noted, three main color types are recognized in Ethiopia:

- White tef (*thaf hagaiz*). This slow-maturing form is grown in the cool season. It is superior for grain. However, it makes higher demands on the soil and can be grown only below 2,500 m altitude. In South Africa this type is being developed as an export grain.
- Red and brown tefs. These are quick maturing and superior for fodder. In Ethiopia they are usually grown above 2,500 m. Elevation seems irrelevant, however, because this is the type being used in South Africa as a fodder crop.

Environmental Requirements

Daylength The exact requirements are unknown. In South Africa the plant seeds freely between 22 and 35°S latitude (average daylength, 12 hours). In Ethiopia, the latitude is between 5°N and 10°N (daylength, 11–13 hours).

Bringing the Dead to Life . . .
. . . Just Add Water

Although the seeds of many flowering plants can survive complete dehydration, all other plant parts die when they dry. Certain plants, however, have the seemingly miraculous ability to recover from desiccation. Within hours of being watered, their leaves, stems, and sometimes even flowers spring back to life. Tissues that were brown and seemingly irreparably damaged take up a healthy green color and resume active growth once again.

No one knows how many species can defy drought in this way, but it is a small number, and at least four of them are African grasses related to tef. This suggests that crossbreeding them with tef might yield hybrids combining the qualities of a good cereal with the ability to withstand the ultimate drought.

This fascinating possibility of a fail-safe crop that can bounce back from complete desiccation is being studied by Australian plant physiologist Don Gaff.* So far, his biggest problem (other than getting funds for such far-out research) has been to get tef to breed with its "resurrection relatives." Fertility barriers between the species are too high for natural pollination, so Gaff has adopted a process known as "somatic hybridization." Using electrical pulses, he induces cells from the leaves to fuse as if they were normal pollen and egg cells. To accomplish this, he must first strip the cells of their cellulose walls. The fused cells resulting from this forced marriage can be regenerated into whole plants using the techniques of tissue culture.

Although only at the beginning of this challenging work, Gaff has already found four eligible partners for tef. These are:

- *Eragrostis paradoxa*. A rare species collected in Zimbabwe, this relatively low-growing grass with very fine leaves has remarkable resilience and has survived growing on soils only 1 cm deep.
- *Eragrostis hispida*. This species, too, was from Zimbabwe and is taller and has broad, hair-covered leaves.
- *Eragrostis nindensis*. A vigorous grower, widely distributed in Namibia and other arid areas of southern Africa, this wild tef is locally valued as sheep fodder.
- *Eragrostis invalida*. Gaff's sample of this perennial was collected in the Tingi Mountains near the Niger River's source in Sierra Leone. Tallest of the four, it is still only 60 cm high; short rhizomes assist its clumps to spread.

* Don F. Gaff, Department of Ecology and Environmental Biology, Monash University, Wellington Road, Clayton, Victoria 3168, Australia.

Rainfall The average annual rainfall in tef-growing areas is 1,000 mm, but the range is from 300 to 2,500 mm. Tef resists moderate drought, but most cultivars require at least three good rains during their early growth and a total of 200 to 300 mm of water. Some rapid-maturing cultivars can obtain the 150 mm they need from water retained in soils at the end of the normal growing season. Most tef in South Africa is planted in the 500–800 mm summer rainfall zone.

Altitude Tef can be grown from near sea level to altitudes over 3,000 m. It is particularly valued for areas too cold for sorghum or maize.[26] It has a wider altitudinal range than any other cereal in Ethiopia. Most is cultivated between 1,100 and 2,950 m.

Low Temperature While tef has some frost tolerance, it will not survive a prolonged freeze.

High Temperature Tef tolerates temperatures (at its lower altitudinal range) well above 35°C.[27]

Soil Type Tef's tolerance of soil types seems to be very wide. As noted, it performs well even on the black cotton soils that are notoriously hostile to crops and farmers. In fact in South Africa it is already very popular on such soils.[28] Soil acidities below pH 5 are apparently no problem for tef.[29]

[26] In Lesotho, for instance, it occurs at altitudes up to 2,000 m, where temperatures drop to -15°C. Information from H. Kreiensiek.

[27] It is, for example, grown with irrigation at Gode on the Wadi Shebele River in the Ogaden where the temperature reaches 50°C.

[28] The farmers, however, use twice the normal seeding rate. Information from N.F.G. Rethman.

[29] Information from H. Kreiensiek.

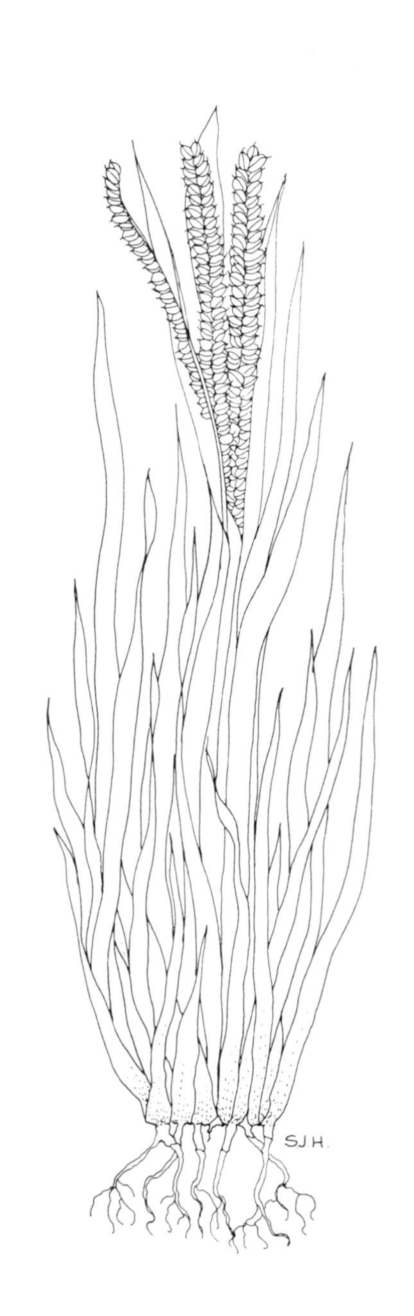
SJH

13

Other Cultivated Grains

Some of the neglected cereals described previously—sorghum, finger millet, and pearl millet, for example—are not, strictly speaking, "lost." But there are a number of African food grains that are indeed truly overlooked by all of modern science. Most of these come from wild grasses (see next chapter), but some are from plants cultivated by farmers on at least a small scale. These last, Africa's least known grain crops, are discussed here.

GUINEA MILLET

Guinea millet (*Brachiaria deflexa*) is perhaps the world's most obscure cereal crop. It is cultivated by farmers only in the Fouta Djallon plateau, a rather remote region of northwestern Guinea. Little, if anything, has been done to improve this crop, yet where it is grown the people value it highly. They grind its soft seeds into a flour, which is used for cakes and fritters.

Although this domesticated plant is grown only in this one area of the Guinea highlands, the wild form is spread throughout the Sahelian zone from Senegal to the Horn of Africa as well as in coastal savannas from Ivory Coast to Cameroon. This wild form is also harvested for food.[1] The main difference between the two is that the cultivated type has much larger grains and is nonshattering (holds its seeds).

This plant grows to about 1 m tall, and looks so much like fonio (see Chapter 3) that for decades it was classified as just a special fonio variety.[2] However, it has botanical differences and bears larger grains.

Although unstudied by agronomists, guinea millet appears to have useful characteristics. For instance, some types mature so quickly

[1] Another wild relative (*Brachiaria stigmatisata*), found from the Gambia to the Sudan, is widely gathered as a cereal as well.

[2] This was previously considered a cultivar of fonio (*Digitaria exilis*, see fonio chapter, page 59). Locally, it is often called "fonio with thick seeds."

Brachiaria species, thought to be guinea millet, growing in Fouta Djallon region of Guinea. (B. Simpson)

they take only 70–75 days from planting to harvest (most, however, require 90–130 days). Commonly, farmers use these fast-maturing guinea millets to fill in any gaps in their fields of sorghum, maize, or other grains. This allows them to get a full harvest from those fields.[3] To achieve truly quick growth, however, a rich and well-drained soil is required.

Guinea millet deserves recognition and attention from scientists and others interested in helping food production and agriculture across West Africa. Despite its current obscurity, it just might have a big future both there and in other regions.

EMMER

Emmer (*Triticum dicoccum*) is not strictly African; it is a wheat that originated in the Near East. Indeed, it was one of the first cereals ever domesticated[4] and was part of the early agriculture of the Fertile Crescent. Farmers had it in fields perhaps as far back as 10,000 years ago. For several thousand years it remained a major cereal throughout the Middle East and North Africa. Then people switched to durum wheat—the type now used worldwide to make spaghetti, macaroni, and other pastas. In fact, durum wheat (*Triticum turgidum* var. *durum*) probably originated from emmer by mutation. Farmers preferred it because its grain was free-threshing (the seed fell out of its husk quite easily), and during the past 2,000 years or so the older form, emmer, became an abandoned waif.

Despite its Middle Eastern origin, emmer nonetheless has an ancient African heritage. It reached Ethiopia probably 5,000 years ago, perhaps more, and it survives there to this day.[5] Whereas it virtually disappeared elsewhere, emmer comprises almost 7 percent of Ethiopia's entire wheat production. Even in what is a major, modern, wheat-growing region, it remains important. Indeed, far from abandoning it, farmers in Ethiopia's highlands have over the last 40 years increased the percentage of emmer that they grow.[6]

Emmer, locally known as *aja*, is used in various ways. Some is ground into a flour and baked into a special bread (*kita*). Some is crushed and cooked with milk or water to make a porridge (*genfo*). And some is mixed with boiling water and butter to produce a gruel. With emmer's high protein content and smooth, easily digested starch, the gruel is especially favored by invalids and nursing mothers.

[3] Portères, 1976.
[4] Together with two-rowed barley (see page 245) and einkorn (*Triticum monococcum*), which, like emmer, is a predecessor of modern wheat.
[5] It also survives as a crop in a small way in Yugoslavia, India, Turkey, Germany (Bavaria), France, and other countries. Information from J. Harlan.
[6] Tesemma, 1986.

Resurrecting Biblical Wheats

Emmer (see previous page) is just one of several ancient wheats that could help the modern world. Two others are being rescued in Europe. The efforts summarized below could be the spur for similar endeavors to bring emmer back as a major crop as well.

Einkorn

Until recently everyone thought that einkorn, perhaps the earliest of all cultivated wheats, was essentially extinct. But in 1989, botanist Jacques Barrau reported the following experience in the south of France.

"In 1971, I decided to look at all the food plants in the mountains of Vaucluse, where my father's family had its origin. From childhood memories, I knew that a kind of porridge was a popular peasant dish there in winter. I started looking for the cereal used for that purpose and found to my surprise that it was the neolithic [Stone Age] einkorn, *Triticum monococcum*. The crop was still being grown there, as well as in some localities in the Southern Alps, as a subsistence cereal of which the unground grain was used to prepare this special porridge. This was unknown to my learned friends in French agricultural research.

"Today, this relict prehistoric wheat is beginning to find markets as a 'natural health-food,' and it sells at a price rather satisfying for the stubborn traditional growers who, through generations, had kept it in cultivation, just to satisfy their lasting taste for this porridge."

Spelt

For the Stone Age inhabitants of what is now south Germany, spelt (*Triticum spelta*) was the main food source. Later, however, this primitive winter cereal was abandoned—not because of inferiority but because farmers found other wheats easier to grow. For one thing, spelt's grain had a close-fitting husk that made it harder to thresh, and its very long straw meant that summer winds could blow the plants down.

Now, spelt (or dinkel as it is usually called in Germany) is coming back as a crop. In this case, the driving forces behind its return are modern consumer preferences—notably the rising appreciation for good nutrition and for protecting the environment.

Nutritionally speaking, spelt is very exciting. Breadmaking wheats in northern Europe generally contain around 11 percent protein. Spelt averages between 14 and 15 percent; some types have even exceeded 17 percent. The grain also has greater concentrations of minerals and vitamins. Even with its lower yield, spelt can produce more protein per hectare than modern breadwheat. And a growing number of consumers are acclaiming the "nutty" taste of products baked from spelt flour.

Spelt's environmental advantages are proving even more important. "The kernel is protected against fungi or insects by the close-fitting husk," explains Christof Kling, head of wheat breeding at Hohenheim University in Stuttgart. "This means the crop is very appropriate for use in environmentally sensitive areas or where farmers want to use less pesticide, or even none at all."

In the old days, when people had to thresh grain by hand, the very attribute that helps to protect the grain against pests and diseases—the close-fitting husk—was an overwhelming disadvantage. But in our mechanized era it is inconsequential.

Like einkorn and emmer, spelt never disappeared entirely, but until recently it was grown in only a few isolated pockets in Germany, Belgium, Switzerland, and Austria. Now all that has changed. In fact, enthusiasm for this long-lost grain is so high that spelt in the early 1990s is being cultivated on over 6,000 hectares in Germany alone. Indeed, a special organization (the Dinkelacker Foundation) has been established to help foster this prodigal son's return from the Stone Age.

Emmer

Recently, researchers in Syria have become excited over emmer. Samples gathered from different parts of the country grew surprisingly well when planted at two ecologically different locations (Tel Hadya and Breda). A wealth of qualities soon became apparent. The researchers concluded that their samples were: "an important genetic reservoir of variability for useful characters such as earliness, short stem, high number of fertile tillers [see picture overpage], long spikes, dense spikes, high number of seeds per spike, weight of kernels per spike, and protein content."

They also noted that most of the emmers exhibited traits suitable for cultivation in the arid areas. "Tolerance to drought is also one of [the] traits, which could be used in breeding wheat for the dry areas," they said.

* This work was conducted by S. Hakim and M. Y. Moualla at Tishreen University, Lattakia, and by A. B. Damania at the International Center for Agricultural Research in the Dry Areas (ICARDA), Aleppo.

Emmer. This sample, with its remarkable number of separate seedheads (tillers), was discovered by researchers in Syria when a small collection of emmer seeds was planted at two ecologically different locations (see previous page). In addition to having more than 20 seedheads, this plant tolerates drought and its grain has a protein content of 18–21 percent. (A.B. Damania)

This "cereal that refuses to die" deserves better treatment from science and commerce. Its economic importance in Ethiopia alone makes it worthy of research attention. However, there might also be worldwide interest. Already, small projects to restore it to widespread modern use are under way in the United States and France (see box). The plant grows in a wide range of environments and can be produced in many parts of the world. The fact that it is the wheat family's "living fossil," little changed from wheat eaten in the times of the Bible and the Koran could give it special consumer appeal. But it can also stand on its own culinary merits. Pliny the Elder (AD 23–79) wrote that emmer wheat makes the "sweetest bread," and even today its virtues are hailed with similar plaudits.

On the face of it, emmer might also benefit the world's wheat-breeding programs. Already, its genes have conferred on the American wheat crop resistance to rust, a virulent fungal disease that in earlier times periodically devastated the nation's food supply.[7] Its other desirable characteristics include early maturity, drought resistance, and a high protein content.

BARLEY

Although barley (*Hordeum vulgare*) is probably not a native of Africa either, it also has been used in Ethiopia for at least 5,000 years. Indeed, Ethiopian barleys have been isolated so long that two of them, irregular barley and deficient barley, were for a time considered distinct species.

Among these two genotypes, as well as among the rest of the diversity of barley forms, can be found a wealth of promising types in addition to genes for use in the world's barley crop. In fact, Ethiopia's assorted barleys are said to be a vital part of its cultural heritage. Under normal circumstances each family sticks tenaciously to its own seed stock. Thus, over thousands of years, each family's stocks have evolved along separate and divergent lines and a vast diversity has resulted. Today, the fields are amazingly rich in different types. In fact, each farmer usually cultivates complex mixtures or even separate plots of quite distinct barleys.

Barley ranks third in terms of area (after tef and sorghum) in Ethiopia. However, its value goes far beyond just economics and nutrition. It is, in fact, deeply rooted in the cultural life. The Oromo people, for instance, consider it the holiest of crops. Their songs and

[7] For example, severe rust epidemics wiped out vast acreages of wheat in 1904, 1918, 1935, and 1953, each time sowing fear and high prices. In 1918, the harvest was so bad that the U.S. government had to declare "wheatless days," on which no wheat products could be sold.

Ethiopian Barley in New Mexico

Although Ethiopia's barley is all but unknown elsewhere, at least one overseas group has attempted to grow it, and with considerable success. In the dry southwestern quarter of the United States, the Ghost Ranch, a facility sponsored by the Presbyterian church, has been growing it as one of its main cereal crops since 1983. Following are comments by the farm's manager. The photograph was taken after the 1991 harvest.

We grow Ethiopian barley at our experimental farm in the northern mountains of New Mexico. We grow it for three main reasons: it matures quickly (about 110 days); it is hull-less; and it is the most drought-tolerant grain we've ever had. In addition, it has been almost trouble free. We've never experienced a problem with lodging. The plant tillers very well and produces good yields in most years. We haven't had any problems with disease, which might be only because our farm is isolated and the nearest barley grower is about 50 km away.

We thresh the dry grain in a small homemade threshing machine or an old combine employed as a stationary thresher. It threshes easily. The seed is then cleaned in a seed-cleaning machine. (Both the threshing machine and the seed cleaner run off our solar electric system.) The grain mills nicely and produces a flour that has good baking and eating qualities.

Lynda S. Prim

sayings often feature this "king of grains." Everyone in the highlands encourages children to consume lots of barley. It makes them brave and courageous, they say.[8]

Ethiopians turn barley into bread, porridge, soup, beer, and many other foods. A favorite snack is roasted unripe barley seed. Several types are made into various barley-water drinks, most of them nonalcoholic.[9] These beverages (made of water infusions of roasted and ground grains) are highly valued. Also, some intoxicating liquors (*areuie*) are home brewed from barley grains.

Ethiopians draw clear associations between each grain type and its use. The white large-grained forms are preferred for porridges. The white, black, or purple large-grained types are made into bread and other baked foods. Partially naked grains are usually roasted or fried. Small-grained types (mainly black and purple) are used for beverages.

Barley is also important to the country's livestock. The grain itself is sometimes fed. (Wealthy farmers, for instance, use it to fatten horses and mules before and after long journeys or to strengthen cattle before the plowing season or going to market.) But more commonly, the animals end up eating the straw. Finely broken barley straw is also employed in constructing mud walls.

For all its importance, however, Ethiopia's barley production can be strengthened. A vast store of indigenous germplasm has yet to be tapped. Indeed, some of it is being lost. (This genetic erosion is happening mainly as farmers switch to crops such a bread wheat, tef, and recently, oats.)

Some of Ethiopia's barley could be made more useful by genes of the barleys developed elsewhere in the world. But the multitude of local types offer great opportunities on their own accounts. Many are unique. Even the number of rows of grains on the seedhead (spike) can be unique. Everywhere else in the world, barleys have exactly two rows or six rows. However, Ethiopia's irregular barley has two full rows as well as parts of other rows. And its deficient barley has two full rows, but the lateral spikelets are greatly reduced or are wanting entirely.

Although essentially unknown elsewhere, irregular barley[10] ranks fourth among Ethiopia's crops, both in quantity produced and area planted.[11] At altitudes above 2,500 m it is usually the only cereal that

[8] The ancients had similar traditions. Greeks, for example, are said to have fed much barley to gladiators. Roman gladiators were called "hordearii" in the belief that barley was the source of their strength.

[9] A popular trail food is roasted and ground barley. The traveler can stop at any stream, stir the powder into a cup or gourd of water, and have "instant barley water."

[10] It is also known as Abyssinian intermediate barley. It occurs also in Yemen, Arabia, and Egypt, but only as a very minor crop.

[11] Hailu and Pinto, 1977.

Thoroughly Modern Millets

Whereas today's most reliable approach to advancing little-known grains is conventional plant breeding, biotechnology might soon be able to leapfrog much of the tedious and time-consuming toil traditionally involved in creating new varieties. Here we identify a few possibilities.

Because they are well known in scientifically advanced countries, wheat, rice, maize, and (to a lesser extent) sorghum have benefited from high-tech research. Millets, however, remain almost exclusively resources of countries with little or no basic research capacity. Millets have therefore barely benefited from the latest instruments and techniques. Given such attention, it seems likely that they can be leapfrogged into the twenty-first century using biotechnology.

This is especially important to Africa, where the needs are so vast and diverse, the resources so few, the time so pressing, the conditions so changeable, and the priorities so uncertain that conventional plant breeding, which can take 10–12 years to perfect a new variety, may not be up to the task. Certainly, its ability to breed for genetically complex attributes such as drought tolerance is limited. Moreover, in environments such as the Sahel, where climatic variables far outweigh genetic ones, plant breeding is all but impossible to do in the normal way in field trials.

When it comes to Africa, then, biotechnology could have a huge impact. For example, breeding can be done more quickly, it can be done indoors in controlled environments, and it can be done with greater precision. Increasingly, biotechnology can deal with genetically complex traits. In sum, technologies such as tissue culture, anther culture, embryo rescue, protoplast fusion, and genetic markers are likely to bring undreamed of breakthroughs that will transform Africa's native grains.

The key to this gene revolution is to develop tissue-culture techniques for each of Africa's grains. If scientists can grow mature, fertile plants from tissues of pearl millet, finger millet, fonio, irregular barley, and tef, they will open doors to the more rapid development of these cereals. Grasses are difficult to culture—so difficult, in fact, that not long ago they were considered impossible—but rice, maize, sorghum, and vetiver have already succumbed and can be grown routinely in tissue culture. Now it seems likely that the right conditions can be discovered for the others.

Once tissue culture has been established, a major challenge will be to "map" the chromosomes using genetic "markers." Knowing the physical location of particular genes will result in many shortcuts to improved strains. This is particularly because thousands of young seedlings can be tested for the presence of specific genes, rather than waiting for the genes to express themselves in the mature plant. It will also allow desirable genes to be more easily transferred, and undesirable ones to be eliminated. The markers could be provided by the restriction-fragment length polymorphism (RFLP) technique (see box, page 34), a process already being applied to maize, barley, and rice.

Following are examples of the gains to be achieved:

Drought Resistance. Breeding drought-resistant varieties has always been difficult because researchers had no way to determine genetic influences on the basic mechanisms of drought injury and tolerance. In basic studies, biotechnology is now helping to show how water stress affects the physiological, biochemical, and molecular organization of plants during their various life stages.

In future, the new techniques could target the genes that govern rooting depth, water extraction, and root penetration of compacted soil layers. Once identified and mapped, the genes for these characteristics (which are extremely difficult to evaluate in the field) could be readily tracked in breeding programs. This would lead to crops with much higher drought tolerance.

Striga. An ability to manipulate the genes that attract or repel the striga parasite could boost cereal yields continent-wide.

Hybrids. Biotechnology would make it much easier to make hybrids within and between species. This might be brought about through chemical hybridizing agents, through clonally propagating sterile seed, or through embryo rescue.

As work progresses on the major crops in the world's most sophisticated laboratories, millets should not be overlooked. Pioneers pushing the frontiers of gene manipulation in wheat, maize, and rice, for example, should not leave the millets trailing so far behind that they will be abandoned willy-nilly. Actually, the high-tech equipment and powerful genetic tools could likely help make major advances in millets and thereby bring more humanitarian benefits than in all the rest of the work.

can be cultivated satisfactorily. It is very important throughout most of the upper highlands, for example, where it accounts for about 60 percent of the population's total plant food. Farmers in that area rely on fast-maturing types to save their families from starving during food shortages.

This is just one example of the genetic wealth to be found among Ethiopia's barleys. Other traits include:

- High yields. Some Ethiopian barleys have big and heavy kernels, some plants tiller (send up multiple shoots and seedheads) very well, and others mature quickly.
- High nutrition. Some have high levels of protein and a few are high in lysine and are thus exceptionally nutritious. They are the only known source of quality-protein barley.[12]
- Disease resistance. Several have resistance to diseases such as powdery mildew, leaf rust, net blotch, *Septoria,* scald, spot blotch, loose smut, barley yellow dwarf virus, and barley stripe mosaic virus.
- Drought resistance. Many have the ability to grow under dry conditions—a feature apparently related to deep and efficient root types.
- Tolerance to marginal soils.
- Resistance to barley shoot fly and aphids.
- Vigorous seedling establishment.

On the other hand, Ethiopia's barleys tend to blow down easily due to weak straw and tall, spindly growth. Some specimens suffer from the condition known as "fragile rachis," in which the seed spike breaks apart and spills the seeds on the ground.

The outside world's barley breeders have not neglected Ethiopia's materials. For example, they employ the accession called Jet (jet-black seeds) to obtain resistance to loose smut, a severe fungal disease. In the United States and several other countries they have employed the genes for resistance to the extremely damaging barley yellow dwarf virus, leading to great savings in grain yields. But many more useful types remain to be employed both at home and abroad.

ETHIOPIAN OATS

Ethiopia also has a native oats, *Avena abyssinica*. Partially domesticated in the distant past, this species is largely nonshattering—that is, it retains most of its grain so farmers can harvest them conveniently.

[12] See companion report, *Quality-Protein Maize*. Quality-protein barleys are rich in amino acids, such as lysine, that are vital to human nutrition and yet normally deficient in cereals. They have been called "Hi-proly" by the Danish food scientists who have studied them most. (For a list of BOSTID publications, see page 377.)

It has long been used in Ethiopia and is well adapted to the high elevations and other conditions there. It is, however, unknown elsewhere. With a rising international interest in oats this little-known species deserves research attention.

Unlike common oats (*Avena sativa*), which is a hexaploid, Ethiopian oats is a tetraploid. It is seldom grown as a solitary crop; it is almost always sown in a mixture with barley. Agriculturists may classify it as a weak-stemmed "weed," but not the farmers. They harvest the two grains together and use them mainly in mixtures. These mixtures generally end up in *injera* (the flat national bread; see last chapter), local beer (*tala*), and other products. Some are roasted and eaten as snacks.

However, some people don't appreciate Ethiopian oats because the plant is not fully domesticated and does shatter somewhat. It is also fully fertile with the weed *Avena vaviloviana,* which creates swarms of weedy hybrids that shatter a lot.[13]

Nonetheless, Ethiopian native oats deserves research attention and a chance to prove itself.

KODO MILLET

Although wild forms of kodo millet (*Paspalum scrobiculatum*) occur in Africa, the plant is not grown as a crop there. However, domesticated forms have been developed in southern India, where they are planted quite widely. This is therefore a plant in the very process of domestication, and the cultivated forms could have an important future in Africa as well.

The wild form is common across tropical Africa (as well as across wetter parts of the Asian tropics from Indonesia to Japan). It is often abundant along paths, ditches, and low spots, especially where the ground is disturbed (which accounts for the reason it is sometimes called ditch millet).

Although kodo millet frequently infests rice fields in West Africa, it is tolerated even there. Many farmers actually take pleasure in seeing it in their plots. Should the rice crop fail or do poorly, they will not have lost everything . . . the field will likely end up choked with kodo millet, which can then be harvested for food. In this sense, the weed becomes a lifesaver for a subsistence-farming family.

All in all, this is another obscure cereal deserving greater modern research and recognition. Two technical problems to evaluate are an ergotlike fungal disease and the probable presence of antinutritional compounds.

[13] Specimens from these two species, as well as the hybrid between them, have also been referred to as the species *Avena barbata* Pott., from which the Ethiopian species may have been derived.

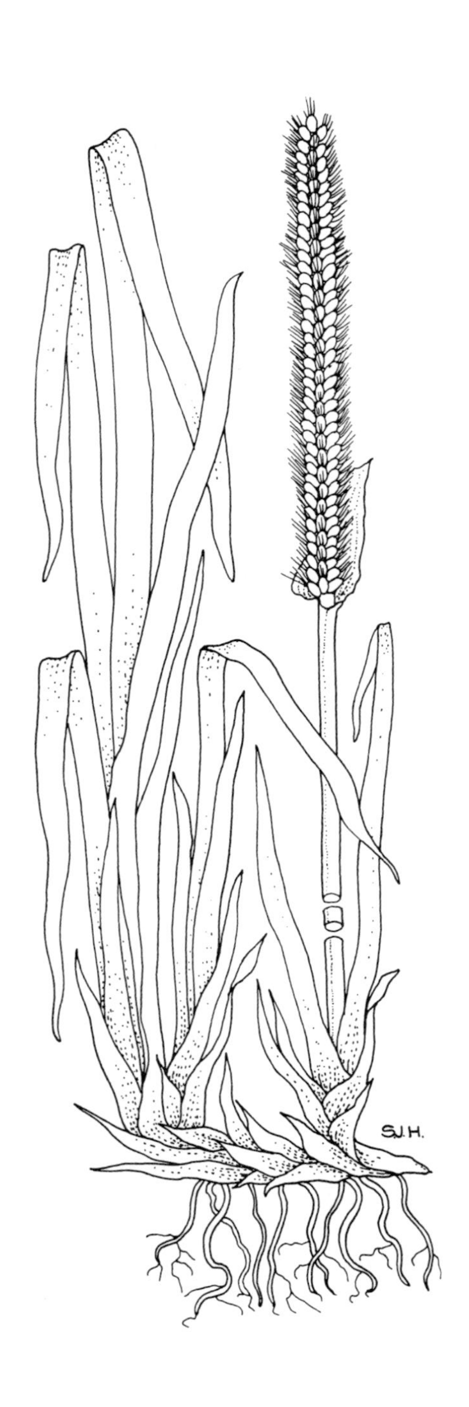
S.J.H.

14

Wild Grains[1]

Over large areas of Africa people once obtained their basic subsistence from wild grasses. In certain places the practice still continues—especially in drought years (see boxes, pages 258 and 264). One survey records more than 60 grass species known to be sources of food grains.[2]

Despite their widespread use and notable value for saving lives during times of distress, these wild cereals have been largely overlooked by both food scientists and plant scientists. They have been written off as "obsolete"—doomed since hunting and gathering started giving way to agriculture thousands of years ago. Certainly there has been little or no thought of developing wild grains as modern foods.

This deserves reconsideration, however. Gathering grains from grasslands is among the most sustainable organized food production systems in the world. It was common in the Stone Age[3] and has been important almost ever since, especially in Africa's drylands. For millennia people living in and about the Sahara, for instance, gathered grass seeds on a grand scale. And they continued to do so until quite recently. Early this century they were still harvesting not insignificant amounts of their food from native grasslands.

However, in previous centuries the grains of the deserts and savannas were harvested in enormous quantities. In the Sahel and Sahara, for example, a single household might collect a thousand kilos during the harvest season.[4] The seeds were piled in warehouses by the ton and shipped out of the region by the caravan-load. It was a major enterprise and a substantial export from an area that now has no equivalent and is often destitute.

[1] Much of this chapter is based on a review by Jack Harlan (Harlan, 1989).
[2] Jardin, 1967.
[3] Many of the stone (Mesolithic) implements found in archeological sites throughout the Sahara were probably created for harvesting wild grass seeds. Some are still used. Modern desert dwellers, for example, find it convenient to employ ancient grindstones they find at archeological sites instead of carrying their own from camp to camp. Information from J. Harlan.
[4] Nicolaisen, 1963.

How the Millets Arose

It is not illogical to think that at least some of the wild grasses in this chapter might be turned into tractable crops for farm fields and household gardens. It has been done in the past . . . by our Stone Age forebears, no less.

Between 12,000 and 6000 B.C., most of the Sahara appears to have been perfectly hospitable to humans. What is today the world's most fearsome desert then enjoyed a mild climate, winter rainfall, and an extensive grass cover. Acacia and tamarisk trees lined the many water courses. Mountainsides were verdant woodlands of myrtle, oak, hackberry, and olive, with juniper and pines at the upper altitudes.*

By 10,000 B.C. people inhabited the area. A scattering of Neolithic (New Stone Age) sites across the central Sahara provide evidence that they were using sickles and grinding equipment, which suggests that they were using the grasses. By 6000 B.C. the central Sahara people were definitely collecting wild grain as well as apparently hunting wildlife and herding livestock. The vast grasslands provided game as well as limitless grazing for cattle, sheep, and goats. Shallow lakes—occupying wide, flat pans—enlarged during the rains and provided plentiful food from fish, hippopotamus, and aquatic plants, including African rice.

But then, after about 4000 B.C., the region began drying out. The desert as we know it today had begun to form. Few archeological sites from this period are found, and the people apparently had been forced to leave.

But before they left, they had time to domesticate some of the grasses around them during the thousands of years the rather sedentary, herding-fishing-hunting people occupied the Sahara. Several cereals seem to have arisen there. African rice, fonio, pearl millet, sorghum, and perhaps finger millet got their start this way.

Those ancients did a miraculous job, considering they had no knowledge of genetics, microorganisms, chemistry, nutrition, or the myriad other sciences we now consider vital for domesticating and developing crops. Nor did they have ready access to the variety of germplasm that any scientist today would demand. If they could do it, surely we can.

* All this is suggested by numerous pollen samples dug up in the Tibesti and Haggar massifs in the heart of the Sahara.

But in modern times these wild grains have been neglected and even much maligned. Various writers repeatedly refer to them as "famine foods." This is obviously wrong. Where the grains were gathered, surplus was often the rule. Wild grains were eaten even when pearl millet was in oversupply, for instance.

Modern writings also imply that the wild-grass grains were eaten only in desperation when nothing else was available. This, too, is apparently false. The harvest was large scale, sophisticated, and commercial: it must have been founded upon a keen and constant demand. Indeed, all evidence suggests that the grains were a delicacy that even the wealthier classes considered a luxury.

Remnants of this once vast and highly organized production still linger. One observer pointed out that harvests of wild grains were still being carried out in 1968, at least 60 years after they had last been major contributors to the local diet.[5] However, despite its former prestige and ancient heritage, the wild-grain harvest has been declining for a century or more.

A major reason for the decline is that the once vast stands of grasses are much reduced. Partly this results from the demise of the nomads. Sedentary life encourages continuous and localized grazing so that the plants never get a chance to form grains. Partly, too, the decline results from the breakdown of traditional authority. Formerly, chieftains banned grazing animals from certain areas while the wild grains were filling out. If camels were caught there during that time, the chieftain could slaughter one of them in recompense; if goats were caught, he could kill as many as 10.

Just because wild grasses no longer contribute greatly to Africa's food does not mean they should be disregarded. Even preliminary study is likely to turn up many fascinating possibilities and perhaps much future potential. Many come from locations where burning temperatures, scant rains, and ravenous insects make the better-known grains impossible to produce. Some can populate and stabilize sand dunes—perhaps even the juggernaut dunes that threaten to bury oases, farms, villages, roads, and towns. Forged upon the unforgiving anvil of survival, these wild grasses are clearly suited to the worst of conditions.

In fact, plants like these—inured to harshness and constantly pressured by pathogens, pests, severe weather, and harsh soils—are just the sort of resources the world needs for overcoming some of its most intractable environmental problems. For example, some of Africa's wild cereals might be especially good weapons for combating desertification. Indeed, resurrecting the ancient grain-gathering industry could well be a way to defeat land degradation across the worst

[5] Gast, 1968.

Northern Aïr, Niger. Relict stand of wild sorghum that is harvested for grain each year. All over Africa, stands like this one are relied upon to provide food, especially during times of crop failure. (John Newby, WWF)

afflicted areas of the Sahel and its neighboring regions. A vast and vigorous grain-gathering enterprise, for instance, would ensure that once again the grass cover is kept in place and that overgrazing is controlled once more.

Such a possibility is not inconceivable. Wild cereals might be made into an everyday food source, a famine reserve, and perhaps even a specialty export crop. This last may seem unlikely, but it should at least be considered. Today, the overall situation is different from that of a century ago. Railroads and airfreight mean that grains can now be shipped from the Sahara with much greater ease than on the backs of camels. Moreover, consumers in affluent nations are increasingly interested in buying and trying "exotic" cuisines. And many people of goodwill are highly motivated and eager to help avoid the horrendous tragedies of Sahelian drought and famine they have witnessed on their television screens in recent decades.

A similar concept is being attempted as a way to combat the destruction of tropical rainforests. In the last few years, for instance, an international trade in special tropical-forest products has begun. The object is to foster an economy based on resources of the rainforest itself. If successful, it will generate powerful local disincentives for destroying the natural environment.

In the case of the rainforest, the products are such things as wild rubber, fruits, nuts, and vegetable-ivory buttons. In the case of Africa's desertifying areas, the product might be kreb.

Kreb is perhaps the most famous food of the Sahara. A complex of a dozen or more different wild grains, it was harvested from natural meadows. Its composition varied from place to place and probably from year to year, depending on the mix of grasses that grew.

These days, given some clever marketing, "kreb from the Sahara" might sell at premium prices in Europe, North Africa, and North America, for example. It would be seen as a gourmet food that provides income to nomads and protects the earth's most fragile lands from further destruction by keeping a cover of wild native grasses on them.

Although this idea is highly speculative, subject to many limitations and uncertainties, it is not beyond reason. Mixed-grain products are not uncommon in Western supermarkets these days. For instance, in the United States a popular breakfast cereal is a grain mixture that people boil in water like rice. (It is made from conventional grains but goes by the trade name "Kashi," another word for kreb.[6]) And some expensive breads are made from as many as 11 different grains.

[6] The pamphlet in each box explains: "Kashi, the breakfast pilaf, is a specially formulated pure blend of whole oats, long grain brown rice, whole rye, triticale, hard red winter wheat, raw buckwheat, slightly hulled barley, and mechanically dehulled sesame seeds; 100 percent quality whole grains that are not cut, cracked, rolled or flaked nor creamy or mushy when cooked."

Resurrecting the production of kreb could provide food, income, and perhaps a protection against famine. It might bring substantial environmental benefits as well. Many of the wild African grains come from perennial grasses that continuously cover the soil and protect it from water and wind erosion. In addition, these plants facilitate the infiltration of rainfall and prevent rapid runoff from desert downpours early in the season, a time when annuals are still getting started and much of the soil around them is exposed and hard. Moreover, perennial crops have long growing seasons and the extra solar energy they collect normally produces good grain yields. (This is why some hybrids, including maize hybrids, have been so productive.)

Native perennials might prove to have economic benefits as well. Perennials save the vast amount of energy and labor that farmers must put in each year to move soil for planting and tilling annual cereals. Also, they save on the often large amount of grains that must be put aside each year for planting—with a perennial, those can be eaten.

Beyond their direct use as cereals, Africa's wild grasses may also have international value as genetic resources. Some are related to species used elsewhere for food or fodder and are likely to have genes of international importance—particularly because many of them have outstanding tolerance and resistance to heat, drought, drifting sand, and disease. On the other hand, some might prove weedy when taken out of the desert and introduced to more salubrious situations.

The nutritional value of wild-grass seeds has seldom been studied in detail, but those analyses that have been made indicate that protein contents are usually considerably higher than that of cultivated cereals. Several Saharan grains, for instance, have protein contents of 17–21 percent, roughly twice that of today's main cultivated cereals.[7]

All cereals are low in vitamins A, D, C, B_{12}, and the amino acids lysine and tryptophan. Wild grass seeds are no exception. However, some may be unusually high in food energy. Certain kram-kram seeds, for instance, apparently have about 9 percent fat and are perhaps higher in energy than any other cereal grain.[8]

Africa's promising wild cereals include those described below. All of these deserve the attention of food and agricultural scientists, as well as of the people involved in battling Sahelian desertification. Even the most basic studies could be extremely valuable. These include the following:

- Tests to determine how best to plant and establish each species (seed treatments, sowing depths, planting times, and so on);

[7] Busson, 1965. Much of the difference may be due to their small seed size. Domesticated grains are usually bigger, and the increase is primarily due to endosperm, which is largely starch.

[8] Busson, 1965.

Harvesting Wild Grasses

To most people, it probably seems inconceivable that in this age of intensive agriculture, wild grasses are still being gathered. The following (adapted from a recent FAO report) gives a sense of the ongoing importance of wild grains in different parts of Africa.

Niger

On their way from the wet- to the dry-season pastures, the Tuareg of Niger regularly harvest wild cereals. The grains, collectively known as *ishiban,* include desert panic (*Panicum laetum*) and shama millet (*Echinochloa colona*). Women do most of the gathering, and around harvest time groups of five or six women often go off for a week or so to gather wild grains (as well as fruits, gum arabic, and other wild products).

They collect the grains in different ways:

- If the seed is ripe and ready to fall, they harvest early in the morning when dew tends to hold the seed in the inflorescence. They swing a deep, cone-shaped basket through the tops of the plants to gather the grain.
- If the seed is not ripe enough to fall, they first cut the grass and then dry, thresh, and winnow the grain as if it were a domesticated cereal.
- If the seed has already ripened and fallen, they cut or burn the stands, and later sweep the seeds up off the ground. (This spoils the taste and adds soil and pebbles, but the harvesters often have no choice.)
- Sometimes the women search for seeds in ant nests and termite mounds. In desperate times, such as the terrible drought of the 1970s, they even dig down to the ants' subterranean storehouses.

Sudan

The Zaghawa of the Sudan and Chad harvest many annual grasses for food and beer. These include Egyptian grass (*Dactyloctenium aegyptium*), desert panic, shama millet, wild tef (*Eragrostis pilosa*), and wild rice (*Oryza breviligulata*). Kram-kram (*Cenchrus biflorus*) and *Tribulus terrestris* seeds are used only during famine. The women generally use the grains for their own families, but they sell some as well. The Zaghawa spend a month or two in the areas where the wild cereals grow, often returning with three or four camel loads of grain. The various sites are visited several times, at intervals of 15–30 days. The earliest harvests usually yield the most. There is much communal cooperation. The women mentally mark off areas for themselves, cut the grass, and pile it up to dry. To foil any goats or wildlife, they cover their piles with thorny branches, and to guard against theft, they leave a symbolic stone representing each woman's clan. Livestock are barred from these areas until after the grain harvest, and herders are fined if any animals get in. It appears that the gathering actually helps maintain a good stand of wild cereals, because less useful plants (especially kram-kram) are taking over the areas where gathering is no longer practiced.

Zambia

The Tonga of Zambia routinely harvest the grains of wild sorghum and Egyptian grass, and during famines they also harvest species of *Brachiaria, Panicum, Echinochloa, Rottboellia,* and *Urochloa.* They supplement these wild cereals with relishes made from leaves, most of which they also usually find in the wild. These two together provide them with sources of starch, proteins, fats, vitamins, and minerals. They also use wild native plants for brooms, building material, fiber, salt, medicine, poisons, and so on.

South Africa

When in the 1930s the Chamber of Mines began asking about edible wild plants, its labor-recruitment offices across South Africa became overwhelmed. "We were inundated with parcels from many parts of the country containing plants or parts of plants," wrote one of the participants recently. "It became clear that a nutritionally significant part of the people's diet was being obtained from the veld."

Among the grains sent in were those from

- *Sporobolus fimbriatus* (matolo-a-maholo)
- *Brachiaria brizantha* (bread grass, long-seed millet)
- *Echinochloa stagnina* (bourgou)
- *Panicum subalbidum* (manna grass)
- *Stenotaphrum dimidiatum* (dogtooth grass)

- Direct seeding trials using rain as the sole source of moisture;
- Searches for elite specimens (those that, for instance, hold onto the ripe seed, that have bigger seed, and that best survive harsh conditions);
- Trials on various sites (from the most favorable locations to moving sand dunes);
- Analyses of food value (physical, chemical, and nutritional) as well as of the foods prepared from them; and
- Multiplication of seeds or other planting materials for distribution to nomads, farmers, governments, and researchers.

DRINN

The grass known in Arabic as *drinn* (*Aristida pungens*) once provided by far the most important wild grain of the northern Sahara.[9] It was extremely abundant, often growing on sand dunes but especially on bottomlands watered by runoff from higher ground. It is a tall (to 1.5 m), tufted perennial with deep roots and long leaves. Its grains are black.

Travelers crossing the Sahara in the past often wrote about drinn's value, both as a food and as forage. Duveyrier (1864) commented: "its grain is often the only food for people." Cortier (1908) referred several times to the abundance of drinn: "The hillocks of sand in all the plain," he wrote, "are embossed by enormous tufts of drinn, whose black grains at the tips of long stems swing and sweep the soil."

Even as recently as 1969, drinn was still a significant part of the diet in the Sahara oases.[10] In earlier times it was an important food from the desert's edge almost to the Ahaggar (southern Algeria). It was, for instance, vital to people living a tenuous existence in the very heart of this fearsome region; the Toubou of Tibesti (northern Chad) are just one example.[11] In fact, this grass was so crucial to life that desert tribes were characterized as those who cultivated cereals (the Mahboud), and those who gathered drinn (the Maloul).

Drinn is extremely drought resistant. It grows, for instance, between Touggourt and El Oued in Algeria on sand dunes where the average rainfall is less than 70 mm per year.[12]

PANIC GRASSES

Various *Panicum* species have been favored by grain gatherers the world over. *Panicum miliaceum* was once so popular in Europe that

[9] It is also known as *toulloult* or *loul*.
[10] Champault, 1969.
[11] Chapelle, 1958.
[12] Information from P. Beckman.

it became a crop that perhaps predates wheat. Today this plant is grown extensively in the Soviet Union and Central Asia under the name proso millet.

At least seven wild *Panicum* species are gathered for food in Africa.[13] The most important are discussed below.

Panicum turgidum Called *afezu* or *merkba* (Arabic), this grass produces seed that closely resembles proso millet. It was once abundant across the Sahara as well as in desert lands as far east as Pakistan. It was widespread, for example, in Senegal, Mauritania, Morocco, Egypt, and Somalia and was the primary wild grass in a vast belt across the southern Sahara. Its grain was formerly gathered in large amounts, and even today it is still harvested, at least to some extent, throughout the plant's range.

This desert species grows where few crops can. It is extremely drought tolerant, thriving in dry sands in semiarid or arid areas with annual rainfalls from 250 mm down to as little as 30 mm. It is also found in semidesert shrublands and is common among the vegetation inhabiting dried-up wadis.

A deep-rooted, clump-forming perennial, this plant forms loose tussocks 1 m or so in diameter. It spreads by long, looping stolons, building up mats of vegetation that are extremely useful for erosion control. (Its stems fall over and root at the nodes, clamping down the soil.) It is known to colonize wind-blown sand dunes (often while they are still moving) and can protect steep slopes. The root system is extensive, penetrating to below 1 m as well as radiating out horizontally more than 3.4 m in plants excavated in Somalia.

Although *afezu's* main nonfood use is as a sand-binder, it provides some grazing for camels, goats, and other animals. Its palatability is generally low, but its ability to grow in virtual desert conditions, together with its perennial nature, gives it great value.

This plant bears its seeds on panicles that rise above the mat. They can be easily collected by holding the seedheads over a bowl and beating them with a stick. Most of the grain collected ends up in a porridge (*tébik*).

Panicum laetum The grain of this particular panic grass is regarded as a special delicacy. It was an important ingredient in kreb. People still collect it for food in many parts of West Africa, sometimes on a large enough scale that it shows up in local markets. The grains are normally crushed and eaten as a porridge.

This plant, which also occurs in massive stands, ranges from Mauritania to the Sudan and Tanzania. It is an annual, often common

[13] Jardin, 1967.

on black-clay soil in areas that are seasonally flooded. Animals like it, and it is especially well suited for making hay or silage. It is not highly drought tolerant, however.

Because it occurs in almost pure stands, the grain is fairly easy to collect. People sweep a small bowl or calabash through the seedheads during the period when the ripe grains are ready to fall.

Panicum anabaptistum Little has been written about this species. However, its grains are also eaten in at least a few parts of Africa. It, too, is liked by animals and can be utilized for hay and silage. The plant prefers heavy soils and is found predominantly on wet sites. It continues producing green shoots well into the dry season, a valuable feature in any desert forage. People weave its long, dried culms (stems) into mats for their houses.

Panicum stagninum This interesting perennial (also known as *Panicum burgii*) is found throughout much of tropical Africa, especially the Sudan and Central Africa. Instead of producing a useful grain, it yields a thick syrup, which is used in confections and sweet beverages that are widely enjoyed in Timbuktu and other places.

KRAM-KRAM

Along the southern fringes of the Sahara the primary wild cereal is kram-kram (*Cenchrus biflorus*).[14] This annual grass builds massive stands over thousands of hectares of sand plains and stabilized dunes. In earlier times, it was the dominant cereal of both the Sahel and the borderland between the Sahel and the Sahara. In those days it was a more important food than pearl millet, and its grains were milled into flour and made into porridge on a vast scale. As noted earlier, some kram-kram seeds contain 9 percent fat and have perhaps the highest food energy of any cereal. They also have a notably high protein content—21 percent in one recent analysis, or about twice the level found in normal wheat or maize.

Kram-kram[15] is now harvested only when other crops fail, but given some attention it might once again become a universal food for the peoples of the northern Sahel. Also, this wild plant might be converted to a useful crop. Domestication could come about quickly, particularly if its grain were enlarged by selection or cross-breeding with other *Cenchrus* species. The plant grows well on sandy soils. It is a reliable

[14] In older literature this is referred to as *Cenchrus catharticus* Delile.

[15] Other common names are "Sahelian sandbur," *chevral*, and *karindja*. Tuareg names include *karengia*, *wujjeg*, and *uzak*.

KRAM-KRAM

Main Components[a]		Essential Amino Acids	
Food energy (Kc)	325	Cystine	1.7
Protein (g)	19.2	Isoleucine	4.8
Carbohydrate (g)	56	Leucine	15.5
Fat (g)	2.9	Lysine	1.1
Fiber (g)	2.3	Methionine	2.2
Ash (g)	10.2	Phenylalanine	5.2
Calcium (mg)	63	Threonine	3.2
Copper (mg)	0.5	Tyrosine	3.2
Iron (mg)	6.4	Valine	5.5
Magnesium (mg)	63		
Manganese (mg)	2.0		
Phosphorus (mg)	162		
Potassium (mg)	153		
Zinc (mg)	5		

[a]Assuming 10 percent moisture.

COMPARATIVE QUALITY

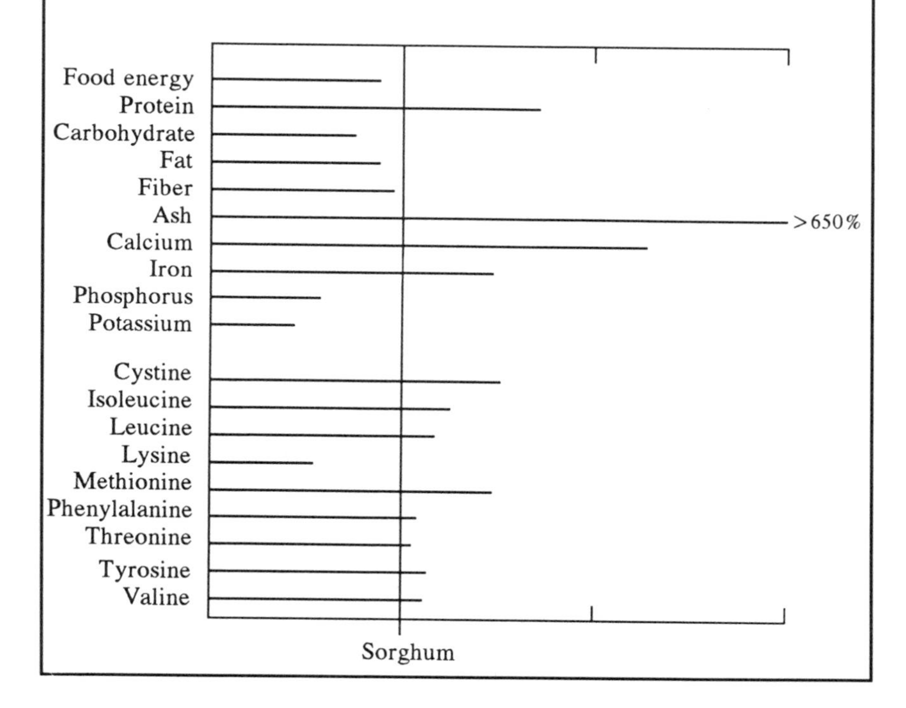

Lakes of Grass

*The following, taken from a 1990 report from the United Nations Sudano-Sahelian Office (UNSO), shows how a farsighted project is restoring one of the formerly important West African wild grasses. Although it emphasizes animal feed, it gives a glimpse of what could be done by developing wild grasses for food**

To farmers and pastoralists in the Inner Delta of Mali, the bourgou floodplains supply a crucial source of fodder. Without these *bourgoutières,* the livestock would die during the dry season. Only bourgou can survive in these bottomlands that go underwater each year for months at a time.

Bourgou is unique in its adaptation to these amazing conditions. As the waters rise around it, the grass grows taller and taller until (after about 3 months) its stems can reach lengths of more than 3 m. At this point bourgou is like an aquatic plant with only its flowers and seedheads sticking above the surface. Once the water level drops, cattle are given access, and as they walk through the shallows, they trample the seeds and runners into the soft ground. This ensures that the crop will survive and grow again. However, when everything has dried out, there remains on the surface a dense mat of grass, half-a-meter thick.

This mat is what is used for fodder. If well managed, bourgou produces nearly 30 tons of dry matter per hectare—a sizable yield even for much more productive locations. When cut and

Bourgou harvest. (UN Sudano-Sahelian Office)

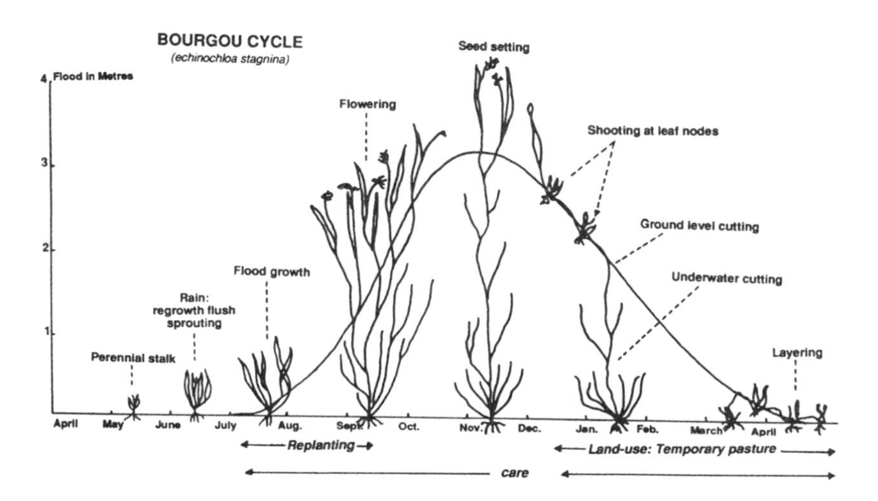

sold in the market, the grass fetches good prices: between 25 and 100 CFA francs per bundle (1–3 kg) in the early 1990s.

The problem, however, is that the period of intense drought, from 1968 to 1985, destroyed many *bourgoutières*. So, in 1982 UNSO and the Malian government began a project to learn how to regenerate bourgou grasslands.

So far, the most effective technique has been to plant rootlings: small, rooted cuttings collected either from existing *bourgoutières* or from nurseries specifically set up for the purpose. The planting (at an average rate of 10,000 plants per hectare) is done by hand. This takes a lot of work, but it has been so successful that this grass has now been re-established on more than 4,000 hectares. And, as bourgou is a perennial, it should continue in those floodplains for decades.

Already, regenerated *bourgoutières* have had a great impact locally. Farmers use the grass both for direct grazing and for making silage and hay. Many have been able to increase their incomes through selling both fodder and milk. Local milk supplies have increased so much that thousands of families have benefited from better nutrition.

UNSO feels that areas all along the Niger River could also be planted with bourgou. It is possible that the grass might even thrive in other river valleys, such as that of the Senegal, where annual floods make better known crops difficult to grow.

* For more information, contact United Nations Sudano-Sahelian Office (UNSO), Avenue Dimdolobsom (section 3), B.P. 366, Ouagadougou, Burkina Faso.

source of forage, since it persists in a dry but palatable state until the next rainy period.[16]

On the other hand, kram-kram is vicious. It is a sandbur whose grains are enclosed in clusters (fascicles) surrounded with many sharp spines. These grab onto the fur of animals and the clothing of people. Indeed, they easily penetrate flesh and have literally been thorns in people's sides for millennia. Travelers have long complained of the plant's "troublesome nature" and "constant inconvenience," but they did admit that it was also very useful. "Many of the Tawarek, from Bornu as far as Timbuktu," wrote Heinrich Barth in the mid-1800s, "subsist more or less upon its seed."

When mature, the burs fall to the sand in great quantities, often clinging together in giant masses that roll along with the wind, growing as they go. People sweep them up with bunches of straw or with giant "combs." They throw them into a wooden mortar and pound and winnow away the troublesome spines, leaving behind the white, flavorful seeds.

Livestock cannot abide the prickly spikelets, but they like grazing on kram-kram both in its juvenile state and after the spiky burs have fallen off. The plant grows vigorously, and during the rainy period it can be cut several times for hay or silage. The hay must be made at times when the burs are absent, but silage can be made at any time because the fermentation softens the bristles, so that animals digest them without difficulty.

Not all forms of this plant are spiky nuisances. At least one has blunt inner spines and no outer spines at all. It has been called *Cenchrus leptacanthus*. If this type breeds true and if it could be developed as a crop, it would make kram-kram easier to handle and perhaps very valuable as a forage for many dry areas.[17]

A related species, also used as a wild cereal, is *Cenchrus prieurii*. It is spread throughout the Sahara from Senegal to Ethiopia (as well as India). People eat the crushed grain, mainly as porridge.

BOURGOU

Of all the grasses of the central delta of the Niger, bourgou (*Echinochloa stagnina*) was once the most prevalent. At one time it covered an estimated 250,000 hectares. (Much of that land, which is

[16] Information from R. Bartha.

[17] A close relative, *Cenchrus ciliaris* (commonly known as "buffel grass"), is a perennial with a very high forage value. It is increasingly used throughout the world's tropics and subtropics.

flooded for part of each year, is now under cultivated rice, see Chapter 1.) The Fulani people, for example, harvested large amounts of bourgou seed for food. They also got sugar from the plant. Some of the sugar produced by photosynthesis is not converted to starch and accumulates in the stems. People used it in beverages, both alcoholic and nonalcoholic. Even today, some sugar is still extracted from bourgou and is utilized especially for making sweetmeats and a liqueur.

This grass is found typically along river banks and other moist areas, especially those of Central Africa and on the central delta of the Niger. Recently, a farsighted UN-sponsored project has begun to restore some of the old bourgou stands in the area (see box, page 264).

Although its seeds are harvested for food, bourgou today is mainly used for fodder. For this purpose, it is notably important at the beginning of the dry season. As the annual floodwaters recede, it provides the vital forage needed to fatten livestock before the dry season sets in and their drastic weight losses begin.

The genus *Echinochloa* is one of the larger ones in the grass family. Two more species used for food in Africa are the following.

Antelope grass (*Echinochloa pyramidalis*) This native of tropical Africa, southern Africa, and Madagascar is primarily used for fodder, but is also used locally as flour.

Shama millet (*Echinochloa colona*) This plant probably originated in Asia, but it has been in Africa a very long time. Today people eat its grain only in dry years, although Egyptians possibly once grew it as a cereal on farms. The plant thrives in wet, clay soils where few grasses do well (in some African languages it is called "waterstraw"). Beyond its use as a food, the plant is suitable for making hay and silage and is relished by livestock.

CROWFOOT GRASSES

At least one *Dactyloctenium* species is eaten in Africa. It is the so-called Egyptian grass (*Dactyloctenium aegyptium*). This annual of the Sahara and the Sudan is now widely naturalized in different parts of the tropics and subtropics, including North America. It has never been considered as a possible cultivated crop, but nomads and others in its homeland (as well as Australian aborigines) gather the grains for food. The plant mostly grows in heavy soils at damp sites below 1,500 m. Livestock enjoy it, and it is also suitable for making hay and silage.

SHAMA MILLET

Main Components[a]		Essential Amino Acids	
Food energy (Kc)	311	Cystine	0.8
Protein (g)	9.5	Isoleucine	4.6
Carbohydrate (g)	56	Leucine	10.8
Fat (g)	5.3	Lysine	2.1
Fiber (g)	11.1	Methionine	1.6
Ash (g)	7.8	Phenylalanine	6.9
Calcium (mg)	45	Threonine	3.5
Copper (mg)	0.4	Tyrosine	4.3
Iron (mg)	9.7	Valine	5.8
Magnesium (mg)	198		
Manganese (mg)	2.5		
Phosphorus (mg)	369		
Potassium (mg)	270		
Sodium (mg)	9		

[a]Assuming 10 percent moisture.

COMPARATIVE QUALITY

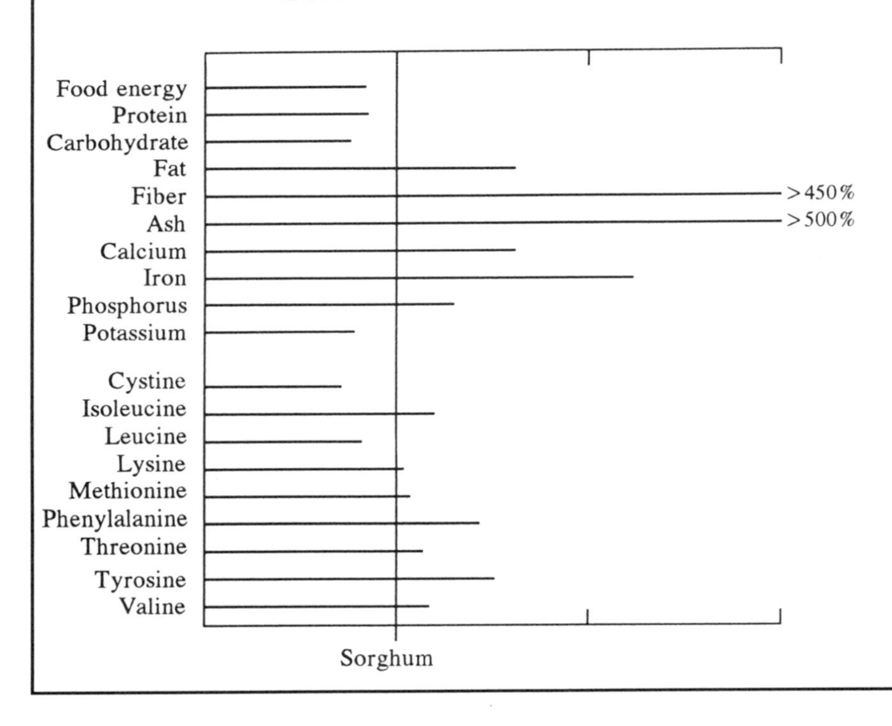

EGYPTIAN GRASS

Main Components[a]		Essential Amino Acids	
Food energy (Kc)	323	Cystine	1.5
Protein (g)	11.8	Isoleucine	4.8
Carbohydrate (g)	65	Leucine	9.9
Fat (g)	1.7	Lysine	2.0
Fiber (g)	4.0	Methionine	3.2
Ash (g)	7.5	Phenylalanine	6.8
Calcium (mg)	963	Threonine	3.5
Copper (mg)	0.6	Tyrosine	3.7
Iron (mg)	10.9	Valine	5.8
Magnesium (mg)	198		
Manganese (mg)	38.3		
Phosphorus (mg)	351		
Potassium (mg)	270		
Zinc (mg)	6		

[a]Assuming 10 percent moisture.

COMPARATIVE QUALITY

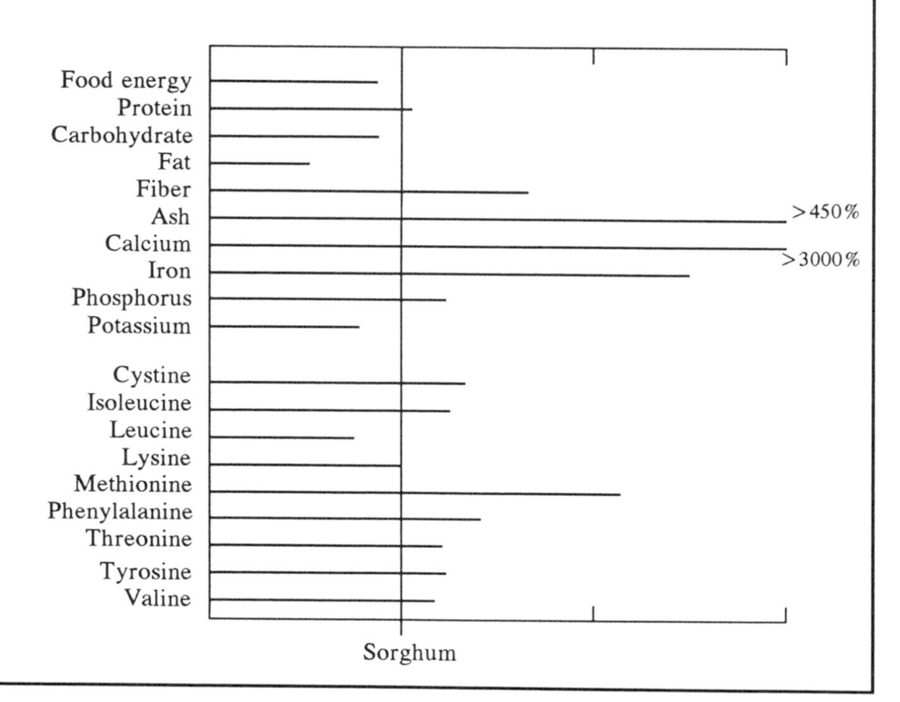

WADI RICE

Main Components[a]		Essential Amino Acids	
Calcium (mg)	36	Cystine	1.5
Copper (mg)	0.6	Isoleucine	4.1
Iron (mg)	15.1	Leucine	8.6
Magnesium (mg)	243	Lysine	3.6
Manganese (mg)	4.4	Methionine	2.2
Phosphorus (mg)	495	Phenylalanine	5.2
Potassium (mg)	333	Threonine	3.4
Sodium (mg)	9	Tyrosine	4.8
Zinc (mg)	4	Valine	5.9

[a]Assuming 10 percent moisture.

COMPARATIVE QUALITY

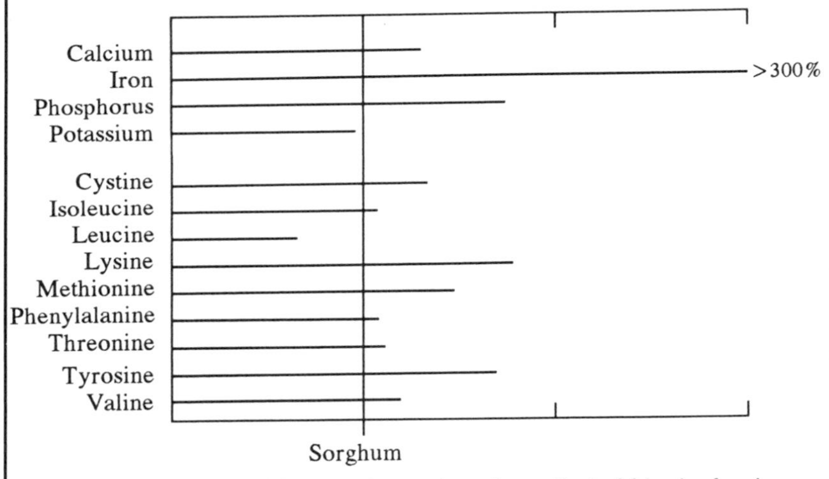

This chapter's tables and graphs show that Africa's famine-food grains can be quite nutritious. They are notably rich in those amino acids that are essential for human health but that are normally deficient in sorghum and the other common staples. Kram-kram, Egyptian grass, and wadi rice, for example, have more of the sulfur-containing amino acids than the FAO reference protein requirement. Egyptian grass and shama millet proteins are also significantly higher in threonine than those usually reported for sorghum protein. Wadi-rice protein (see above) is notably better than sorghum, but it closely resembles that of common cultivated rice in its amino-acid composition.

WILD RICES

Cereals of the West and Central African savannas include two wild rices. One, *Oryza barthii*, is the wild progenitor of the African domesticated rice (see African rice chapter, page 17, and especially the map, page 23). An annual, it tends to grow in shallow depressions that fill with water during the rains but later dry up. It produces abundant seed and is still harvested on a considerable scale.

The second species, *Oryza longistaminata*, is perennial and thus requires a more continuous supply of moisture. It is a relatively shy seeder, but its grain is sometimes harvested in sufficient quantities to reach the local markets.

A third wild rice (*Oryza punctata*) is indigenous to eastern Africa. This so-called "wadi rice" is a freely tillering annual that grows up to 1.5 m tall, and it, too, commonly occurs in rain-flooded depressions. Its seeds are relatively large and resemble those of cultivated rice except that they have a reddish husk. In Central Sudan, where wadi rice is widespread, the grains are boiled with water or milk and eaten as a staple.

OTHER WILD GRAINS

Among other wild African grasses that are, at least on a few occasions, used as food are the following. Little or nothing is known about these or their food uses, but certain botanical tomes contain the following cryptic comments.

Urochloa mosambicensis. Central and East Africa. Grains boiled.

Urochloa trichopus. Tropical Africa. Grains sometimes eaten.

Themeda triandra. Tropical and southern Africa. Perennial grass. Grain eaten during times of famine. Forms principal cover in fire-climax savanna areas. Used as fodder for livestock. Possibly of use in papermaking. Used a lot for thatching; bundles are sold in Ethiopian markets for the purpose.

Latipes senegalensis. Tropical Africa. Annual grass. Seeds are eaten by desert tribes.

Eragrostis ciliaris.[18] Widespread in tropics. Grains used as famine food.

Eragrostis gangetica. Tropical Africa and Asia. Grains used as famine food.

[18] This and the following *Eragrostis* species are related to tef (see chapter 12, page 215).

Eragrostis pilosa. Grains harvested regularly in East Africa.

Eragrostis tremula. Tropical Africa and South Asia. Grains used as famine food.

Setaria sphacelata. Eastern South Africa, South Cape, Botswana, Namibia. Perennial, robust, usually tufted grass. Of much economic importance. Different varieties or ecotypes have various uses: for hay and silage; for silage only; or just for grazing. Seeds eaten as famine food.

Appendix A
Potential Breakthroughs for Grain Farmers

This book was intended to be solely a survey of Africa's promising grains. However, in drafting it the staff became aware of certain nonbotanical developments that could bring enormous benefit to the use and productivity of Africa's indigenous grains. Some of these promising developments that deal with farming methods are presented here; others dealing with food preparation are given in Appendixes B, C, and D. It should be understood that the innovations described are not the only ones. Indeed, there may be dozens of alternatives for helping to solve the problems described. Nor is it our intention to suggest that these are panaceas. It should be understood further that the novel subjects described here are largely unproved or even undeveloped. Each incorporates a sound and seemingly powerful concept, but whether any will become truly practical in the harsh reality of rural practice and poverty is uncertain. We present them to encourage scientists and administrators to explore these unappreciated topics that just might become vital to Africa's future.

CONQUERING QUELEA

A tiny bird is perhaps the greatest biological limit to African cereal production. The most numerous and most destructive bird on earth, the seed-eating quelea (*Quelea quelea*) can descend on a farm in such numbers as to consume the entire grain crop in a matter of hours.

Quelea occurs only in Africa, but there its population is estimated to be at least 1.5 billion.[1] Although it holds much of the continent's agriculture hostage, its worst outbreaks are in parts of the eastern and southern regions, where its plagues are worse than those of any locust.

[1] One estimate puts the number as high as 100 billion.

The fields of ripe grain lying in the path of quelea migrations are essentially doomed. And it is unlikely that the consequences will diminish. Indeed, marginal lands are increasingly employed to grow grains, and future destruction is likely to be even greater.

The quelea's influence is insidious. This bird not only eats enough farm grain to feed millions of people, it destroys the farmers' morale and drains all interest in planting more land. Where quelea occurs, family members must patrol the ripening fields for weeks, disrupting their lives and restricting all outside activities such as jobs or schooling. Its preferences even dictate what is planted—millions of families now grow dark-seeded, tannin-rich, poorly digestible sorghums, at least in part because the birds, quite naturally, dislike them (see Chapter 10).

Trying to scare away hordes of ravenous birds is clearly futile in all but the smallest plots. Efforts to control quelea with poisons, napalm, dynamite, pathogens, and electronic devices have failed. Dynamiting the densest concentrations can achieve temporary local control, but a single flock may contain more than two million pairs and spread over an area far too wide for an explosion to have much effect. However, one line of research is now showing some promise.

At sunset each day queleas congregate in patches of tall grasses or trees. As the sky darkens they crowd together, until thousands are packed side by side in a small space. Researchers at the Zimbabwe Department of National Parks and Wildlife Management have observed that (provided the night is dark and the roost is isolated and fairly homogeneous, such as a patch of bulrushes) the birds are loath to leave. When disturbed, the chattering flock flutters forward a meter or two and only reluctantly decamps into the soundless darkness beyond. Indeed the scientists found that, once the flock had settled in, they could "herd" it around in the roost on moonless nights. By blowing whistles, beating on metal, or making some other disturbance, they could hustle the birds from one end to the other at will.

This was the key. If a barrier (a sheet of glass or transparent plastic, for example) was placed across the middle of the roost, thousands of queleas could be forced to fly into it each night (at least, for three consecutive nights, after which the birds became more cautious). If a holding cage was placed beneath the barrier, at least some of the half-stunned birds tumbled in. They could then be dispatched humanely, or, even better, could be trucked directly to a slaughtering facility and processed like poultry.[2]

[2] These grain-fed fowl make good food and traditionally have brought high prices in Zimbabwe. However, Zimbabwe law now prohibits eating them because 16–65 million quelea are killed each year by spraying bird toxins onto their roosts and nests. It has been noted, however, that people follow the spray teams and few dead birds remain on the ground for long.

In a second step, the Zimbabwe researchers tested tailor-made roosts. In isolated locations (and on sites quelea should find irresistible), they planted plots of napier grass and shaped them with slightly narrowed waists where the barriers and traps could be easily erected.

This seemed like an excellent way to turn a pest into profit, or at least into food, but it proved to have operational difficulties. The biggest problem was that only a few birds ended up in the cages. Those coming in from the fields flew fast enough to stun themselves on the glass, but most of those herded within the roost recovered too fast to fall.

Actually, because of such disappointing results the Zimbabwean authorities dropped the whole idea. They do, however, still use trap roosts to concentrate the birds so that workers with backpack sprayers can get to them with avicides (bird-killing chemicals).[3] This is much cheaper than using aircraft.

For most parts of rural Africa, killing birds with chemicals is unlikely to be nearly as practical or as appealing as capturing them for food. Thus, even though not yet perfected, the trap-roost concept seems to have promise. Indeed, it might in the end prove ideal for much of rural Africa because it offers the hungry poor both food and source of income. In principle, the operation is simple, cheap, and easy to understand and replicate. Given a new burst of innovation, today's limitations might well be overcome. Nets might be devised or the cages raised so that the chattering flocks would fly right in during the dark of the night and not have to stun themselves at all. Certainly, there seems to be much scope for improvement.

Of course, at this early stage there are many uncertainties even if the method can be made operational. Will it work in locations where the birds normally roost in trees? Could it be modified for use in trees? Are there grasses better than napier?[4] Will the birds learn, over time, to avoid the seductive patches of grass?

These issues are of course unresolved. However, if this approach can be made to succeed even partially, its effects could be far-reaching. And if it can be brought to perfection, it might transform the production of cereals throughout the quelea combat zone. Relieved of this feathered scourge, farmers could grow the best-adapted, best-tasting, and most nutritious grains. They could plant more land, their children could stay in school during bird season, and they themselves could keep their outside jobs.

[3] Information from C. Packenham.

[4] Vetiver grass seems likely to be a more practical choice. It is a perennial that neither spreads nor is attractive to grazing animals. Trap roosts made of vetiver would stay in place, perhaps for decades. Whether the birds will roost in blocks of vetiver grass should be quickly tested. This plant is described later in the chapter.

Although the trap-roost technique will never be a panacea,[5] it appears to have advantages over other approaches on several grounds:

- Environmental. The method requires no bird-killing chemicals.
- Economic. Trap roosts need no imported materials and farmers can build them with their own labor and materials so the technique could be employed by subsistence farmers, who have no cash to spare for bird control.
- Conservation. Although the fact that other species roost with quelea is a concern that needs to be evaluated,[6] techniques such as use of chemicals and explosions, for instance—are as indiscriminate or more so.
- Logistical. The method is independent of supplies, government, consultants, or high-level training.
- Adaptability. Catching birds in trap roosts seems infinitely adaptable to various locations and to the differing needs of users from subsistence farmers to large-property owners. For instance, a village farmer might install a small trap roost to get a little "poultry" for a party or a corporate farmer may establish many large ones to maximize a crop worth millions.

EXORCISING WITCHWEED

A small plant is the second largest biological constraint on Africa's cereal production. Usually called striga or witchweed, it is a parasite that lives off other plants during its first few weeks of life. Its roots bore into neighboring roots and suck out the fluids, leaving the victims dried out and drained of life.[7]

Unfortunately, striga (there are two main species, *Striga indica* and *Striga hermonthica*) loves maize, sorghum, millet, cowpeas, and other crops. Millions of hectares of African farmland are continually threatened; hundreds of thousands are annually infested. The traditional defense was long, idle fallow—now impossible because of population pressure.

[5] Even were it possible, eradicating quelea would not solve all bird problems because sparrows and other grain-eating species also occur.

[6] It is possible that desirable or endangered birds might inadvertently get caught, but so far experience shows that the tailor-made quelea roosts invariably contain few or no other species.

[7] Actually, striga seedlings are so small that the "drain" they put on the host is probably only moderate. However, the victim does dry out and die. It is suspected, though not proven, that striga somehow modifies the host's metabolism to interrupt its resistance to drought (thus the drying-out effect) and to increase its production of roots (at the expense of its leaf growth). Obviously both are processes that greatly reduce grain yields. Information from J.L. Riopel.

And today when striga breaks out severely, nothing can be done. Farmers usually abandon their land. Some of the most productive sites now lie idle—victims of this abominable sapsucker.

And the problem is worsening. Striga is most damaging when crops are stressed by drought or lack of nutrients—phenomena that are increasingly common. Changes in farming practices are also helping striga to conquer ever more countryside. The continuous cropping of cereals, for example, contributes more and more striga seed to the soil.

At present, the only way to keep this weed in check is by carefully crafted farming practices: crop rotations, fertilization, and skillful use of herbicides, for instance. But this is impractical for the millions of subsistence farmers who have no surplus land for crop rotations and can afford neither fertilizer nor herbicides. Also, it would be nearly impossible to train millions of farmers to modify their farming practices, especially in the impoverished zones where striga is most threatening.

A "technological fix" to take care of the problem easily, universally, and permanently has never been found, but there is a possibility that it might be just around the corner. A crack in the plant's biological armor has been discovered, and through it researchers see exciting new prospects.

The excitement is based on the recognition that striga relies heavily on "chemical signals" to locate its victims. The mechanisms of this signaling have now been defined. In addition, approaches have been designed to cut striga's "lines of communication" or to provide misinformation. And control methods are proving successful in laboratory trials and even early field experiments.

Striga seeds refuse to germinate until they receive a chemical signal from the root of a potential host. The signal telegraphs the fact that a victim is nearby and that moisture is adequate for successful germination. The seed may lie dormant for decades awaiting this chemical confirmation that it is safe to come out.

But striga's elegant adaptation provides a window of opportunity. Farmers could, at least in theory, block the signals. Better still, they could supply false signals and trigger striga seeds into suicidal germination. Striga depends so much on the lifeblood of other plants that unless its seedlings can latch onto a root within four days, they die. Each striga plant produces millions of tiny seeds, but a chemical trigger could perhaps fool all of them into germinating. If the land had been newly plowed, the parasite would find no victims and four days later farmers could safely plant their crops.

Recently, scientists have identified chemical signals that trigger striga's germination as well as others that inhibit it. Apparently, the balance between stimulation and inhibition is what determines whether the seed will germinate. Both chemical types are extremely active. The stimulants, for instance, can be diluted 10,000-fold or more and still cause striga seed to germinate.[8]

If compounds like these can be synthesized, mimicked, or economically extracted from plant roots, they could be (at least in humanitarian terms) among the most valuable of all organic chemicals. For example, it may be possible to produce striga-suicide sprays, perhaps even in the regions that require the most help. This approach has been exploited by Robert Eplee of the U.S. Department of Agriculture to dramatically reduce striga attachment in greenhouse tests.

Also, another striga signal has been identified. This compound (2,6-dimethoxybenzoquinone) "tells" the germinating striga seedling to form the organ (haustorium) that pierces the victim's root. This, too, may offer a way to overcome striga. For instance, an antagonist chemical might blunt striga's underground weapon. If the pest can find no host, it never develops a growing shoot (apical meristem), it never becomes photosynthetic, and it dies.[9]

Recently, scientists have found that nature is ahead of them. At least one strain of sorghum can already foil striga by producing water-soluble compounds that are striga inhibitors. This sorghum, SRN-39, both resists the parasite and has desirable agronomic characteristics and good-quality grain. Its striga resistance appears to be simply inherited (only one or two genes). Crosses with other cultivars have already been made and promising progeny obtained. Moreover, an assay has been developed to screen breeding material for this resistant characteristic. These results suggest that sorghum breeders may soon be able to breed for striga resistance rapidly and efficiently.[10] Similar progress has been achieved in maize.

It has also been found that some leguminous plants—*Crotolaria* species are examples—excrete their own striga-stimulating signals but do not serve as hosts. Although the striga germinates, it immediately dies. Thus, plants like these could be employed to deplete the striga seed bank in the soil. They may prove extremely valuable species for fallow crops or alley crops. *Crotolaria* species (rattleboxes) are le-

[8] Information from L. Butler.

[9] New results suggest that striga uses a "chemical radar" approach to host detection. The striga itself releases enzymes that remove the stimulants from the root's surface. This is a novel, and very effective, means of detecting the presence of a potential host. Disruption of this enzymatic function is also being effectively exploited by the U.S. Department of Agriculture.

[10] In fact, about 10 years ago a series of SAR (*Striga asiatica* resistant) varieties that have a very high level of resistance to the white-flowered *asiatica* were developed in India. More recently, in southern Africa, five SAR lines have been found to have reasonable resistance to the red-flowered *asiatica* found in Africa. As with SRN-39, inheritance is simple. (Information from L. House.)

gumes, so they not only knock out the parasitic pest, they also enrich the soil with nitrogen and organic matter.

All these approaches to the striga problem should be top research priorities, and not only in Africa. This parasite already affects India and has broken out in a small part of the United States. It could easily come to infect much of the world's farmland. Solving the problem now would lift from African agriculture a burden so big that the result might compare with a "Green Revolution." It would also help insulate the rest of the world from the heartbreak of this herbaceous horror. All countries have a stake in the outcome of this challenging research.

LIQUIDATING LOCUSTS

Numerous African countries, but especially those in the Sahel, are victimized by the desert locust (*Schistocerca gregaria*). Controlling this one pest soaks up vast amounts of money, time, and insecticides—700,000 liters of concentrate were sprayed over 14.5 million hectares in 1988, for instance. It has generally been effective, but in recent years some of the locust's relatives have risen up to become equally menacing. In 1989, for example, grasshoppers—in particular the Senegal grasshopper (*Oedalus senegalensis*)—arrived just at harvest time, causing 10 times more damage than the locusts had the previous year.

For nearly 30 years Dieldrin was the pesticide of choice. Applied in strips across the desert terrain where locust larvae hatch, it seemed an ideal way to stop the insects before they reached their damaging migratory stage. It worked, it needed no repeated spraying, it was cheap, and it could be stored without degrading even in the scorching heat of the Sahara. But in the late 1980s, even while locust swarms were swelling to worrisome levels, people began protesting because of Dieldrin's potential toxicity to humans and animals.

On environmental grounds, organophosphorus chemicals and pyrethroids seemed preferable but they remain effective for a few days only and must be reapplied over and over. This means higher costs, more work, and the destruction of all insect life—even beneficial species.

Now, a new approach to chemical control seems to offer some hope. Research in Germany has shown that oil from the seed of the neem tree (*Azadirachta indica*) stops locust nymphs from clustering.[11] After exposure to even tiny doses, the juvenile locusts fail to form the

[11] Information from H. Schmutterer. This tree and its promise for controlling insects and other pests is described in the companion report *Neem: A Tree for Solving Global Problems*. (For a list of BOSTID publications, see page 377.)

massive, moving plagues. They remain alive but solitary and lethargic; they sit on the ground, almost motionless, and are thus very susceptible to insectivorous birds. Grasshopper nymphs are affected in the same way.

This is very different from the earlier applications of neem against locusts. Those first attempts used alcoholic extracts of the seed kernel, and were aimed at disrupting metamorphosis or at stopping the adults from feeding on crops. Although highly promising in experiments, they proved less successful in practice.

The new approach uses neem oil rather than neem-kernel extracts. Experiments have shown that at very low concentrations (2.5 liters per hectare) this oil, like Dieldrin, prevents locusts from developing into their migratory swarms. It doesn't kill them but it keeps them in the harmless, solitary (green) form. It apparently disrupts the formation of hormones necessary for the transformation into the yellow-and-black gregarious stage whose plagues are the bane of arid Africa and Arabia.

The neem tree grows throughout West Africa, and thus the locust-control agent could, in principle, be locally produced. To press the oil out of the neem kernels and to spray it over the areas where locusts breed and gather requires neither particularly high-technology equipment nor unexpected expense. The oil itself is neither toxic to mammals nor to birds and is biodegradable.

Another approach that may have some localized merit is to provide nesting sites for insectivorous birds. In western China, where another plague locust occurs, farmers have reportedly met with success by protecting, and even building, nesting sites for the feathered locust eaters of the area.

EASING EROSION

The effects of soil erosion are well known: it devastates farms and forests; worsens the effects of flooding; shortens the useful lifetimes of dams, canals, harbors, and irrigation projects; and pollutes wetlands and coral reefs where myriad valuable organisms breed. But there could now be a way to slow or even stop it.

Hedges of a strong, coarse grass called vetiver have restrained erodible soils for decades in Fiji and several other tropical locations. The hedges are only one plant wide and the land between them is left free for farming, forestry, or other purposes. This persistent grass has neither spread nor become a nuisance. If current experience is applicable elsewhere, vetiver offers a practical and inexpensive solution to the problem of soil losses in many locations. It could become an exceptionally important component of land use, at least in the hot parts of the world.

This deeply rooted perennial can already be found throughout Africa, but in most places the idea of using it as a vegetative barrier to erosion is new and untested. However, it is not farfetched. Strips of vetiver certainly are able to catch and hold back soil. The stiff lower stems act as a filter that slows the movement of water enough that it drops its load of soil.

Equally important, the dense, narrow bands of grass cause the runoff water to spread out and slow down so that much of it can soak into the soil before it can rush down the slopes. This captured moisture allows crops to flourish when those in unprotected neighboring fields are lost to desiccation.

So far, all the international attention has focused on an Indian vetiver (*Vetiveria zizanioides*). This is already widespread in Africa and has shown promise for controlling erosion in Nigeria, Ethiopia, Tanzania, Malawi, and South Africa, and appears to be a blessing for many countries. However, Africa has its own native *Vetiveria* species. These are entirely untested, but they may confer similar benefits. One (*Vetiveria nigritana*) has long been used to mark out boundaries of properties in northern Nigeria, for instance,[12] and it has been employed for the same purpose in Malawi and Zambia as well.

Vetiver has many interesting and unexpected uses. Tobacco farmers in Zimbabwe report that putting a vetiver hedge around their fields keeps out creeping-grass weeds, such as kikuyu and couch. It even seems to be a good barrier to ground fires.[13]

In the Sahel, vetiver hedges may prove extremely useful as sand barriers. Winds off the Sahara often blow sand with such power that it scythes across the landscape at ankle level, cutting off young crops before they are barely beyond the seedling stage. Rows of vetiver planted on the windward side of fields could be an answer. The stiff stalks would doubtless halt the scurrying sand, providing both a windbreak and a sand trap.

Rows of vetiver planted across wadis may also make excellent water-harvesting barriers. Once planted, the barriers would be essentially permanent. The deep-rooted grass is likely to find enough soil moisture to survive even the driest seasons in most arable locations. Although the upper foliage may die back, the stiff, strong lower stalks that block the sand, soil, and water will remain. These are so coarse that not even goats will graze them to the ground.

[12] It spreads so little that in legal disputes vetiver hedges have been officially accepted as valid property lines. At one documented site in northern Zambia, vetiver still exists in the same narrow lines that were planted 60 years ago.

[13] Vetiver and its promise are described in the companion report *Vetiver: A Thin Green Line Against Erosion*. (For a list of BOSTID publications, see page 377.)

HANDLING SMALL SEEDS

As has been noted several times, a major problem with many of Africa's grains—finger millet, fonio, and tef, for example—is that they have tiny seeds. Size alone is holding these crops back. Small seeds create many difficulties. They are hard to store and hard to handle because they pour uncontrollably through even the smallest holes. They also make the crop difficult to plant because the soil must be very finely textured (clods or clumps can overwhelm the seeds' puny energy reserves), and the seeds must be placed precisely at just the right depth. Moreover, because the emerging seedlings are small and weak, they are easily smothered by weeds.

Many innovations could probably be devised to overcome these problems; here we present several examples of seeding devices newly developed in four Third World countries. These are undoubtedly not the only innovations for planting small-seeded crops, but we present them here as guides to those who wish to help Africa's lost crops.

Cameroon In the late 1980s, the Cameroonian Agricultural Tools Manufacturing Industry (CATMI) in Bamenda produced a seeder that, compared to traditional planting by hand, reduces planting time by 60 percent and seed requirements by 33 percent. It is not specifically for small-seeded crops but includes a simple distributer mechanism that can be adjusted to accept seeds of different sizes.[14] It is said to reliably plant the desired number of seeds at the right depth and distance apart. It is simple to handle, suitable for planting both on ridges and on flat land, durable, easy to maintain, and cheap.

In 1988, 30 prototypes were distributed to farmers and research stations for field testing. After further improvements, 300 more were produced and sent out. Various agricultural services ran information and demonstration campaigns to promote the planter. A line of credit was set up in the Northwest Province to enable small farmers to purchase one. In addition, other provinces were contacted and provided with demonstrators and seed planters.

A survey after the first planting season (1989) indicated that 97 percent of the farmers who tried the implement bought it. Not only did it make the work easier (no back pain) and speeded up planting, but it also reduced the need for hired labor and helped increase both the area farmed and the yields achieved.

[14] This work was done in cooperation with the Departments of Agricultural Engineering and Rural Socio-Economics of the University of Dschang.

Peru[15] In the Andean city of Cuzco, Luis Sumar Kalinowski has created a seeder capable of handling kiwicha,[16] whose seeds are as small as sand grains. It is a simple, almost cost-free device that can sow large areas evenly and in uniform rows. It may also work well with Africa's small seeds.

One version of the Sumar seeder uses a scrap piece of plastic pipe with a foam-plastic cup taped to the end.[17] A nail is pushed gently through the bottom of the cup to leave a hole of known diameter. Another version employs a commercially available plastic end piece, which is drilled to provide the hole. In either case, seed placed in the pipe trickles out at a constant rate, and the farmer can vary the seeding density by walking faster or slower.

Indeed, by measuring the flow of seed through the hole, it is easy to calculate how fast to walk (in paces per minute, for example) to sow the desired density of seed. With a little practice, the farmer can attain an accuracy rivaling that of mechanical drills. For the method to work, however, it is important that the seeds be clean and free of straw, small stones, or other debris that could block the hole.

Tanzania[18] Engineers at Morogoro have designed and developed a low-cost, hand-operated device known as the Magulu hand planter. It includes an attachment that can be fastened to a hand hoe and can be used to plant both maize and beans in a straight row. It is said that to plant a hectare of land using the Magulu hand planter takes between 18 and 27 man-hours as compared with 80 man-hours using the conventional method of planting by hand hoe.

Thailand The Asian Institute of Technology (AIT), which is located near Bangkok, has developed a mechanical seeder that is now being popularized in many Asian countries. In one stroke, this so-called "jab seeder" makes a hole, drops a seed, and covers the site, without the operator ever having to bend over.

The seeder weighs only about 1.5 kg and costs about US$10.00 (including labor, materials, and mark-up). In Thailand, a farmer can

[15] For more information, contact Luis Sumar Kalinowski, Centro de Investigaciones de Cultivos Andinos, Universidad Nacional Técnica del Altiplano, Avenida de la Infancia N° 440, Huanchac, Cuzco, Peru.

[16] This crop (*Amaranthus caudatus*), a species of amaranth, is discussed in the companion volume *Lost Crops of the Incas*. (For a list of BOSTID publications, see page 377.)

[17] It is not necessary for the pipe to be plastic. Any tube—bamboo, cardboard, or other material—will do. However, standard household water pipe and the disposable coffee cups common in many countries fit together well. Also, the foam-type cups are easy to pierce with a nail, and they leave a clean, smooth hole. Caps with different-sized holes can be kept on hand for use with different crops.

[18] T.E. Simalenga and N. Hatibu, Department of Agricultural Engineering, Sokoine University of Agriculture, Morogoro, Tanzania.

recover the cost, in terms of labor saved, in only 5 days and on an area as little as one-fifth of a hectare. Mass production is expected to reduce the cost even further.

In Thailand's northern province of Chiang Mai, the idea has already caught on: a number of local manufacturers are producing mechanical seeders based on the AIT model.

At present, this machine is not intended specifically for small seeds. It is used mainly with soybean, rice, maize, and mungbean. But even with these crops, it brings big advantages in labor saving and yield.

In Nepal, field tests have found that—at wages of 25 rupees (US$1) a day—a farmer can recover the cost of a jab seeder by planting maize or soybean in just 1 hectare of land. Fifty seeders made locally by the Agricultural Tools Factory in Birganj cost US$13.50 each.

By making a less onerous and more systematic operation, the jab seeder could well increase grain-crop productivity and thereby benefit millions of Africa's grain farmers.

OTHER INNOVATIONS

Seed planters are probably the main need for small-seeded crops, but they are not the only need. Various appropriate technologies are required also for harvesting, storing, shipping, and handling tiny cereal grains. Some of these might come from techniques devised to produce ornamentals, forages, and vegetable crops, many of which also have minute seeds.[19]

Also, it is not impossible that the size of the seeds could be increased through selection and breeding. Luis Sumar has already created a simple machine for doing this in the case of kiwicha. The Sumar sorter uses a small blower and a sloping plastic pipe. The seeds are blown up the pipe and drop into different containers, depending on their weight. With it, Sumar has increased the grain size in kiwicha. He keeps only the heaviest for planting, so that over the years the crops produce seeds that are ever larger, on average. The use of such a simple, inexpensive device in Africa might dramatically benefit fonio, finger millet, and tef, to mention just three cereals.

[19] A reviewer from Oklahoma wrote us: "We have been handling small seeds in the Southern Great Plains with precision for half a century. I worked with native grass seeds myself for 25 years. Some of the seeds are smaller than tef, fonio, or finger millet. We had equipment that would mete out seed at low seeding rates very accurately and plant them with precision. Our planters, processors, and cleaners are, perhaps, too sophisticated for subsistence farmers, but modified versions are well within the capabilities of most village mechanics and blacksmiths. The technology has been available for a long time. Suggest you contact Chet Dewald, Southern Great Plains Range Research Station, 2000 18th Street, Woodward, Oklahoma 73801."

Appendix B

Potential Breakthroughs in Grain Handling

Appendix A identified technological advances that might boost the production of indigenous African grains. Here we identify other advances that might similarly influence the methods of milling and storing those grains. These, too, are innovations that, in principle, could bring outstanding benefits continent-wide. Again, however, it should be realized that they are just a smattering of examples that caught our attention as the book was being prepared. Other cutting-edge technologies may be as good, or better.

NO MORE POUNDING

Every day of the year, perhaps 50 million Africans—most of them women and children—spend hours preparing the grain that their families will eat that day. They usually soak the grain in water, pound it with the butt end of a heavy wooden pole (pestle) to knock off the outer seed coat, winnow the beaten mixture to separate the bran, moisten the grain a second time, and finally pound it yet again to break it up into flour.

This is always a hot and disagreeable task. It limits both cereal use and life itself. Decorticating enough pearl millet for a family meal (about 2.5 kg) takes two women about 1.5 hours; converting the product into flour with a mortar and pestle requires an additional 2 hours, sometimes more. Moreover, because the flour spoils quickly and cannot be put aside for later use, it has to be done day after day, in fair weather and foul, and regardless of sickness or other indisposition.

Probably no single development could help rural Africa more than relief from this never-ending drudgery. It would recover millions of "lost" hours every year, it would improve health and family welfare,

Sorghum and Women

Sorghum is a women's crop in Africa. To a large extent, they are its planters, cultivators, and harvesters. Through the accumulated wisdom of centuries, women have amassed information about the crop and its handling. Many are expert in distinguishing closely related varieties . . . a knowledge which men—even professional scientists—seldom attain. Only now, however, are researchers beginning to pay attention to this knowledge.

Joyce Kanyangwa is one of those. Working under the auspices of Texas Tech University, she traveled to three sorghum-growing areas of Lesotho, visiting selected households to gain a perspective on attitudes about the use of sorghum. "I was interested in finding out what might be done to expand the use of sorghum in the diet to give women more income for their labor, as well as a cheaper staple for their tables," she explains.

Her research indicates that improving sorghum use can do much to help Africa's women. "Sorghum is a woman's crop, but the market for the product is limited primarily to brewing beer for men," she notes.

Better processing methods are particularly needed. The processing and cooking of sorghum and millet takes more time than rice. Women going to work, either in the fields or in the community, have less and less time available for processing and cooking. Small-scale rural sorghum and millet processing mills, like the rice mills already available in India, could help promote the consumption of sorghum and millet.

"When sorghum is processed using a special machine, people like it," Kanyangwa says. "I'm optimistic that the crop has the potential for helping female-headed households feed their families better and for helping women make more money."

The introduction of suitable dehullers and flour mills will:

- Reduce the drudgery of women in the sorghum eating areas.
- Convert sorghum into a much more convenient grain.
- Improve the quality of sorghum products.
- Check the tendency of shifting from sorghum to other grains.
- Help develop composite flours and commercialize sorghum products.

OPPOSITE:
Ethiopia. Danakil Depression. At the entrance to her hut, over a goatskin, an Adoimara Danakil woman grinds sorghum between two stones. (Victor Englebert)

and it would make the whole continent more productive. Perhaps most important in the long run, it would secure the future of the local grains. At present, the burden of the terrible toil is causing a silent rebellion against sorghum, millet, and the other indigenous cereals.

Now an option is emerging. Small power mills can, in just a few minutes, perform the task that now absorbs so much human energy and time. Some of the most successful consist of a series of 8 or 12 grinding stones of the type used for sharpening tools. The essential component, the dehuller, was originally designed at the Prairie Regional Laboratory in Saskatoon, Canada. A small version specially sized for rural Africa has been built, field-tested, and improved at The Rural Industries Innovation Centre in Kanye, Botswana. It is powered by a small diesel engine.

Reportedly, the machines waste no more grain than hand pounding does. (Recovery rates of 85 percent have been achieved, which is 10 percent better than is normal in the village.) Also the machine-dehulled grains apparently make no detectable changes in local foods. Since they use dry grain, the dehullers are more flexible than traditional methods, and the resulting flour can be stored.

The dehuller does only half of what African women do: it takes off the seed's outer layer, leaving white, ricelike grain. A further grinding is needed to make flour or grits. To do this mechanically, a hammer mill is employed. In some cases, the dehuller and hammer mill are combined into a single unit.

Although these mechanical systems were designed primarily for processing sorghum and pearl millet, they have also proved satisfactory for fonio and food legumes such as cowpeas and pigeon peas.[1] One of the main attractions is their capacity to handle (without major adjustment) grains of widely different size.

Under a Canadian-sponsored program, different models are currently being developed or distributed for use in Senegal, the Gambia, and Zimbabwe. Mali and Niger, following Botswana's lead, are creating designs suitable for local toolmakers to build.

Mechanized processing probably has its most immediate use in cities and towns. In rural areas, people must carry their grain to the mill and then carry home their flour and bran. For them, the chore of carrying several kilograms several kilometers may be just as onerous as staying home and pounding the grain with a pole. However, there are ways to circumvent this. In Botswana, for instance, a donkey cart is being made available without charge to carry the grain and flour back and forth. (The donkey is fed on the customer's bran waste.) Also, the milling unit could conceivably be mounted on a cart and

[1] Finger millet is difficult to mill mechanically, but in India a suitable device has been perfected (see page 48).

wheeled to the customers. Thus, for example, a mobile mill might stop at various villages once a week and process the grain on the consumer's own doorsteps. A hammer mill perhaps might not work on such a system, but the dehuller alone would relieve the major and most unpleasant part of the drudgery.

All of this opens the possibility of substantially lessening the burdens that at present fall so heavily on millions of people. It will probably widen the mix of crops they grow. It could increase lifestyle options and employment opportunities by freeing women from the daily morning and evening chore of pounding grain. It may contribute dramatically to better health among women and children, provide time for more productive pursuits, create better markets for farmers, and lead to a more stable food situation for many countries.[2]

Despite the fact that people must pay to have their cereal mechanically milled, this mini-milling industry is already starting to take hold in parts of Africa. Several nations have introduced the Canadian-type mills, and support for their maintenance has quickly spread, even into remote areas. Moreover, merchants and consumers throughout Africa are showing increasing interest in buying and using flours instead of unprocessed grain. A grain revolution seems to be arising, bringing new options for farmers and consumers, as well as new possibilities for a better life in the rural areas.

GRAIN DRAIN

To worry only about grain production is not enough; what counts is the amount and quality of the food that gets into people's bodies. Today, unfortunately, much of Africa's cereal crop never gets that far—it spoils or is lost sometime after the harvest. Estimates suggest perhaps 25 percent of each year's food production is either lost or rendered unfit.[3] The reasons are clear. During handling and storage, heat and humidity foster molds and rots that ruin much grain. Insects, rodents, and birds steal enormous amounts. Most subsistence farmers store their harvest in small granaries (capacity 1.5 tons or so) and 10–20 percent usually deteriorates or disappears before it can be eaten.

An obvious answer is better storage, and these days pest-proof silos built of several materials are showing much promise. Examples follow.

[2] This topic seems especially well suited to the U.S. Agency for International Development's current programs on "family issues."

[3] The exact figure is not known for certain. Some writers claim that postharvest grain losses in Africa sometimes top 40 percent, mainly due to poor processing and storage. Others say that it is less than 10 percent. Certain types of traditional stores are very effective, but a 25-percent loss is very common in government stores (partly because farmers often contribute only their poorest materials). And the losses on the farm can rise dramatically if a new variety of grain is produced.

Brick

A Zimbabwean engineer, Campbell D. Kagoro, has for years been developing a granary built of local brick.[4] His structures—known as ENDA granaries—have been installed in dry, poverty-prone areas of Zimbabwe. People there (as elsewhere in Africa) know how to manufacture baked-clay bricks. To build the silos, they lay the bricks directly on gravel-covered soil or on rock. (In some instances, wooden joists and masonry footings are used.) They cover the final structure with a waterproof thatch roof. The silos have a capacity of about 2.5 tons and may include up to five compartments for storing different products. They are equipped with air vents and are said to offer excellent protection against dampness, insects, and rodents.

Ferrocement

A form of reinforced concrete, ferrocement utilizes materials that are normally readily available—wire mesh, sand, water, and cement. It does not corrode easily and can last a lifetime.

Experience in Thailand and Ethiopia has demonstrated that ferrocement silos can be built on site relatively inexpensively, using unskilled labor and only one supervisor. In such silos, losses are less than 1 percent per year. Rodents, birds, insects, and dampness cannot get in.[5] If the bin is well constructed and its lid tightly sealed (tubing from a bicycle tire makes a useful gasket), even air cannot get in. Inside an airtight silo, the respiring grain quickly uses up the oxygen. Insects (eggs, larvae, pupae, or adults), as well as any other air-breathing organisms introduced with the grain, are then destroyed.

The possibility of putting ferrocement silos on every farm is demonstrated by a remarkable program in Thailand. There, where the concern is storing pure water rather than grain, the government has provided three ferrocement jars (each two cubic meters in size) for every family of six in rural areas. The project involved three million families and nine million jars. Each jar costs $20, but the per-capita cost—because a revolving fund of $13 million is recoverable—can be as low as 42 cents.

Heat is a basic problem with ferrocement (and most other) silos. Bins in the burning sun can warm up so much that moisture evaporates

[4] This work has been done at the Agricultural Technical and Extension Service (Agritex) in Zimbabwe. The design was developed by the Institute of Agricultural Engineering (IAE) and Agritex. ENDA-Zimbabwe (Environment, Development, Activities) has joined with Agritex to undertake a major study of the subject, with financial assistance from IDRC.

[5] Hundreds of ferrocement boats floating on the world's waterways demonstrate that this material can be watertight, but the construction must be top quality because the ferrocement is usually only a centimeter or so thick.

from the grain, collects at the top, and fosters molds or sprouting. For this reason, silos should always be located in the shade of trees or houses, sunk in the ground, or surrounded with some rough-and-ready sun shield (thatch or scraps of foamed plastic comes to mind).

Although much of the promise is for small bins for household use, ferrocement can also be used to construct large storage facilities for town or regional use. One of the most intriguing is the horizontal "sleeping silo" pioneered in Argentina (where they are used mostly for storing potatoes). These large structures are shaped like the hull of an upside-down ship half buried in the ground. Bulkheads give strength and also create separate compartments in which different products or different owners' products can be stored. Compared to the towering grain elevators now used in much of the world, the horizontal counterparts lie on the ground and require little in the way of engineering, footings, or structural reinforcing.

Mud

Recently, an airtight grain store made from clay and straw has been introduced to Sierra Leone. The silo, demonstrated by Chinese instructors brought by the UN's Food and Agriculture Organization (FAO), is simple in construction, low in cost, and has potential to significantly decrease postharvest grain losses.

The raw materials in this case consist of mud and straw, and the finished silo is roofed with boards, straw, reeds, or other waterproof materials. Its inventors are the peasants of northeast China who, from time immemorial, have built tiny mud turrets to store their household food reserves. In recent years, a national campaign to decentralize grain storage has led to this very simple and economical technique being used throughout the Chinese countryside. In fact, mud silos are now being built as large as 8 m in height and diameter, to hold 200 tons.

Ghana, too, has been testing improved mud silos.[6] Instead of ordinary mud, however, sun-dried molded mud bricks, from a locally made mold, are used for the circular wall. The top is a separately molded mud slab. The whole unit is sealed with a mixture of mud and clay, and the wall is whitewashed to maintain coolness.

Neither of these two silos requires any great expertise to construct or use.

Plastic

Researchers in Australia and the Philippines in recent years have jointly developed sealed plastic enclosures for storing grain in ware-

[6] Adopted by the Ghana German Agricultural Development Project (GGADP).

houses located in the humid tropics. In 1992, a new project was begun to design a counterpart suited to the smaller-scale and outdoor needs of cooperatives, small millers, and merchants. The scientists have developed a plastic container that is rodent- and insect-proof and protects grain against the extremes of the tropical environment. It is also simple to fumigate and suitable for storing damp grain before drying. The plastic silos have been designed using the general principles already employed for storing bulk grains in Australia. Although conducted in and for the Philippines, this work seems suitable for application throughout the humid tropics.[7]

Rubber

Israel's agricultural research organization, familiarly known as the Volcani Institute, has pioneered development of simple, cheap, and easily movable grain stores with capacities up to 1,000 tons. These collapsible, tentlike structures can be taken down, trucked to a new site, and quickly reassembled—a novel feature that makes them especially useful for handling emergency food supplies and for storing excess grain from unexpectedly bountiful crops. The walls are constructed of rolls of strong wire mesh (actually weldmesh fencing material), but the grain is held within UV-resistant plastic liners. These silos are sufficiently airtight to control insect infestation without requiring pesticides. They are primarily for use in drier areas.

DRYING GRAIN

Insects and rodents are not the only grain despoilers. Insufficient drying also leads to vast amounts of damage. Dampness fosters molding, sprouting, and decay that renders grain inedible. Drying the grains before storing them is therefore vital. Techniques for doing this under Third World conditions are being devised in several parts of the world.

Sierra Leone

Farmers in six districts of Sierra Leone are replacing traditional mud floors, used for drying freshly harvested rice, with improved drying yards. This cheap and simple change keeps the grain clean, lessens the drying time, and reduces postharvest losses by more than half.

[7] This research was sponsored by the Australian Centre for International Agricultural Research (ACIAR), G.P.O. Box 1571, Canberra, A.C.T. 2601, Australia.

United States

The Food and Feed Grains Institute of Kansas State University has designed a new kind of dryer for developing country use. It has no fan or other moving parts and uses heat generated by burning weeds, rice husks, agricultural by-products, or other wastes.

This natural-convection, hot-air drying could open up new options in many areas of Africa where today the only cereals that can be grown are those that mature after the rains have ceased (when grains can be dried in the sun). In 1990, Kansas State tested its dryers under conditions of very high rainfall in Peru and Belize. Sun-drying was impractical, even impossible, but the new dryer proved very effective: rough rice was reduced from a level of 20 percent moisture to 14 percent in only about an hour. While this is too fast for everyday practice with rice, it clearly demonstrated that the dryer would perform well in the dampness of the tropical rainy season.

Thailand

The Asian Institute of Technology (AIT), near Bangkok, has developed a simple solar dryer, constructed of bamboo poles and clear plastic sheeting.[8] It can process up to one ton of rice at a time and even in the wet season can reduce the moisture content from 22 percent down to 14 percent in about 2 days. It is said to cost only around US$150 to build.

In this device, sunlight passes through a clear plastic sheet and strikes a layer of black ash (burnt rice husk) or black plastic sheet. This absorbs the solar energy, converting it into warm air. The heated air rises by natural convection through the slatted floor of the rice box, up through the grain (contained in fine wire mesh), and out a tall chimney (again fabricated from bamboo and plastic sheet).

Korea

In the early 1980s rice farmers in South Korea faced postharvest losses of about 10 percent. But now those losses have been halved, thanks to a new technology.[9] The system has been so successful that just 8 years after the project was launched, 70,000 dryers had been purchased. By 1995, half a million are expected to have been built.

[8] This work is part of a project sponsored by the International Development Research Centre of Canada.

[9] The dryer was developed jointly by the University of Hohenheim in Germany and the Korea Advanced Institute of Science and Technology (KAIST). Funds for the program were provided by the Deutsche Gesellschaft für Technische Zusammenarbeit (GTZ) GmbH.

With this method, the grain is dried using a low-temperature process that mainly exploits the drying potential of ambient air. Basically, a fan blows air through grain in a silo. The process is cheap, requires little capital investment, and the silo can subsequently be used for storage purposes. To enable drying in humid weather and during the night, a small electric heater is used to heat the ambient air a few degrees.

In practice, the dryer is a room-sized brick structure, with a false floor to prevent soil moisture from seeping up. The air is uniformly distributed using wood or sheet metal air ducts, laid on this false floor. The air is pushed through the piled-up grain by a small 400-watt electric fan.

KILLING STORAGE INSECTS

The need to protect Africa's stored food from insects is particularly important these days. The larger grain borer, a Central American beetle introduced accidentally into Tanzania and West Africa, is relentlessly spreading through maize-growing areas. This voracious pest feeds on stored maize, cassava, wheat, sorghum, sweet potato, peanuts, and other foods. The destruction it causes can be devastating; in tests in Tanzania up to 34 percent of cob maize in a crib has been destroyed after only 3 months, and up to 70 percent of dried cassava after only 4 months.

Insects get into even the best silos when the grain is added. Previously, there were no cheap and effective controls for subsistence farmers to use. However, some innovations follow that might help overcome the problem.

Sunshine

Researchers in India have found that farm produce can be disinfested by "roasting" the bugs in the sun. They first wrap a square sheet of black polyethylene around two slats of wood, leaving a "mouth" at either end. After filling the resulting pouch with produce to a depth of 3 or 4 cm, they seal the ends by weighing them down with slats of wood or bags of earth. Finally, they add a covering of transparent polyethylene. This transmits sunlight through to the black inner pouch and traps the heat inside.

The inventors, T.S. Krishnamurthy and colleagues,[10] report that insects, at all stages of the life cycle, die when kept at 60°C for 10

[10] At the Central Food Technological Research Institute in Mysore, India. This work is described in *Appropriate Technology*, December 1991, p. 15.

minutes. They tested pouches of varying sizes containing several kinds of produce, including wheat, rice, pulses, and semolina. A pouch containing 40 kg of peanuts, for example, reached an internal temperature of 67°C in just 4 hours. Wheat took 6 hours to reach 61°C. No insects survived.

Neem Products

Neem (see Appendix A, page 279) is an Indian tree that has been introduced widely in Africa and now can be found from Mauritania to Mauritius. People in neem's homeland have long known that ingredients in its leaves and seeds can disrupt the lives of storage insects. For thousands of years, Indians, for example, have placed neem leaves in their grain bins to keep away troublesome bugs. Now, scientists are finding that there is technical justification for this process and commercial neem-based pesticides are already being employed in the United States.[11]

With all the neems in Africa (not to mention the new ones being planted because of the rising international enthusiasm for this tree), neem-based methods for controlling insects in grain stores are soon likely to be widely available.

Some German-sponsored research has already pioneered one approach that employs the oil extracted from neem seeds.[12] In this project, neem oil has proved effective against bruchid beetles—the prime pest of Africa's stored products. Amounts as small as 2–3 ml per kg of stored food will protect grains and legume seeds up to 6 months—long enough to overcome the critical period when bruchids and other storage insect pests are active.

In Togo, a program for teaching farmers how to protect seeds with neem has been under way for the past 15 years.[13] Now Niger, Senegal, and other nations are following suit. Neem oil imparts no bitterness to the food. In trials, people could not distinguish the seeds protected by it.

Probably in the long run, however, it will be neem leaves that are used most. This is the simple technique employed since ancient times in India. The leaves are merely added to the grain at various levels in the bin. The leaves eventually dry out, turn to powder, and (for all intents and purposes) disappear. The important thing is that bruchids, weevils, and flour beetles disappear also.

[11] See the companion report *Neem: A Tree for Solving Global Problems* for more information. For a list of BOSTID publications, see page 377.
[12] See also Appendix A, page 279.
[13] Deutsche Gesellschaft für Technische Zusammenarbeit (GTZ).

Mineral Dusts

For some time researchers have known that certain powdery minerals can kill insects. The sharp-edged dust particles "spear" through the thin joints between the horny plates of the animal's exoskeleton. This was first recognized with diatomaceous earth, a widely available, completely safe powder that kills cockroaches almost on contact. Now scientists in Nigeria have found that a common local mineral called "trona" also works in the same way—at least on certain storage pests.

In experiments, powdered trona proved lethal to the maize weevil (*Sitophilus zeamais*), causing almost 100 percent mortality after 15 days of exposure. It also reduced the maize weevil's fecundity in grains treated with the dust.[14]

Trona, $Na_2CO_3 \cdot NaHCO_3 \cdot 2H_2O$, is a crystalline carbonate/bicarbonate that occurs naturally in several parts of Africa. It is apparently not toxic to humans and livestock. Indeed, in most African countries, rural people use it as a food additive.[15] For example, they commonly drop it into okra soups to increase the mucilaginous quality or into boiling cowpeas to reduce the cooking time. In northern Nigeria, farmers add trona to their cattle's drinking water.

Mixing trona dust with maize grains (at 1.5 percent by weight or more) killed or inhibited the biological activities of the most ubiquitous pest of stored maize, the maize weevil; but another noxious pest, the red flour beetle (*Tribolium castaneum*), was unaffected.[16]

Mineral dusts may never be fully reliable in grain-store insect control, but their permanence, low toxicity, and ready availability make them attractive possibilities for a simple, cheap, and ubiquitous answer to at least part of the massive and widespread storage losses.

[14] L.C. Emebiri and M.I. Nwufo. 1990. Effect of Trona (Urao) on the survival and reproduction of *Sitophilus zeamais* and *Tribolium castaneum* on stored maize. *Agriculture, Ecosystems and Environment* 32:69–75. (L.C. Emebiri and M.I. Nwufo, Department of Crop Production, S.A.A.T., Federal University of Technology, Owerri, P.M.B. 1526, Nigeria.)

[15] It is known locally as *kaun* in Nigeria and *kanwe* in Ghana.

[16] A similar finding has been reported in respect to an inert diatomaceous earth, which was lethal to eight pests of stored products but was harmless to the red flour beetle. S.D. Carlson and H.J. Ball. 1962. Mode of action and insecticidal value of a diatomaceous earth as a grain protectant. *Journal of Economic Entomology* 55:964–970.

Appendix C
Potential Breakthroughs in Convenience Foods

Most people have never considered (or perhaps have abandoned) the idea of sorghum, millet, and the other African grains becoming prestigious foods for up-scale mass consumption. Everyone accepts that wheat is sold as bread, pastries, and baked goods; rice comes in all sorts of precooked forms; and maize is routinely available in convenient flour or grits. However, almost no one thinks of sorghum and millets in the same light. These African cereals are relegated to the limbo of foods suited only for personal use in rural regions by individual families who have to prepare their own food from raw grain.

But possible ways to upgrade Africa's own grains are on the horizon, and these deserve thorough investigation and development. Such processing breakthroughs can break the malicious mind-set, diversify the uses, improve the nutritive value, and boost the acceptability among consumers. Their success will create convenient-to-use foods, open vast new markets for Africa's farmers, and improve both rural economies and the balance of payments of many nations. In this particular sense, food technologists hold the key to the future of the lost grains of Africa.

This topic is far too broad to be covered adequately here. (It actually deserves a major international research endeavor.) Nonetheless, a few possible innovations—encountered while compiling this report—are mentioned below, just to provide perspective on some opportunities that are now languishing through lack of initiative.

POPPING

Popping is a simple technique that produces light, attractive, ready-to-eat products. It improves taste and flavor and it yields a crunchy, convenient food. Most people think of it as a process only for maize, and no wonder: popcorn is wildly popular among Americans and others who know it well. What has hardly been appreciated, however, is that most of Africa's grains also pop. While less spectacular than popcorn, they do expand dramatically and they, too, take on an agreeable toasty flavor. In the future, popped forms of sorghum, pearl millet, finger millet, fonio, and perhaps other grains could find extensive usage.

As has been mentioned previously (pages 43 and 177), people in India already pop sorghum and finger millet on a large, and sometimes commercial, scale. They often mix together milk, brown sugar (*jaggery*), and popped finger millet to create a very pleasant dessert. Popped finger millet is also used in brewing.

For finger millet, as well as for Africa's other cereals, popping seems to offer many benefits. It is a promising way to increase the grain size, create ready-to-eat foods, and add flavor to what are often bland dishes. Something similar is happening in the United States with amaranth. This former staple of the Aztecs and Incas is making a comeback, largely as a popped snack food. Recently, a continuous popper designed to handle amaranth's extremely small seeds was patented.[1] Such a device may well be the key to commercially popping Africa's small-grain cereals as well.

Once the popped grains are available, many new foods are likely to be created. Indian food scientists have blended popped finger millet with legumes such as puffed chick pea or toasted green gram to form nutritious and very tasty new foods.[2] In Africa, something similar might be done using legumes such as peanut, cowpea, or bambara groundnut.

PUFFING

The process of puffing, a variant on popping, was discovered almost a century ago. Since then, cereals made from puffed rice and puffed wheat have been breakfast staples worldwide. Puffed oats and maize are now also produced.

In the puffing process the grain is placed in a sealed chamber and heated until the pressure rises. Then the chamber, or puffing "gun," is suddenly opened. Relieved of the pressure, the water vapor expands, blowing up the grains to many times their original size (for wheat, 8–16 times; for rice, 6–8 times). Finally, they are toasted and dried until crisp.

Puffing has probably never been attempted with African rice, fonio, tef, or the other African grains, but it is another possible way to

[1] The machine was developed by Edward S. Hubbard, American Amaranth, Inc., Bricelyn, Minnesota.
[2] Information from H.S.R. Desikachar.

boost the size of these small grains, add flavor, and produce quality convenience foods with high consumer demand.

MALTING

Germination also upgrades the quality and taste of cereals. The sprouting process, known as malting, releases amylase enzymes that break starches down into more digestible forms including sugars. The result is to liquefy, sweeten, and raise the nutritional value.

Malting is particularly good for children because they can better assimilate the partially digested nutrients.[3] During World War II, government authorities in Great Britain (to mention just one country) seized on malting as a way to prevent childhood malnutrition brought on by wartime food shortages. Malt extract was produced in large amounts and distributed for daily use by children. This thick, dark, pasty material may have looked awful, but children loved its sweet and pleasant taste. It is in fact still sold in parts of the world, not so much as a nutritional supplement but as an everyday food that people buy for its flavor. It is also the key flavoring ingredient in famous foods such as malted milk and Ovaltine®.

Why malting is not more widely used in these days of mass malnutrition is a puzzlement. Perhaps the process is so associated with barley that the two have become almost synonymous, and, because barley will not grow where malnutrition mostly occurs, it is never considered. What has been overlooked, however, is that finger millet and some sorghums are almost as good at malting as barley. Their amylase activity is also high. And they *will* grow where the malnutrition is rife.

It is perhaps the ultimate irony that malting is practiced every day in many African homes, but the fact that malted grains make fine foods is overlooked.[4] Finger millet malt, for example, is great tasting, easily digested, rich in both calcium and sulfur-containing amino acids, and an ideal base for foods for everyone, from the very young to the very old.[5] But most of what is made these days is used in fermentations that produce beer (see box, page 168).

[3] For more on this topic, see Appendix D, where the use of malts in preparing weaning foods is discussed.

[4] Villagers are not the only ones who misunderstand malting. Missionaries in more than one country have preached against it in the mistaken belief that malted foods are alcoholic.

[5] Malleshi and Desikachar, 1986a.

FERMENTING

Lactic acid fermentations are used worldwide to produce foods such as sour cream, yogurt, sauerkraut, kimchee, soy sauce, and pickled vegetables of all kinds. Except for making sourdough bread, it is so far not used widely to "sour" cereal products. But in Africa it is traditionally used to flavor and preserve porridges and to produce popular foods such as *bogobe* (sour sorghum porridge) in Botswana, *nasha* (sour sorghum and millet porridge) in the Sudan, and *obusera* (sour millet porridge) in Uganda. People in many parts of the continent prefer the sharp flavor of these fermented porridges.[6]

Despite its almost complete neglect by cereal science, acid fermentation is yet another process for upgrading a grain's taste and nutritive value. For the food supply of Africa, it is particularly promising. The lactic acid fermentation process is well known. It is generally inexpensive and requires little or no heating, making it fuel efficient. It yields highly acceptable and diversified flavors. And it usually improves nutritive value.

It is commonly used in households (at least throughout eastern and southern Africa) and remains one of the most practical ways to preserve food for hundreds of millions of hungry people who cannot obtain or afford canned or frozen foods.

Lactic acid fermentations make foods resistant to spoilage, thereby performing an essential role in preserving wholesomeness. The bacteria rapidly acidify the food to a pH so low that dangerous organisms are no longer able to grow. They also produce hydrogen peroxide, which kills organisms that cause food spoilage (the lactobacilli themselves are relatively resistant to hydrogen peroxide). Certain lactic bacteria (notably, *Streptococcus lactis*) produce the antibiotic nisin, active against gram-positive organisms. Others produce carbon dioxide, which also helps preserve foods, notably by displacing oxygen (if the substrate is properly protected).

The course of the fermentation can be controlled by adding salt. Salting limits the amount of pectinolytic and proteolytic hydrolysis that occurs, thereby controlling softening (as well as preventing putrefaction).

Although fermented porridges were once extremely popular in rural Africa and are still widely consumed, their popularity appears to be declining. Some consumers are turning to alien alternatives that are widely advertised, such as tea or carbonated drinks. In many districts, farmers (as we have noted earlier) are giving up sorghum and millet and are growing maize. And in others, people are said "to lack the will and the interest" to prepare traditional fermented porridges.

[6] The use of such fermentations to make baby food is discussed in Appendix D.

But for all that, fermentations have a future and deserve recognition and attention. For one thing, they are very promising for creating weaning foods that may overcome mass malnutrition (see next appendix). For another, lactic acid fermentations are promising as commercial methods of processing and preserving food as well as for creating business enterprises.

PRECOOKING

To help meet the demands of an ever hungrier Africa (not to mention the world), the partial cooking of grains looks particularly promising. When dropped into boiling water, most (perhaps all) of the grains described in the earlier chapters soften within 5 or 10 minutes. The hot water partially gelatinizes the starch so that the dough sticks together and can be rolled into sheets or squeezed into noodles.

Some food technologists have already begun applying such processes to sorghum and pearl millet.[7] In the future, precooking might be applied to most of Africa's native cereals to produce top-quality, ready-to-cook foods that are stable, more nutritious, and easy to store.

Below we highlight three techniques—parboiling, flaking, and extruding.

Parboiling[8]

Parboiling is basically the process of partially cooking grain while it is still in the husk (that is, before any milling). The raw grain is briefly boiled or steamed. (Generally, it is merely soaked in water, drained, and then heated.) Only after the resulting product is dried is it dehusked and decorticated.

What results is very different from the normal milled grain. Sorghum kernels, for instance, come out looking like rice: light-colored, translucent, firm, and intact—attractive in both appearance and aroma and much less sticky than normal. Of course, they still must be cooked to become edible.

Parboiling not only gelatinizes the starch in the grains, it also does the following:

- Makes the milling process more efficient. (In a recent trial

[7] In this regard, notable work is being done at the Central Food Technological Research Institute (CFTRI), Mysore, Karnataka 570 013, India. There, N.G. Malleshi and his colleagues, although thousands of kilometers from Africa, have been doing work of great possible significance to the future of African grains.

[8] This section is based largely on the paper by R. Young, M. Haidara, L.W. Rooney, and R.D. Waniska. 1990. Parboiled sorghum: development of a novel decorticated product. *Journal of Cereal Science* 11:277–289.

with soft-kernel sorghum, parboiling more than doubled the yield of decorticated grain.)

- Inactivates enzymes and thereby greatly extends shelf life. (It even improves the storability of pearl-millet flour, a material notorious for turning smelly during storage.)
- Kills insects and their eggs so that it reduces storage losses.
- Improves the grain's cooking characteristics. (Boiling parboiled sorghum, for instance, doesn't produce mush; instead, the kernels remain separate, whole, and very much like pilaf or rice.)
- Improves nutritional values. (This is notably because it helps retain water-soluble constituents—such as the B-vitamins and certain minerals—that otherwise are thrown out with the cooking water.)
- Upgrades certain grains that have poor processing characteristics (the soft endosperm in finger millet, for example).

Given its now widespread use in the rice industry, parboiling is a surprisingly recent newcomer to commerce. Until the 1930s, it was hardly known outside South Asia where it was a village technology employed by poor people in their cottages. In the last 60 years, however, parboiled rice has rocketed into extensive worldwide use, and parboiling is now done on a giant commercial scale in countries such as the United States.

Parboiling is still good for village-level use, however. For example, field trials in Mali, using local sorghum and pearl millet, showed that it was practical, satisfactory, and boosted the yields from milling. Malian families tested the parboiled grains in local dishes and condiments (such as peanut sauce) and rated them very acceptable.[9]

At first sight, the extra energy and effort needed to parboil grains would seem to be a major disadvantage. However, the increases in yield and quality provide both the processor and consumer with substantial benefits.[10] Rice is already parboiled in the villages of some parts of Mali (not to mention half of India), which certainly suggests that the product is good enough so that people will find the fuel and put in the extra effort to prepare it.

Flaking

In this process, decorticated (pearled) grains are soaked, heated, partially dried (to about 18 percent moisture), pressed between rollers,

[9] These tests were run at Sotuba. The whole grain was washed, placed in cast-iron pots (covered) and heated in tap water over an open fire until the boiling point was reached. The pots were then taken from the fire and allowed to cool overnight. The next morning they were heated again and drained immediately after once more reaching the boil. The moist grain was next spread out in the shade to dry (24 hours for pearl millet and 48 hours for sorghum). The final product was decorticated with a mechanical mill.

[10] An increase of only 1–2 percent in the milled yield of commercially parboiled rice gives the processor enough profit to offset the extra energy costs.

The Power of Processed Foods

Despite the reliance on sorghum and millet in some countries, and despite consumer preference for flour made from them, the industrial production and commercialization of local flour has barely been established in Africa. Sorghum and millet flours are still mainly produced by each individual household. On the other hand, the introduced grains—wheat, rice, and maize—are more commonly milled at commercial facilities.

This makes the foreign grains look superior and it holds back the local cereals. And the situation is worsening. Soon, the rural labor force could be insufficient. Thus, even if production is increased there won't be the people to process it. For example, in most regions it is the young women who process most of the grain, but increasingly they are going to school, getting jobs, or abandoning the countryside to seek opportunity in the cities.

In a sense, then, it is imperative to find and develop good profitable uses for millet, sorghum, and the others. And the time is ripe. With increasing urbanization and rising disposable incomes, the demand for preprocessed and convenience foods is accelerating. This is one reason why commercially milled wheat and maize flour are increasingly preferred. Sorghum and millet are much cheaper, but they are unprocessed and therefore less convenient to use. As a result, markets for locally grown sorghum and millet are diminishing, incentives for local production are deteriorating, and foreign exchange reserves are dwindling to meet ever-rising demands for preprocessed flours.*

In dry regions, processing facilities are particularly vital to the future of local cereal farming. There, sorghum and millet are essential for a viable agricultural community. Both crops are so drought tolerant they can grow where other cereals cannot. When imported flour crushes the demand for them, the farmers are left with no outlet for their grain in years of good rainfall when they have a surplus. And when market prices fall, farmers cannot afford the inputs, such as fertilizer, that can keep their yields up.

If, as has been noted, markets for local flour and processed foods are developed, a large and healthy trade between a country's own sorghum, millet, and fonio farmers and its cities could operate to everyone's benefit.

Success with processing would likely transform Africa's native cereals into big-time, high-value worldwide foods.

* FAO reports that between 1961 and 1977, imports of wheat, rice, and maize to Africa increased between 5 and 10 percent a year, whereas the production of sorghum and millet increased 0.2–1 percent.

and, finally, completely dried into flakes.[11] The resulting product is a convenience food of many potential uses. The flakes store well and hydrate quickly when dropped into warm water or milk. They can be used in many types of sweet or savory dishes. When deep fried, they burst into light and crispy products.

African grains are particularly suitable for flaking because they are small and soak up water quickly. But although the process is simple, it is seldom used today. The holdup seems to be purely technological: grain-flaking machines are large, expensive, and inappropriate for Third World use. Now, however, a simple, inexpensive machine capable of flaking cereals in villages has been developed in India.[12] A unit has been installed in a village near Bhopal, and the people took to it and were able to operate it without supervision.

This type of invention could open up a new world for sorghum, millet, fonio, and other grains. More than 30 years ago, South African researchers mixed sorghum flour with water, then passed the slurry through a hot roller that both cooked and dried it. The resulting ready-to-eat flour proved very palatable and would keep for at least 3 months without deteriorating. Whole milk or skim milk (used in place of the water) produced a similar flour that was not only tasty but rich in protein, calcium, and phosphorus. Processing costs were reportedly low.[13]

Extruding

Extruding is a variant of the flaking process. The moistened and half-cooked grains are squeezed out through small holes. It is how noodles and pastas of all kinds are prepared. It, too, improves water absorption and cooking quality. Noodlelike products can probably be made from all the grains highlighted in this report. The pearled grains are first soaked for a day or two, then drained, mashed, cooked, extruded, and dried.

Noodles prepared from blends of finger millet and legume flours are already being used in India to form nutritionally balanced foods that can be used as supplementary foods for malnourished children.[14] When deep fried, they make excellent crispy products—said to equal those

[11] This is pretty much the basic process worked out by J.H. Kellogg in his kitchen in 1906. What resulted was the famous Kellogg's Corn Flakes.

[12] At the Central Institute of Agricultural Engineering in Bhopal. The machine consists of a hopper, four "large" rollers (112 mm x 230 mm), a smaller roller (88 mm x 230 mm), and a gear train to provide the differential speed needed to squeeze out the flakes. The whole unit is powered by a 2 hp electric motor, and the power required is only 150 W. The rollers are hollow and made of nylon to reduce weight and noise. The flakes come out as thin as 0.35 mm.

[13] Coetzee and Perold, 1958.

[14] Kumate, 1983.

Success Brewing in South Africa*

Mohale Mahanyele's story exemplifies the immense business opportunities to be found in commercializing the traditional foods made from African grains.

In the late 1980s South Africa's government set out to privatize the sorghum-beer industry. For at least 20 years, sales had been dropping, as workers migrated to the cities and left the rural villages where the low-alcohol, high-protein drink is embedded in the culture. The government hired a management consultant, Mohale Mahanyele, to advise it on how to get rid of the business. His task seemed like a thankless one; the sales decline seemed inexorable. One analyst said the authorities were merely unloading "an old Third World product doomed to die."

Mahanyele did not agree. "There were a lot of leaders in the African community who thought we were being set up to fail," he says. "But I thought differently. Here was a drink that had always been associated with our festive occasions, and it had been taken away from us and tainted. It was humiliating, degrading. I wanted to restore the dignity of sorghum."

Armed with that vision, Mahanyele himself set out to buy the business from the government-run monopoly in 1990. It seemed like a foolish notion. He had to raise $20 million to purchase the corporation and its 21 factories, but he had no access to white capital. So he did something never before attempted in his country: he sold shares to fellow Africans, building on the centuries-old custom of *stokvels*—small, informal savings societies—in traditional communities.

National Sorghum Breweries ended up with 10,000 shareholders, more than 90 percent of whom are black—a novel arrangement in a country where few blacks own the roof over their heads.

But Mahanyale's problems were far from over. In addition to the dropping sales, he had to overcome sorghum beer's political stigma, created during the 80 years when the white-minority government ran the business. To his own people, "Kaffir beer," as it was known, had become a symbol of white oppression.

But Mahanyele succeeded. Today, National Sorghum Breweries is by far South Africa's most successful black-owned business. It has nearly doubled its volume in the past three years, while

* This vignette is adapted from an article by Paul Taylor (*The Washington Post*, July 21, 1993).

paying annual dividends of 20 percent or better. "We understand the product," he says. "We have a color fit and a culture fit with our customers."

Through the development of the sorghum-brewing business, Mohale Mahanyele has become South Africa's foremost apostle of black economic empowerment. The company's board and management team, once all white, is now nearly all black. Most of its contractors are black, including 500,000 small businessmen who distribute the beer to stores throughout the country. It employs a quarter of South Africa's black accountants, and is putting more than 100 of its executives through an MBA program that it runs on the premises.

Today, National Sorghum Breweries is beginning to diversify into other products—food, soft drinks, computers and, most daunting of all, conventional beer, a market in which a giant white-owned brewery currently has a 98-percent share. Can more success be far behind?

Sorghum beer has a rather thick consistency with a refreshing acid flavor; the alcohol content is only 3–4 percent by volume, but large amounts are apt to be consumed on festive occasions. Women have brewed it in Africa's villages for centuries (see page 168).

No one has ever written a definitive work on African beers and their nutritional or social roles. This could be a major project for African scholars. These beers are more important than most people realize. A special quality is their safety. Because they are highly acidic (ranging between 3 and 4 on the pH level), they are free of bacterial contamination. So far, however, science has shied away from investigating such beers. Anthropologists and nutritionists refer to them, but that is about all. This is surprising because sorghum beers are an important part of life throughout most of Africa below the Sahara.

Opposite: In his executive suite in a suburban office tower in Johannesburg, Mohale Mahanyele merrily lowers himself on his haunches onto the plush carpet. He is demonstrating the traditional way to consume sorghum beer. "You gather around in a circle, and everyone squats," he says. "Then you pass around the calabash, and everyone takes a drink. If you're standing up, it's a sign of disrespect." (Louise Gubb, courtesy *Washington Post*)

prepared from rice. Noodles from finger millet and other African grains could probably be economically produced in small-scale industries, as the equipment needed is not overly complicated and the capital investment is modest.

LEAVENING LOAVES

Raised bread has become what is perhaps the world's premier food. Wherever it is introduced people eagerly adopt it and clamor for more. Unfortunately, however, leavened breads can be made only from wheat or rye, neither of which grows well in the tropical zone where the neediest people are concentrated.[15]

For at least 30 years scientists worldwide have searched for ways to make raised bread without using wheat and rye. Such work could have profound implications for Africa (see box, page 310) but, despite the theoretical promise, nowhere has there been much practical success so far. Local staples tend to make unattractive, short-lasting, poor-rising breads that the public shuns. Dough strengtheners and other modifiers (such as emulsifiers, pentosans, xanthan gum, and wheat gluten) can be added. They make acceptable breads, but usually they must be imported and are expensive.

Now, however, there is a possibility of a breakthrough: research has shown that it is possible to prepare loose-structured bread from local grains using a swelling and binding agent. Different types have been tested. Dried pregelatinized cereal or tuber starches have shown some success. Glyceryl monostearate is said to be effective. Locust bean gum, egg white, and lard are also fairly good. These compounds act to bind the starch granules together, making it possible for the dough to hold carbon dioxide gas and thereby to rise. Baked products obtained this way have greater volume, softer crumb, and a more regular texture.

FAO Bread

Although none of the techniques has yet yielded light, high breads like those from wheat, there has been partial success. Perhaps the most advanced is a project operated by the Food and Agricultural Organization of the United Nations (FAO). The FAO method involves

[15] Gluten gives bread its light texture, and this elastic protein is unique to wheat and rye. When the dough is fermented, gluten's network of protein strands traps carbon dioxide released by the yeast. As the gas bubbles up, it raises the dough into the light, open texture of leavened bread. Triticale, a man-made hybrid between wheat and rye, not unexpectedly, can also produce raised breads. (Triticale is described in a companion report, *Triticale: A Promising Addition to the World's Cereal Grains*. For a list of BOSTID publications, see page 377.)

Avoiding the Wheat Trap

Researchers in several southern African nations have banded together to produce a white sorghum that can be locally grown to make flour for bread and mealies (cornmeal). They seem to be already on the verge of success. If so, they will have developed the first truly African bread grain. The following is a recent announcement from PANOS, an international organization that specializes in disseminating Third World news.

Fifty scientists from Angola, Botswana, Lesotho, Malawi, Mozambique, Namibia, Swaziland, Tanzania, Zambia, and Zimbabwe, the 10 countries grouped in the Southern African Development Coordination Conference (SADCC), are now being trained to breed and produce sorghum hybrids. Soon, that number of trainees is expected to double. Why all the excitement?

To help reduce the region's dependence on imported wheat, researchers in Zimbabwe have developed hybrid strains of sorghum and millet that are designed for use in making flour and bread. The work at the Matopos Research Station near Bulawayo forms part of a drive to reduce food shortages in the SADCC countries.

For most people in the region maize is the staple, but the crop does not grow well in the drier areas. Researchers are trying to develop substitutes that can be grown there and mixed with wheat for bread or maize for mealies. Any surplus could be sold to make high-quality malt.

In farm tests, the new hybrids have produced bigger yields than existing varieties. The researchers expect to have white-grained hybrid sorghum for milling very soon. It is hoped that the white sorghum will satisfy a popular preference for white maize meal. A local milling company is already working with a nongovernmental organization called Enda-Zimbabwe to set up pilot mills in rural areas to grind the hybrid grains for bread.

Before people in areas of low rainfall can be persuaded to abandon their often futile efforts to grow maize, good varieties of the new hybrids must be available in large quantities of the seed.

boiling part of the flour from a local cereal (or root) until it thickens into a gel strong enough to hold the gas released during breadmaking. When added to local flour, yeast, sugar, and salt, this starchy substitute for gluten produces a puffy bread of acceptable texture, taste, and color.

Reportedly, this new technology is simple, inexpensive, and uses nothing but local ingredients. It can, for example, produce leavened loaves using sorghum, millets, and other African grains.

Leavening with Fungus

Recently, food scientists in India have found that fermenting a mixture of grain and pulse (legume seed) can produce a gum thick enough to act like gluten. This special process, locally known as *idli* or *dosai* fermentation, involves the microorganisms *Leuonostoc mesenteroides,* which is used in other parts of the world for producing dextran gums from sucrose. Using this fermentation, a mixture of rice and *dahl* (made with black gram or other legume) can be turned into a dough that will produce breadlike products without employing any gluten. Either the legume, the microorganisms, or the combination produces a gum that holds the carbon dioxide gas, thereby leavening the products. It is a fermentation that enables raised breads to be

The Wheat Trap

Africa is finding itself more and more caught up in what is being termed the "wheat trap." During the past 20 or 30 years, certain governments as well as private companies have responded to consumer demand by establishing wheat mills. As a result, various countries now spend large amounts of foreign exchange importing wheat to feed those mills. The bulk of the flour produced is used to make bread for the working population, as well as for the small expatriate population living in the towns and cities.

Bread is a convenient food because it is ready to eat, easily carried around, and not messy like porridges and gruels. Its taste is highly acceptable, it gives a feeling of bulk and fullness, and it is relatively cheap. With large numbers of people migrating from rural areas to the cities, the demand for bread has increased.

However, the population is being fed on food the country does not grow, with scarce foreign exchange being used to import wheat to produce flour. More foreign exchange is also spent on spare parts and foreign managers to maintain and run the flour mills. The process not only damages the economy but the indigenous African cereals as well. They are being left in a state of underdevelopment and inadequate processing.

J. Maud Kordylas

made without any wheat or rye.[16] Perhaps other fermentations or other substrates for this fermentation to do this job can also be found.

Biotechnology

With all the advances in biotechnology these days, it seems likely that the genes that cause gluten to form in wheat will soon be isolated. Inserting them into the chromosomes of Africa's native grains could bring profound changes. Suddenly, sorghum or pearl millet would (at least in theory) produce bread that rises without any extra help. This is not a far-fetched idea. Indeed, it will be surprising if it does not come about within the next decade or two.

[16] Information from N.G. Malleshi and H.S.R. Desikachar.

Appendix D

Potential Breakthroughs in Child Nutrition

As in the three previous appendixes, we report here innovations relating only indirectly to Africa's cereals. Once again, these seem of notable significance to the continent as well as to the future of the traditional grains. In this case, the potential breakthroughs are of great humanitarian significance—no less than a means by which Africa may at last put behind it the horrors and heartbreak of childhood malnutrition.[1]

WEANING FOODS

In most parts of the world, baby foods are commonplace. In North America, for example, supermarkets may carry whole aisles of liquefied and semisolid concoctions carefully created from cereals, vegetables, and fruits. Through these foods, a child gets a diet that is easily digested, rich in energy, and balanced in protein, vitamins, and minerals. Such foods help the child make the complex and otherwise life-threatening transition from mother's milk to adult fare.

The tragedy for Africa's millions of malnourished children is that comparable bridging foods are unavailable to, or at least far beyond, a family's financial reach. A child in Africa, therefore, faces a cataclysmic change from a balanced and hygienic liquid diet of mother's milk to an unbalanced solid adult food that is often very unwholesome. Although the young milk-fed bodies are basically unprepared for such

[1] As before, the coverage is far from exhaustive. In fact it is based on a single report. This report, a quite technical 400-page document full of data and diagrams, includes information contributed by several dozen nutritionists and food technologists from all parts of Africa. In this appendix we can only skim some highlights. The details can be found in the full report: D. Alnwick, S. Moses, and O.G. Schmidt, eds. 1988. *Improving Young Child Feeding in Eastern and Southern Africa: Household-Level Food Technology*. Report No. IDRC-265e. International Development Research Centre, Ottawa, Canada.

a switch, they must start digesting foods of alien consistency and inferior quality. Moreover, they often must do this while battling new and numerous intestinal infections introduced through unclean hands and utensils as well as through inadequate cooking.

This situation constitutes the gravest emergency facing children today. As UNICEF's executive director, James P. Grant, has pointed out: "The period of weaning, during which a young child becomes accustomed to the change from a diet consisting solely of his or her mother's milk, to one totally devoid of it, may take a year or more, and in much of the world this is perhaps the most dangerous period of the child's life. Many will not survive it. Of those that do, too many will be stunted in body, and perhaps in mind, and never be able to attain the full promise of their birth."

Today this hazard falls heaviest on Africa's children. Perhaps in the future centrally processed weaning foods will, as in North America, serve the children's needs. However, at present the cost of such products and the inability to distribute them throughout the rural regions makes this impractical. The only answer for the moment, then, is weaning foods that can be prepared either in the home itself or at least in nearby locations in the rural districts.

Given the extent of present malnutrition, one could be forgiven for concluding that household weaning foods are an impossibility for rural Africa—that appropriate ingredients must be unavailable, or that the people cannot make foods appropriate for children. But a number of knowledgeable nutritionists and food technologists believe that bridging foods for the critical nutritional years of each new generation can indeed be produced locally and cheaply. And, in their view, it is the traditional native grains—sorghum and finger millet, in particular—that are the key to this vital and life-saving possibility.

The reason for this is unexpected but understandable.

Those who, in the past, blamed malnutrition exclusively on the lack of certain nutrients in the foods were largely wrong. The local cereal products are not as poor in nutrient quality as was (and is) generally claimed. Today's nutritionists increasingly blame the low quantity of solids (what they call the "nutrient density") in the foods used for feeding the very young.

Africa's traditional weaning foods are watery gruels based on boiled cereal. These may have the right consistency for a child whose sole diet has been milk, but they are just too dilute. A gruel whose consistency is acceptable to a one-year-old contains merely one-third the food energy of a typical Western weaning diet. A child simply cannot consume enough to meet its energy and other nutrient requirements. Even when stuffed with gruel to its limit, a small stomach contains too little solid to keep its owner fed for very long. And most of the children must get by on only two feedings a day because mothers who work in the fields have no time to boil batches of gruel throughout the day. The children therefore get fed only in the morning and evening when the rest of the family's food is prepared.

A tragic irony is thus becoming apparent. Although the gruels are too thin, the porridges the mothers are cooking for the rest of the family would be satisfactory except for one fact: they are too thick to be swallowed by an infant. A stiff porridge is useless to anyone who cannot eat solids.

What can be done? The answer, the nutritionists now say, is to take a small part of the adults' thick porridge and change its consistency so any child can "drink" it. How? By the age-old African methods of malting or fermenting (see Appendix C). Both procedures break up boiled starch so that it collapses into smaller saccharides, including sugars, and releases the water that keeps it thick.

For the rest of the world, malting and fermenting are not everyday household operations, but in Africa they are. Indeed, these two processes are probably better known at the household level in Africa than anywhere else in the world. Both techniques require only a minimum of equipment and appear to be good ways to turn stiff starchy porridges into liquid weaning foods.[2]

MALTED FOODS

Given what is currently available in an African village, probably nothing can compare with malting as a means for carrying rural babies across the nutritional abyss between mother's milk and adult foods. The previous appendix discussed malted grains and the potential they offer in and of themselves. Here, however, we discuss another side of these versatile materials: their use as culinary catalysts for modifying starchy foodstuffs. This is a process all but unknown to most people, but it is by far the biggest use of malted grains and is conducted all over the world. It is, in short, the vital first step in making beer and whisky.

Perhaps because of this association, malting has been saddled with a somewhat seedy reputation. But it is a simple, safe process that produces no alcohol and should be more widely used and better known to cooks everywhere.

In Africa, malting has a special promise. Two of the native staples—finger millet and certain sorghums—are rich in the malting enzymes (amylases) that break down complex starches. To liquefy even the

[2] Although germination of cereals is mainly associated with the preparation of local beers, there are a few examples of this procedure being used to prepare local weaning foods with low dietary bulk.

thickest cereal porridges takes only a small quantity of flour from germinated sorghum or finger millet. When this flour and the porridge are heated slowly, the amylase enzymes hydrolyze the gel-like starch in the porridge so that it collapses and can no longer hold water. In this way, sprouted sorghum and finger millet can turn a pasty porridge semiliquid in minutes.

Moreover, the food not only thins down, it becomes, to a certain extent, predigested so that it is easier for the body to absorb. In addition, the enzymes hydrolyze not only the starches but some of the proteins as well. They also reduce antinutritional and flatus-producing factors, improve the availability of minerals, and enhance some of the food's vitamin content. Further, the malting process imparts sweetness and flavor that makes for a tasty end product.

Considering the extent of malnutrition, it is more than ironic that individuals throughout Africa know more about this process than people anywhere else in the world. Indeed, throughout sub-Saharan Africa, millions of homes have a crock in the corner that contains malted grain. A small sample of the contents would transform thick porridges into baby foods sufficiently liquid for children to consume and sufficiently nutrient-dense to keep them healthy. Tests have shown that adding a little germinated cereal while a porridge is being prepared doubles the amount of food energy and nutrients a child can ingest. However, at present the malt is used only to make beer, almost never to prepare weaning foods.

Experiences in Tanzania suggest that the concept of liquefying porridges for baby food is not an impractical dream. In the early 1980s, scientists at the Tanzania Food and Nutrition Centre found that small quantities of flour from germinated sorghum or finger millet could be used to thin the traditional viscous porridges.[3] They called their product "Power Flour." When a spoonful was added during cooking, porridges thick enough to hold up a spoon turned liquid within 10 minutes.

The researchers found that mothers in Tanzania's villages were only too willing to use Power Flour. Most of the mothers knew how to prepare germinated cereals for brewing but knew nothing about making foods for their children from them. However, because the procedure was already so well known, they quickly adopted it.[4]

Although it is ironic (even tragic) that malting is so well known across Africa, it is also an advantage. Using germinated cereal to improve weaning foods is simply a variation on an already widespread

[3] Mosha and Svanberg, 1983.

[4] In fact, it is hard to avoid the conclusion that Power Flour (or some barley-based counterpart) could find a place in kitchens worldwide, including the most sophisticated. With an aging world population and in some wealthy countries an intense interest in dieting, liquid diets and highly digestible foods of all kinds are now much in vogue and are the bases of billion-dollar industries.

technology—not a strange foreign food or technique to be imposed by an outside authority. Local, national, and international efforts to stimulate appreciation of this could see a new level of weaning foods sweep across Africa with little outside involvement. The key in many areas may be to educate village brewmasters to the potential of a second product from their ongoing malting operations.

Sorghum is the most widely available malting grain in Africa, and it has been used in most of the nutritional experiments so far. However, finger millet is a better choice: it has a higher amylase activity; it has no tannins; it develops no potentially toxic materials on germination;[5] it is rich in calcium and methionine, both of which are needed for child growth; its malt has a pleasant aroma and taste; and, finally, it does not mold or deteriorate during germination.

Considering the fact that the technology and raw materials are common in most village situations, why has this immensely beneficial practice not been more widely used? For one thing, the process of germinating grain does take some time; mothers, already weighed down with burdensome work loads, tend to reject anything that takes up more of their day. However, germinated flour need not be produced daily. Indeed, small portions can be set aside whenever a fresh batch of beer is begun. In addition, as in the case of Tanzania's Power Flour, the malt could be made centrally and sold widely. Unlike the weaning foods themselves, it is a stable, concentrated material that is used only a pinch at a time.

FERMENTED FOODS

The fermentation of cereals by lactic-acid-producing bacteria has been discussed in the previous appendix. It, too, appears to be a way to prepare weaning foods. Like malting, fermentation is a household-level food technology that reduces the viscosity of stiff porridges (although not as much and not in minutes). It raises the levels and bioavailability of proteins, vitamins, and minerals. It enriches the foods through the synthesis of some B vitamins, and it adds flavor. On top of all that, it helps protect the foods from diarrhea-causing microorganisms.

As has been noted in Appendix C, lactic fermentation is practiced throughout the world to make pickles, sauerkraut, soy sauce, sourdough bread, and other popular foods, but it is especially well known in Africa. From Senegal to South Africa "sour" porridges are popular.

[5] Certain sorghums on sprouting show a marked increase in hydrocyanic acid. This is worrisome, especially when the product is to be fed to a very small child. However, it seems probable that the normal cooking of porridges quickly drives off the cyanide.

However, although still widely consumed, they are often overlooked as weaning foods.

But sour porridges seem to fulfill many of the characteristics required, and they also reduce the risk of pathogenic diarrhea—Africa's leading cause of infant death. They save time and energy as well, and might be very suitable for use during the day when a working mother has no time to cook.

A few fermented foods are already employed as weaning preparations. One example is *ogi,* a blancmange-like product that is one of Nigeria's most important foods. *Ogi* is created by fermenting a slurry of sorghum, millet, or maize. Adults eat it for breakfast, but some is kept aside and used as a weaning food.

There are possibilities, too, of combining fermentation and malting. Thus, fermented doughs, such as *ogi* or *ugi* (a similar product widely eaten in East Africa), might be liquefied with Power Flour into forms that weanlings can "drink." In that way children could ingest more, and the double processing would likely produce highly digestible foods, easy for any young, old, or sick bodies to assimilate.

Appendix E
After Words

While compiling this book, we were in contact with several hundred researchers who specialize in the various crops described. Along with their technical advice, some sent in provocative quotes, valuable for their pith and perceptiveness. In addition, during the four years that have gone into this book, we came across a number of equally intriguing quotes in the published literature. All in all, there were too many to include in the body of the text, so a selection of them is appended here. Some contradict each other, a reflection of the contributors' different visions and of the complexity of the issues. Each, however, contains insights that complement the earlier parts this book, which perforce had to be focused exclusively on the plants and their promise.

Philosophical Overview

The negative trends in Africa are not solely due to lack of knowledge. We shall claim too much if we say "give us money, we will do research, and we will solve the African food problem."

A.H. Bunting

The resources of farmers are not confined, let us remind ourselves, to the classical factors of land, labor and capital, although by suitable definitions we can fit all resources into one or other of those omnibus packages. We have to think also of seed, equipment, knowledge, chemicals, credit and many other things, as well as of external encouragement, services and support, particularly from the policy of governments. Development in Africa might well take a different course if governments were able to be more effective. Many African governments and government services are inexperienced and some are unstable. Many of them have great difficulty in forming and executing development plans.

A.H. Bunting

Farmers are rightly suspicious of the counsel of anyone who does not himself have to live by the results.

John Kenneth Galbraith

African farmers are not a bunch of village idiots; far from it. They can squeeze more out of a hectare than you or I could, and under difficult circumstances.

Jack Harlan

At least eleven hundred million people do not have enough to eat. Many of them live in countries that cannot afford to import food and where per capita domestic food production has declined since 1980. Most of these countries are in Africa, where the gap between food production and demand is expected to quadruple by the year 2000.

Inji Islam

What Africa needs is more agricultural research conducted by well-trained scientists with good support. It should include—at the least—plant breeding, pathology, agronomy, biotechnology, entomology, and soil science.

Arthur Klatt

The right technology—be it genetic or agronomic—will be put to use. If it increases yields economically, Africa's farmers will adopt it.

Arthur Klatt

Unless we satisfy the basic needs of four billion poor, life for the rest of use will be extremely risky and uncomfortable. Struggling farmers . . . threaten environmental stability, while the growing masses of urban poor are a menace to political stability.

Klaus Lampe

These "old" plants are neglected mostly because both local and foreign "experts" are prejudiced against them, but also because of the experts' own preference for anything that is new!

James M. Lock

The promotion of any indigenous crop must be done within local constraints of labor availability, gender relations, cultural constructs, and environmental stress. If local constraints, practices, and beliefs are not realized, promotion of the crop will not succeed.

Clare Madge

Of the two billion persons living in our developing member countries, nearly two-thirds, or some 1.3 billion, are members of farm families, and of these are some 900 million whose annual incomes average less that $100 . . . . for hundreds of millions of these subsistence farmers life is neither satisfying nor decent. Hunger and malnutrition menace their families. Illiteracy forecloses their futures. Disease and death visit their villages too often, stay too long and return too soon.

The miracle of the Green Revolution may have arrived, but, for the most part, the poor farmer has not been able to participate in it. He cannot afford to pay for the irrigation, the pesticide, the fertiliser, or perhaps for the land itself, on which his title may be vulnerable and his tenancy uncertain.

Robert McNamara
President, World Bank [1973]

The persistence of child malnutrition in Rwanda is attributed largely to a lack of time and money on the part of the mothers. In the northern parts of the country, women spend nearly 10 hours in the field and so can prepare the family food only once or twice each day; this food is usually high in bulk but low in nutritional value and is, therefore, inadequate for feeding young children.

M. Ramakavelo

One of the problems that makes the task of the prevention of famines and hunger particularly difficult is the general sense of pessimism and defeatism that characterizes so much of the discussion on poverty and hunger in the modern world. while pictures of misery and starvation arouse sympathy and pity across the world, it is often taken for granted that nothing much can be done to remedy these desperate situations, at least in the short run.

There is, in fact, little factual basis for such pessimism and no grounds at all for assuming the immutability of hunger and deprivation. Yet those unreasoned feelings dominate a good deal of public reaction to misery in the world today. In fact, pessimism is not new in this field, and has had a major role over the centuries in dampening hearts and in forestalling preventive public action.

Amartya Sen

Instead of running away from these traditional products, we should be encouraging their use as quality foods that are as good or maybe even better than some of the foods people are presently substituting for them.

S. Vogel and M. Graham

Cereals in General

There is no doubt that cereals selected and cultivated by man are the basis for a stationary human culture as in the cities and villages of the world. The apparent value of the cereals was high convenience in storage and in cooking quality as well as a pleasant smell and bland taste of the final product combined with a high level of satiety after consumption.

Lars Munck

One of the possible reasons of the lack of research on native grains is that many African postgraduates go abroad either to USA or Europe and do their higher university degrees on wheat or maize. When they return, it is quite natural for them to continue their studies. (I have seen this happening in the past here in Australia but this is now changing.) It would be a step in the right direction if these postgraduates work on the crops of their own country for these degrees. (As a bonus it might even broaden the thinking of their supervisors.)

Donald F. Beech

There is no doubt that the human body was designed mainly to get calories from carbohydrates—starches and sugars—and since most starchy foods are fairly bulky it can be actually quite difficult for children to consume enough carbohydrates in a day if they come entirely from starchy foods like bread and potatoes and root vegetables.

John Birkbeck

Some 80–85 percent of the population in many African countries subsists on farming, and this large segment needs to be helped in improving itself. As improvements occur in agriculture and as it becomes less marginal and less subsistence-oriented, opportunities will need to be created for people to move to other sectors of activity.

Norman E. Borlaug

Although starchy fruits, roots, and tubers will continue to be important in the diets of African people in many countries and regions, much of the extra food needed will consist of cereals.

A.H. Bunting

There are many weaknesses in the output delivery systems such as physical infrastructure, transport, markets, storage, processing, wholesaling and retailing, and prices. These components determine the extent to which farmers can sell off the farm, which is the essential nexus in the whole business of agricultural and rural development.

A.H. Bunting

These cereal grains supply man with 60 percent of his energy and 50 percent of his daily protein requirements . . . the volume of grain required each year to satisfy man's needs can be calculated to be a highway of grain 2 meters high by 23.5 meters wide, that circles the earth at the equator. Approximately 1000 meters of new highway must be added each year to satisfy population increases.

Vernon D. Burrows

In Africa in the 1970s, the total area under all three cereals [sorghum, maize, and millet] increased by 8 percent, while mean yield declined by 1.5 percent and the human population increased by 29 percent. Unless this trend can be reversed, there is real trouble ahead.

Hugh Doggett

Often, a new variety fails to enter the traditional agricultural setup because no one checked if it will make the preferred foods at an acceptable quality. In Ethiopia, for example, bread-wheat varieties have been identified, but the farmers only grow them for cash as they cannot make good bread or grits using the traditional food-making techniques.

Sue Edwards

An essential feature of African diet is that the staple food—either maize, sorghum, millet, rice, cassava or wheaten bread—supplies about 80 percent of the people's calories, compared with approximately 30 percent eaten by Europeans in the form of bread. For Africans, the staple food is not merely the main source of carbohydrates, but also of proteins, minerals and vitamins.

M. Gelfand

Politics is probably the biggest "stumbling block" in Africa. In one country, they told me that the farmer could double the grain yield of pearl millet with existing agronomic practices but when the farmer did this, the government cut the price in half.

Wayne W. Hanna

The colonial literature is full of nonsense about "scarcity foods." They [the colonials] thought people harvested wild grass seeds because they were hungry and did not know that these were staples and gourmet foods.

Jack Harlan

A major widespread constraint to increased production that remains in Africa, in contrast to Southeast Asia, is that of unstable grain markets. In consequence, rural families grow sorghum and pearl millet by the most reliable methods to meet their own needs and produce relatively little surplus to market. When there is a good year, everyone has a surplus and the market price falls catastrophically. Very rationally, farmers invest their efforts into cash crops or some other enterprise where returns are more assured.

R.C. Hoseney, D.J. Andrews, Helen Clark

Since the most ancient of days, the destiny of humanity has been inseparable from grain. Even today in the age of the microchip processor, humanity's affairs remain closely linked to the Fates attending cereal grains.

KUSA

African cereal production has two great weaknesses: there are no facilities for producing top-quality seed and there are no conduits for conditioning, storing and distributing it. Africa is full of entrepreneurs and there is a tremendous opportunity for them to start businesses selling quality seed. India started its own seed-trade that way: entrepreneurs began selling locally produced elite seed to their neighbors. Gradually, an entire distribution system developed.

A. Bruce Maunder

Nowhere in Africa are grains traditionally grown for "yield per hectare." Rather, they are grown for basic ingredients of specific foods such as ugali, injera, couscous, or beer.

J.F. Scheuring and M. Haïdara

I suggest that researchers are now avoiding many of these traditional cereals because they consider it *infra dig* to use simple breeding and selection technology. The crops' status suffers from solely because there are no high tech (genetic engineering, etc.) papers in the literature.

Gerald E. Wickens

In cereal production, Africa's greatest weakness is that there is little local storage. At harvest time farmers must sell their grain, regardless of price. Even in the United States, the drop in grain prices can be startling at harvest time, but most American farmers have their own storage and any farmer can rent storage, either locally or near the markets (which may be thousands of miles away). This allows the farmers the chance to wait and benefit from price rises after the harvest. It also buffers price swings, which benefits everybody except the speculators.

In Africa, the situation will change when a large demand for sorghum and millet flour develops. That will create a need for year-round supplies, and storage capacity will have to be created to provide millers with grain during the off-season. This will serve to draw off grain stocks during flush seasons while maintaining grain stocks during periods of shortage. In turn, it will allow farmers to hold their grain until they're happy with the price. It will also give the farmers an incentive to use superior seedstock, especially because prices won't fall as much during good years.

John Yohe

Plant Breeding

New variety types have to complement a farmer's food security strategy. Farmers in southern Mali have related to me that pearl millet and maize have expected storage times of three years, sorghum up to seven years, and fonio of well over seven years.

D.J. Andrews

I am sure that breeding for multiple objectives is essential if we are to attain our objectives sufficiently rapidly to benefit hundreds of millions of farmers and consumers by the year 2000.

S.C. Harland transformed Tanguis, the main cotton of Peru, by what he named the mass pedigree system of selection. By setting standards for six characters which could be measured on single plants, rejecting plants or small bulks in which these characters were below the norm or the arithmetic mean, and by advancing the standards in successive years, he soon produced populations of improved quality which yielded very much more than before. Starting from preliminary observations in 1940, the first wave of about 500,000 kg of improved seed was issued in 1943; and by 1949 yields around 1 ton of lint per hectare were being harvested on a field scale by some farmers. In respect of characters other than those for which they had been selected, the new populations were genetically heterogeneous and further improvement in them was evidently feasible.

A.H. Bunting

There are still abundant examples of major plant breeding programs which do not take account of the real constraints faced by many farmers. This is equally applicable to national and to international programs. The importance of this is vividly highlighted by the fact that after forty years of breeding on sorghum and millet at internationally supported research stations in West Africa, less than five percent of the crop is planted to such material. The products simply do not meet most farmers' needs.

Stephen Carr

There has been a tendency to so under-rate the value of traditional cultivars that the extension staff ignore them. In so doing they miss the opportunity to provide a well-worthwhile service to their clients.

Stephen Carr

The germplasm story requires a whiff of skepticism. While the collections may not have *everything* (do they ever?), the real problem is to use what we have. We need more *real breeders* and fewer people pontificating about germplasm.

Geoffrey P. Chapman

Time has come when our breeding strategy has to change from the one where land is tailored to suit the requirements of a high-yielding cultivar, to where we tailor the cultivars to suit the harsh and ordinarily inhospitable habitats where the small farmers have to grow their crops.

T.N. Khoshoo

Above all, it is the imagination and ingenuity of the breeder that will be the decisive element in producing any new cereal crop in the future.

C.N. Law

Much progress has been made in the training of African scientists, such as by the Title 12, Sorghum-Millet Collaborative Support Research Program, INTSORMIL. Whereas vehicles and computers have been supplied to thier in-country projects, little or no input has been given to adequate seed storage. Therefore, the maintenance of lands races, varieties, and breeding lines requires frequent re-increases; inefficient activity with risks of losing the original genetic composition.

A. Bruce Maunder

Simple harvesting and processing machines could greatly increase the effectiveness of seed production, and at minimal cost. Even on research stations in Africa, it is common to see sorghum and millet being pounded with wooden clubs. This is just too inefficient: even working night and day, there's no way they can handle the quantities required.

In fact, many suitable small machines are lying around the developed world, having been superseded by newer and more sophisticated models.

A. Bruce Maunder

Traditional grain varieties have been selected over the centuries to fit the constellation of agronomic adaptability in diverse environments, and at the same time have optimum milling, food quality, and storage properties. Most of the recent improved varieties from breeding programs in Africa yield grain that is poorly developed, headbug damaged, and chaffy when harvested from stressed environments. Such grain lends itself to high storage losses, low decortication yields, poor food quality, and poor seedling vigor. That the farmers don't adopt those varieties should not be a surprise. Cereal grain yield in Africa is the amount of nutrient per hectare that finally makes its way to the human stomach as food and to the animal stomach as feed. It is our challenge to start measuring that.

J.F. Scheuring and M. Haïdara

Everybody wants to help the poorest of the poor. However, when it comes the reality of applying modern knowledge it is often logistically impossible. To create a new variety—even of a well-understood crop like wheat—can easily take a decade of dedication and perhaps a million dollars of support. It is therefore clearly impractical to reach, individually, the thousands of different subsistence regions, each with its likes and dislikes, needs and desires, climates and conditions.

Noel Vietmeyer

There is a need to strengthen the links between sorghum and millet breeders and the food scientists, home economists, and other scientists involved in postproduction systems and the commercialization of sorghum and millet end products.

S. Vogel and M. Graham

Agronomy

When the aim is to improve a crop, one has also to improve the cropping system and the management of the fields (in terms of plant population, plant protection, soil fertility, etc.). The yield of any crop is very often related to the degree of intensification of the farming system. Therefore if we remain within the context of a traditional farming system or a slightly improved farming system, the agronomists and the breeders should not aim at achieving high dry-seeds yield; rather they should define the adaptive potentialities of the local varieties and try to utilize these to their maximum.

J.P. Baudoin

Despite the tremendous increases in food production in Asia, the Middle East, and parts of Latin America in recent years, agriculturalists today face even greater production challenges to feed future generations. New Green Revolutions must occur in the more marginal production areas of Asia, sub-Sahara Africa, and parts of Latin America. These areas are generally rain-fed environments that suffer from moisture and temperature stresses, soil fertility problems, diseases and pests, and other difficult production conditions.

Norman E. Borlaug

For arid and semiarid regions with their variable and unpredictable climate breeders should select cultivars that can yield moderately well over a wide climatic spectrum and low agricultural inputs. Maybe the local farmers growing a mixture of cultivars in a field have the right idea!

Gerald E. Wickens

Sorghum

Sorghum is an excellent example of a low-input grain crop that has tremendous potential to meet the needs of an increasing demand for lower input, sustainable solutions to the world's agricultural production problems. Its present adaptation to marginal production areas and its lack of research input to increase its response to external inputs guarantees its better fit into any future agricultural production systems. Its wide, untapped genetic variability found in landraces and its wild an weedy relatives lend tremendous genetic wealth to increase its productivity in these more sustainable systems.

Paula J. Bramel-Cox

Far more attention needs to be paid to sorghum as a human food. In temperate zones the staple grain is wheat, but many of the developing tropical countries cannot grow wheat, and the strain on their financial resources of importing this grain on any scale would be great. They must, of necessity, grow most of their own food grains. Rice is a good grain type in areas where it can be grown. Maize is a valuable grain, but it shows a narrower range of variation in grain type than does sorghum, and cannot be grown everywhere. Of the tropical grains, the one most likely to repay research is sorghum, because it has so much variation in which to work. It should prove possible to develop sorghum grains of a better standard than any present-day tropical grains.

Hugh Doggett

Our responsibility is to develop even more stable and higher yielding [sorghum] cultivars from this wealth of diversity by making the appropriate collections from dissimilar climates and recombining them into more widely adapted improved types useful to the world's people.

Fred R. Miller

The profuse branching and wide distribution of the root system is one of the main reasons why the sorghums are so markedly drought resistant. Other factors are however of importance. In the first place the plant above ground grows slowly until the root system is well established. Secondly, the system has to supply a leaf area which is approximately half the leaf area of maize. Thirdly, the low transpiration rate must influence the water demands. Finally, the plant can remain dormant during a prolonged drought and thereafter recontinue its development.

Hector (1936)

Appendix F
References and Selected Readings

AFRICAN RICE

Buddenhagen, I.W. and G.J. Persley, eds. 1978. *Rice in Africa: Proceedings of a Conference Held at the International Institute of Tropical Agriculture (IITA), Ibadan, Nigeria, 7–11 March 1977.* Academic Press, New York. 356 pp.

Carney, J.A. 1993. From hands to tutors: African expertise in the South Carolina rice economy. *Agricultural History* 67(3):1–30.

Clayton, W.D. 1968. West African wild rice. *Kew Bulletin* 21:487–488.

Clement, G. and K. Goli. 1987. Yield capacity of *Oryza glaberrima* as an upland crop. *L'Agronomie Tropicale* 42(4):275–279.

Fatokun, C.A., A.F. Attere, and H.R. Chheda. 1986. Variation in inflorescence characteristics of African rice (*Oryza glaberrima* Steud.). *Beitraefe zür Tropischen Landwirtschaft und Veterinaermedizin* 24(2):153–159.

Ghesquiere, A. 1985. Evolution of *Oryza longistaminata*. Pp. 15–25 in *Rice Genetics*. International Rice Research Institute (IRRI), Los Baños.

Jusu, M.S. and S.S. Monde. 1990. *Panicle and grain characters of some glaberrima cultivars in Sierra Leone*. International Rice Research Newsletter 15(3):5–6.

Leung, Woot-tseun, W., F. Busson, and C. Jardin. 1968. *Food Composition Table for Use in Africa*. U.S. Department of Health, Education, and Welfare and the Food and Agriculture Organization of the United Nations (FAO). Bethesda, Maryland.

Miezan, K., M.A. Choudhury, A. Ghesquiere, and G. Koffi. 1986. The use of *Oryza sativa* and *Oryza glaberrima* in West African farming systems. Pp. 213–218 in *Progress in Upland Rice Research*. Second International Upland Rice Conference, Jakarta, Indonesia, March 4–8, 1985. IRRI, Manila.

Morishima, H. and H.I. Oka. 1979. Genetic diversity in rice populations of Nigeria: influence of community structure. *Agro-Ecosystems* 5:263–269.

Netting, R. McC., D. Cleveland, and F. Stier. 1980. The conditions of agricultural intensification in the West African Savannah. Pages 187–505 in S.P. Reyna, ed., *Sahelian Social Development*. USAID, Abidjan. (especially the section entitled "Interior Niger Delta of Mali")

Ogbe, F.M.D. 1993. Lost crops of Nigeria: West African rice. Pp. 71–94 in J.A. Okojie and D.U.U. Okali, eds., *Lost Crops of Nigeria: Implications for Food Security*. Conference Proceedings Series No. 3. University of Agriculture, Abeokuta, Nigeria.

Ogbe, F.M.D. and J.T. Williams. 1978. Evolution in indigenous West African rice. *Economic Botany* 32(1):59–64.

Oka, H.I. 1975. Mortality and adaptive mechanisms of *Oryza perennis* strains. *Evolution* 30:380–392.

Oka, H.I. 1977. Genetic variations of *Oryza glaberrima*: their survey and evaluation. Pp. 77–86 in *IRAT [Institut de Recherches Agronomiques Tropicales et des cultures vivrières] -ORSTOM [Institut Français de Recherche Scientifique pour le Développement en Coopération de Montpellier] Meeting on Africa Rice Species*. Paris, 25–26 January 1977. IRAT-ORSTOM, Paris.

Oka, H.I. 1988. Origin of cultivated rice. In *Developments in Crop Science 14*. Japan Scientific Societies Press. Tokyo Elsevier, Amsterdam. 254 pages.

Pernes, J. 1984. Les riz. In *Gestion des Ressources Génétiques des Plantes*. Volume 1, Monographies (212 pp.). Volume 2, Manuel (346 pp.). Technique et documentation. Available from Lavoisier, 11 rue Lavoisier, 75384 Paris cedex 08, France.

Portères, R. 1956. Taxonomie agrobotanique des riz cultivés *O. sativa* Linné et *O. glaberrima* Steudel. *Compilations d'articles du JATBA*. Museum National d'Histoire Naturelle, Paris.

Portères, R. 1956. *Journal d'Agriculture Tropical et de Botanique Appliquées* 3:341, 541, 627, 821.
Richards, P. 1986. *Coping with Hunger: Hazard and Experiment in a West African Rice-Farming System*. Allen and Unwin, London.
Sano, Y., T. Fujii, S. Iyama, Y. Hirota, and K. Komagata. 1981. Nitrogen fixation bacteria in the rhizosphere of cultivated and wild rice strains *Oryza glaberrima*, *Oryza perennis*, *Oryza punctata*. *Crop Science* 21(5):758–761.
Sano, Y., R. Sano, and H. Morishima. 1984. Neighbour effects between two co-occurring rice species, *Oryza sativa* and *O. glaberrima*. *Journal of Applied Ecology* 21:245–254.
Schreurs, W. 1988. Les experiences en riz flottant dur project FENU a Tombouctou. FENU MLI/83/006, FAO, PNUD MLI/84/009, Tombouctou. (Copies available from author, see Appendix G.)
Thom, D.J. and J.C. Wells. 1987. Farming systems in the Niger inland delta, Mali. *The Geographic Review* 77(3):328–342.
Toure, A.I., M.A. Choudhury, M. Goita, S. Koli, and G.A. Paku. 1982. Grain yield and yield components of deep water rice in West Africa. Pp. 103–112 in *Proceedings of the 1981 International Deepwater Rice Workshop, Bangkhen, Thailand*. IRRI, Laguna, Philippines.
Treca, B. 1987. Bird damage in floating rice in Mali. *Journal d'Agriculture Traditionnelle et de Botanique Appliquées* 34:153–170.
Vallee, G. and H.H. Vuong. 1978. Floating rice in Mali. In I.W. Buddenhagen and G.J. Persley, eds., *Rice in Africa*. Academic Press, London.
Yabuno, T. 1981. The transfer of a gene for glutinous endosperm to *Oryza glaberrima* Steud. from a japonica variety of *Oryza sativa* L. *Euphytica* 30(3):867–873.

FINGER MILLET

Appa Rao, S., L.R. House, and S.C. Gupta. 1989. *Review of Sorghum, Pearl Millet, and Finger Millet Improvement in SADCC [Southern Africa Development Coordination Council] Countries*. SACCAR/International Crops Research Institute for the Semi-Arid Tropics (ICRISAT). Bulawayo, Zimbabwe. 170 pp.
Barbeau, W.E. and K.W. Hilu. 1993. Protein, calcium, iron, and amino acid content of selected wild and domesticated cultivars of finger millet. *Plant Foods for Human Nutrition* 43(2):97–104.
Engels, J.M.M., J.G. Hawkes, and M. Worede, eds. 1991. *Plant Genetic Resources of Ethiopia*. Cambridge University Press, Cambridge, UK. 383 pp.
Flack, E.N., W. Quak, and A. von Diest. 1987. A comparison of the rock phosphate mobilizing capacities of various crop species. *Tropical Agriculture (UK)* 64(4):347–352.
Gupta, S.C. and J.N. Mushonga. 1994. Registration of 'SDEY 87001' finger millet. *Crop Science* 34(2):536.
Hiremath, S.C. and S.S. Salimath. 1992. The 'A' genome donor of *Eleusine coracana* (L.) Gaertn. (Gramineae). *Theoretical and Applied Genetics* 84(5/6):747–754.
Kono, Y., A. Yamauchi, T. Nonoyama, and J. Tatsumi. 1988. Comparison of growth responses of summer cereals with special reference to waterlogging and rooting ability. *Japanese Journal of Crop Science* 57(2):321–331.
Marathee, J.P. Structure & characteristics of world millet economy. 1993. Pp. 159–178 in K.W. Riley, S.C. Gupta, A. Seetharam, and J.N. Mushonga, eds., *Advances in Small Millets*. Proceedings of an International Development Research Centre/ICRISAT conference. Oxford International Book House, New Delhi.
Mtebe, K., B.K. Ndabikunze, N.T.A. Bangu, and E. Mwemezi. 1993. Effect of cereal germination on the energy density of togwa. *International Journal of Food Sciences and Nutrition* 44(3):175–180.
Pius, J., L. George, S. Eapen, and P.S. Rao. 1994. Evaluation of somaclonal and mutagen induced variation in finger millet. *Plant Breeding* 112(3):239–243.
Purseglove, J.W. 1972. *Tropical Crops: Monocotyledons. Volumes I and II*. Longman Scientific and Technical, Harlow, Essex, UK; John Wiley, New York. 607 pp.
Rachie, K.O. and L.V. Peters. 1977. *The Eleusines: A Review of the World Literature*. ICRISAT, Hyderabad, India. 179 pp.
Rai, R. 1991. Isolation, characterization and associative N-fixation of acid-tolerant *Azospirillum brasilense* strains associated with *Eleusine coracana* in low pH-Al-rich acid soil. *Developments in Plant and Soil Sciences* 45:663–671.
Riley, K.W., S.C. Gupta, A. Seetharam, and J.N. Mushonga, eds. 1993. *Advances in Small Millets*. Proceedings of an International Development Research Centre (IDRC)/ICRISAT conference. Oxford International Book House, New Delhi.

Seetharam, A., K.W. Riley, and G. Harinarayana, eds. 1989. *Small Millets in Global Agriculture: Proceedings of the First International Small Millets Workshop, Bangalore, India, October 29–November 2, 1986*. Oxford & IBH Publishing Company, New Delhi. 392 pp.

Serna-Saldivar, S.O., C.M. McDonough, and L.W. Rooney. 1990. The millets. Pp. 271–300 in K.J. Lorenz and K. Kulp, eds., *Handbook of Cereal Science and Technology*. Dekker, New York.

Shankara, R., N.G. Malleshi, H. Krishnamurthy, M.N. Narayana, and H.S.R. Desikachar. 1985. Development of mini grain mill for dehusking and grinding of cereals. *Journal of Food Science Technology* 22:91.

Somasekhara, Y.M., S. Viswanath, T.B. Anilkumar. 1991. Evaluation of finger millet [*Eleusine coracana* (L.) Gaertn.] cultivars for their reactions to blast (*Pyricularia grisea* Sacc.). *Tropical Agriculture* 68(3):231–234.

Somasekhara, Y.M., S. Viswanath, and T.B. Anilkumar. 1992. Screening of finger millet (*Eleusine coracana*) cultivars for multiple disease resistance. *Tropical Agriculture* 69(3):293–295.

Tewari, L., B.N. Johri, and S.M. Tandon. 1993. Host genotype dependency and growth enhancing ability of VA-mycorrhizal fungi for *Eleusine coracana* (finger millet). *[World] Journal of Microbial Biotechnology* 9(2):191–195.

Vittal, K.P.R., K. Vijayalakshmi, and U.M. Bhaskara Rao. 1991. Effects of topsoil, rainfall, and fertilizer on finger millet on an alfisol in India. *Soil Science* 152(1):3–6.

FONIO (Acha)

Baptist, N.G. and B.P.M. Perera. 1956. Essential amino-acids of some tropical cereal millets. *British Journal of Nutrition* 10:334–337.

Busson, F. 1965. *Plantes Alimentaires de l'Ouest Africain: Etude Botanique, Biologique et Chimique*. Les Presses de l'Imprimerie Leconte, Marseilles.

Carbiener, R., P. Jaeger, and F. Busson. 1960. Study of the protein fraction of fonio grain, *Digitaria exilis* (Kippist), Stapf., a protein exceptionally rich in methionine. *Ann. Nutrit. Alimentation Mem.* 14:165–169.

De Lumen, B.O., R. Becker, and P.S. Reyes. 1986. Legumes and a cereal with high methionine/cysteine contents. *Journal of Agricultural and Food Chemistry* 34(2):361–364.

De Lumen, B.O., S. Thompson, and W.J. Odegard. 1993. Sulfur amino acid-rich proteins in acha (*Digitaria exilis*), a promising underutilized African cereal. *Journal of Agricultural and Food Chemistry* 41(7):1045–1047.

Gbodi, T.A., N. Nwude, Y.O. Aliu, and C.O. Ikediobi. 1986. The mycoflora and some mycotoxins found in acha (*Digitaria exilis* Stapf.) in Plateau State, Nigeria. *Food and Chemical Toxicology* 24(4):339–342.

Harlan, J.R., J.M.J. de Wet, and A.B.L. Stemler. 1976. *Origins of African Plant Domestication*. Mouton Publishers, The Hague. 498 pp.

Jideani, I.A. 1990. Acha, *Digitaria exilis*, the neglected cereal. *Agriculture International* 42(5):132–134.

Jideani, I.A. and J.O. Akingbala. 1993. Some physicochemical properties of acha (*Digitaria exilis* Stapf) and iburu (*Digitaria iburua* Stapf) grains. *Journal of the Science of Food and Agriculture* 63(3):369–374.

Oyenuga, V.A. 1968. *Nigeria's Foods and Feeding Stuffs (Their Chemistry and Nutritive Value* (third ed., rev.). Ibadan University Press, Ibadan, Nigeria. 99 pp.

Oyewole, O.B. and J.O. Akingbala. 1993. "Acha" (*Digitaria exilis*): a little known cereal with high potential. Pp. 323–326 in J.A. Okojie and D.U.U. Okali, eds., *Lost Crops of Nigeria: Implications for Food Security*. Conference Proceedings Series No. 3. University of Agriculture, Abeokuta, Nigeria.

Portères, R. 1955. Les céréales mineures du genre *Digitaria* en Afrique et en Europe. *Journal d'Agriculture Tropical et de Botanique Appliquée* 2:349–386, 477–510, 620–675.

Riley, K.W., S.C. Gupta, A. Seetharam, J.N. Mushonga, eds. 1993. *Advances in Small Millets*. Proceedings of an IDRC/ICRISAT conference. Oxford International Book House, New Delhi.

Temple, V.J. and J.D. Bassa. 1991. Proximate chemical composition of Acha (*Digitaria exilis*) grain. *Journal of the Science of Food and Agriculture* 56(4):561–563.

PEARL MILLET

Andrews, D.J. 1974. Responses of sorghum varieties to intercropping. *Experimental Agriculture* 10:57–63.

Andrews, D.J. and P.J. Bramel-Cox. In press. Breeding varieties for sustainable crop production in low input dryland agriculture in the tropics. Pages 211–223 in D.R. Buxton et al., eds. *International Crop Science I*. Proceedings of a Conference, Ames, Iowa, July 1993. CSSA, Madison, Wisconsin.

Andrews, D.J. and K.A. Kumar. 1992. Pearl millet for food, feed and forage. *Advances in Agronomy* 48:90–139.

Andrews, D.J., J.F. Rajewski, and K.A. Kumar. 1993. Pearl millet: a new feed grain crop. Pp. 198–208 in J. Janick and J.E. Simon, eds., *New Crops*. Purdue University Press, Lafayette, Indiana.

Annegers, J.F. 1973. Ecology of dietary patterns and nutritional status in West Africa 1: distribution of starchy staples. *Ecology of Food Nutrition* 2(2):107–119.

Annegers, J.F. 1973. The protein-calorie ratio of West African diets and their relationship to protein calorie malnutrition. *Ecology of Food Nutrition* 2(3):225–235.

Annegers, J.F. 1973. Seasonal foods shortages in West Africa. *Ecology of Food Nutrition* 2(4):251–257.

Annegers, J.F. 1974. Protein quality of West African foods. *Ecology of Food Nutrition* 3(2):125–130.

Appa Rao, S. 1987. Traditional food preparations of pearl millet in Asia and Africa. In J.R. Witcombe and S.R. Beckerman, eds., *Proceedings of the International Pearl Millet Workshop, 7–11 April, 1986*. ICRISAT, Patancheru.

Appa Rao, S., M.H. Mengesha, and V. Subramanian. 1982. Collection and preliminary evaluation of sweet-stalk pearl millet (*Pennisetum*). *Economic Botany* 36:286–290.

Badi, S., B. Pedersen, L. Manowar, and B.O. Eggum. 1990. The nutritive value of new and traditional sorghum and millet foods from Sudan. *Plant Foods for Human Nutrition* 40:5–19.

Bationo, A., C.B. Christianson, and W.E. Baethgen. 1990. Plant density and nitrogen fertilizer effects on pearl millet production in Niger. *Agronomy Journal* 82(2):290–295.

Bationo, A., C.B. Christianson, and M.C. Klaij. 1993. The effect of crop residue and fertilizer use on pearl millet yields on Niger. *Fertilizer Research* 34(3):251–258.

Brunken, J.N., J.M.J. de Wet, and J.R. Harlan. 1977. The morphology and domestication of pearl millet. *Economic Botany* 31:163–174.

Bunting, A.H. 1984. Advancing agricultural production in Africa: a personal review. Pages 435–445 in D.L. Hawksworth, ed., *Advancing Agricultural Production in Africa*. Commonwealth Agricultural Bureaux, Farnham Royal, UK.

Burton, G.W. and J.B. Powell. 1968. Pearl millet breeding and cytogenetics. *Advances in Agronomy* 20:49–89.

Burton, G.W., W.W. Hanna, J.C. Johnson, Jr., D.B. Lenck, W.G. Monson, J.B. Powell, H.D. Wells, and N.W. Widstrom. 1977. Pleiotropic effects of the *tr* trichomology gene in pearl millet on transpiration, forage quality and pest resistance. *Crop Science* 17:613–616.

Burton, G.W. and B.K. Werner. 1991. Genetic markers to locate and transfer heterotic chromosome blocks for increased pearl millet yields. *Crop Science* 31(3):576–579.

Buxton, D.R. et al., eds. In press. *International Crop Science I*. Proceedings of a Conference, Ames, Iowa, July 1993. CSSA, Madison, Wisconsin.

Council of Scientific and Industrial Research (CSIR). 1966. *The Wealth of India (volume 8)*. Publications & Information Directorate, CSIR, New Delhi.

Curtis, D.L., G.W. Burton, and O.J. Webster. 1966. Carotenoids in pearl millet seed. *Crop Science* 6:300–301. (vitamin-A millets)

Geiger, S.C., A. Manu, and A. Bationo. 1992. Changes in a sandy Sahelian soil following crop residue and fertilizer additions. *Soil Science Society of America Journal* 56(1):172–177.

Hamon, S. 1993. *Le mil en Afrique*. Proceedings of a meeting on millet (*Pennisetum glaucum* L.), Montpellier, France, 24–26 November, 1992. Institut Français de Recherche Scientifique pour le Développe en Coopération (ORSTOM), Montpellier. 283 pp.

Hanna, W.W. 1987. Utilization of wild relatives of pearl millet. In J.R. Witcombe and S.R. Beckerman, eds., *Proceedings of the International Pearl Millet Workshop, 7–11 April, 1986*. ICRISAT, Patancheru.

Hanna, W.W. 1990. Transfer of germ plasm from the secondary to the primary gene pool in *Pennisetum*. *Theoretical and Applied Genetics* 80(2):200–204.

Hill, G.M. and W.W. Hanna. 1990. Nutritive characteristics of pearl millet grain in beef cattle diets. *Journal of Animal Science* 68:2061–2066.

Hoseney, R.C. and E. Varriano-Marston. 1979. Pearl millet: its chemistry and utilization. Pages 461–494 in *Cereals for Food and Beverages*, G.E. Inglett and L. Munck, eds. Academic Press, New York.

Hoseney, R.C., D.J. Andrews, and H. Clark. 1987. Sorghum and pearl millet. Pages 397–456 in *Nutritional Quality of Cereal Grains: Genetic and Agronomic Improvement*. Agronomy Monograph No. 28. CSSA, 677 South Segoe Road, Madison, Wisconsin 53711.

Hoseney, R.C., E. Varriano-Marston, and D.A.V. Dendy. 1981. Sorghum and millets: anatomy, composition, milling. Pp. 71–144 in Y. Pomeranz, ed., *Advances in Cereal Science and Technology, Volume IV*. American Association of Cereal Chemists (AACC), St. Paul, Minnesota.

Howarth, C.J. 1990. Heat shock proteins in sorghum and pearl millet; ethanol, sodium arsenite, sodium malonate and the development of thermotolerance. *Journal of Experimental Botany* 41(228):877–883.

Hulse, J.H., E.M. Laing, and O.E. Pearson. 1980. *Sorghum and Millets: Their Composition and Nutritive Value*. Academic Press, New York. 997 pp.

International Sorghum/Millet (INTSORMIL) Collaborative Research Support Program/ICRISAT. 1987. *Proceedings of the International Pearl Millet Workshop*. ICRISAT, Patancheru. 278 pp.

INTSORMIL. 1990. *Annual Report 1990*. INTSORMIL, University of Nebraska, Lincoln. 246 pp.

INTSORMIL. 1991. *Annual Report 1991*. INTSORMIL, University of Nebraska, Lincoln. 256 pp.

Kumar, K.A. 1989. Pearl millet: current status and future potential. *Outlook on Agriculture* 18(2): 46–53.

Lambert, C. 1982. IRAT *[Institut de Recherches Agronomiques Tropicales et des cultures vivrières]* et l'amelioration de mil. *Agronomie Tropical*. 38:78–88.

Mahalakshmi, V., F.R. Bidinger, K.P. Rao, and D.S. Raju. 1992. Performance and stability of pearl millet topcross hybrids and their variety pollinators. *Crop Science* 32(4):928–932.

Marathee, J.P. Structure & characteristics of world millet economy. 1993. Pp. 159–178 in K.W. Riley, S.C. Gupta, A. Seetharam, and J.N. Mushonga, eds., *Advances in Small Millets*. Proceedings of an IDRC/ICRISAT conference. Oxford International Book House, New Delhi.

Matlon, P.J. 1990. Improving productivity in sorghum and pearl millet in semi-arid Africa. *Food Research Institute Studies* 22(1):1–43.

Mohammed, M.S. and M.D. Clegg. 1993. Pearl millet-soybean rotation and nitrogen fertilizer effects on millet productivity. *Agronomy Journal* 85(5):1009–1013.

Niangodo, O. and B. Ouendeba. 1987. Varietal improvement of pearl millet in West Africa. Pp. 95–105 in J.R. Witcombe and S.R. Beckerman, eds., *Proceedings of the International Pearl Millet Workshop, 7–11 April, 1986*. ICRISAT, Patancheru.

Obizoba, I.C. and J.V. Atii. 1994. Evaluation of the effect of processing techniques on the nutrient and antinutrient contents of pearl millet (*Pennisetum glaucum*) seeds. *Plant Foods for Human Nutrition* 45(1):23–34.

Oendeba, B., G. Ejeta, W.E. Nyquist, W.W. Hanna, and A. Kumar. 1993. Heterosis and combining ability among African pearl millet landraces. *Crop Science* 33(4):735–739.

Pantulu, J.V. and M. Krishna Rao. 1982. Cytogenetics of pearl millet. *Theoretical and Applied Genetics* 61:1–17.

Rachie, K.O. and J.V. Majmudar. 1980. *Pearl Millet*. Pennsylvania State University Press, University Park, Pennsylvania. 307 pp.

Read, J.C. and E.C. Bashaw. 1974. Intergeneric hybrid between pearl millet and buffelgrass. *Crop Science* 14:401–403.

Serna-Saldivar, S.O., C.M. McDonough, and L.W. Rooney. 1990. The millets. Pp. 271–300 in K.J. Lorenz and K. Kulp, eds., *Handbook of Cereal Science and Technology*. Dekker, New York.

Singh, P., Y. Singh, B.O. Eggum, K.A. Kumar, and D.J. Andrews. 1987. Nutritional evaluation of high protein genotypes in pearl millet. *Journal of the Science of Food and Agriculture* 38:41–48.

Smith, R.L., L.S. Jensen, C.S. Hoveland, and W.W. Hanna. 1989. Use of pearl millet, sorghum, and triticale grain in broiler diets. *Journal of Production Agriculture* 2:78–82.

Tostain, S. 1992. Enzyme diversity in pearl millet (*Pennisetum glaucum* L.). 3. Wild millet. *Theoretical and Applied Genetics* 83(6/7):733–742.

PEARL MILLET, Subsistence Types

International Development Research Centre (IDRC). 1979. *Sorghum and Millet: Food Production and Use*. Report of a workshop held in Nairobi, Kenya, July 4–7, 1978. S. Vogel and M. Graham, eds. Report No. IDRC-123e. IDRC, Ottawa, Canada

Wendt, J.W., A. Berrada, M.G. Gaoh, and D.G. Schulze. 1993. Phosphorus sorption characteristics of productive and unproductive Niger soils. *Soil Science Society of America Journal* 57(3):766–773.

Zaongo, C.G.L., L.R. Hossner, and C.W. Wendt. 1994. Root distribution, water use, and nutrient uptake of millet and grain sorghum on West Africa soils. *Soil Science* 157(6):379–388.

PEARL MILLET, Commercial Types

Bramel-Cox, P.J., K. Anandkumar, J.H. Hancock, and D.J. Andrews. In press. Sorghum and millets for forage and feed. In D.A.V. Dendy, ed., *Sorghum and Millets: Chemistry and Technology*. AACC, St. Paul, Minnesota.

D.A.V. Dendy, ed., In press. *Sorghum and Millets: Chemistry and Technology*. AACC, St. Paul, Minnesota.

Hoseney, R.C. and E. Varriano-Marston. 1980. Pearl millet: its chemistry and utilization. Pp. 462–494 in G.E. Inglett and L. Munck, eds. *Cereals for Food and Beverages: Recent Progress in Cereal Chemistry and Technology*. Proceedings of an international conference, August 13–17, 1979, Copenhagen, Denmark. Academic Press, New York.

Hoseney, R.C., E. Varriano-Marston, and D.A.V. Dendy. 1981. Sorghum and millets: anatomy, composition, milling. Pp. 71–144 in Y. Pomeranz, ed., *Advances in Cereal Science and Technology, Volume IV*. AACC, St. Paul, Minnesota.

Hulse, J.H., E.M. Laing, and O.E. Pearson. 1980. *Sorghum and Millets: Their Composition and Nutritive Value*. Academic Press, New York. 997 pp.

Kumar, K.A., S.C. Gupta, and D.J. Andrews. 1983. Relationship between nutritional quality characters and grain yield in pearl millet. *Crop Science* 23:232–235.

Rachie, K.O. and J.C. Majmudar. 1980. *Pearl Millet*. Pennsylvania State University Press, University Park, Pennsylvania. 307 pp.

SORGHUM

A sorghum newsletter is sponsored by the Sorghum Improvement Conference of North America. For information on ordering, contact Ronny R. Duncan, Editor, *Sorghum Newsletter*, Department of Agronomy, University of Georgia, Georgia Station, Griffin, Georgia 30223–1797, USA.

Andrews, D.J. 1972. Intercropping with sorghum in Nigeria. *Experimental Agriculture* 8:139–150.

Andrews, D.J. 1974. Responses of sorghum varieties to intercropping. *Experimental Agriculture* 10:57–63.

Andrews, D.J. and P.J. Bramel-Cox. 1993. Breeding varieties for sustainable crop production in low input dryland agriculture in the tropics. Pages 211–223 in D.R. Buxton et al., eds. *International Crop Science I*. Proceedings of a Conference, Ames, Iowa, July 1993. CSSA, Madison, Wisconsin.

Bramel-Cox, P.J., K. Anandkumar, J.H. Hancock, and D.J. Andrews. In press. Sorghum and millets for forage and feed. In D.A.V. Dendy, ed., *Sorghum and Millets: Chemistry and Technology*. AACC, St. Paul, Minnesota.

Brand, T.S., H.A. Badenhorst, and F.K. Siebrits. 1990. The use of pigs both intact and with ileo-rectal anastomosis to estimate the apparent and true digestibility of amino acids in untreated, heat-treated and thermal-ammoniated high-tannin grain sorghum. *South African Journal of Animals Science* 20(4):223–228.

Bunting, A.H. 1984. Advancing agricultural production in Africa: a personal review. Pages 435–445 in D.L. Hawksworth, ed., *Advancing Agricultural Production in Africa*. Commonwealth Agricultural Bureaux, Farnham Royal, UK.

Buxton, D.R. et al., eds. In press. *International Crop Science I*. Proceedings of a Conference, Ames, Iowa, July 1993. CSSA, Madison, Wisconsin.

de Milliano, W.A.J., R.A. Frederiksen, and G.D. Bengston, eds. 1992. *Sorghum and Millet Diseases: A Second World Review*. ICRISAT, Patancheru. 370 pp.

Dendy, D.A.V., ed. In press. *Sorghum and Millets: Chemistry and Technology*. AACC, St. Paul, Minnesota.

DeWalt, B.R. 1985. Mexico's second Green Revolution: food for feed. *Mexican Studies/Estudios Mexicanos* 1(1):29–60.

Doggett, H. 1988. *Sorghum*. Longmans, London. 515 pp.

Doumbia, M.D., L.R. Hossner, and A.B. Onken. 1993. Variable sorghum growth in acid soils of subhumid West Africa. *Arid Soil Research and Rehabilitation* 7(4):335–346.

Eggum, B.O. 1990. Importance of sorghum as a food in Africa. Pp. 222–228 in G. Ejeta, E.T. Mertz, L. Rooney, R. Schaffert, and J. Yohe, eds., *Sorghum Nutritional Quality: proceedings of an international conference held February 26–March 1, 1990*. Purdue University, West Lafayette, Indiana.

Einhellig, F.A. and I.F. Souza. 1992. Phytotoxicity of sorgoleone found in grain sorghum root exudates. *Journal of Chemical Ecology* 18(1):1–11.

Ejeta, G. 1983. *Hybrid Sorghum Seed for Sudan: Proceedings of a Workshop, November 5–8, 1993, Gezira Research Station, Sudan*. Agricultural Research Corporation, Wad Medani, Sudan.

Ejeta, G., E.T. Mertz, L. Rooney, R. Schaffert, and J. Yohe, eds. 1990. *Sorghum Nutritional Quality: proceedings of an international conference held February 26–March 1, 1990*. Purdue University, West Lafayette, Indiana.

Engels, J.M.M., J.G. Hawkes, and M. Worede, eds. 1991. *Plant Genetic Resources of Ethiopia*. Cambridge University Press, Cambridge, UK. 383 pp.

House, L.R. 1985. *A Guide to Sorghum Breeding*. 2nd edition. ICRISAT, India. 238 pp.

Howarth, C.J. 1990. Heat shock proteins in sorghum and pearl millet; ethanol, sodium arsenite, sodium malonate and the development of thermotolerance. *Journal of Experimental Botany* 41(228):877–883.

Hulse, J.H., E.M. Laing, and O.E. Pearson. 1980. *Sorghum and the Millets: Their Composition and Nutritive Value*. Academic Press, New York. 997 pp.

ICRISAT. 1982. *Sorghum in the Eighties: Proceedings of the International Symposium on Sorghum*. November 2–7 1981. Patancheru, A.P. India. ICRISAT, Patancheru.

International Board for Plant Genetic Resources (IBPGR). 1976. *Proceedings of the Meeting of the Advisory Committee on Sorghum and Millets Germplasm*. Rome. Mimeo.

IDRC. 1979. *Sorghum and Millet: Food Production and Use*. Report of a workshop held in Nairobi, Kenya, 4–7 July 1978. S. Vogel and M. Graham, eds. IDRC, Ontario, Canada.

IRAT. 1978. L'IRAT et l'amelioration due sorgho. *Agronomie Tropicale* 32:279–318.

INTSORMIL/ICRISAT. 1987. *Proceedings of the International Pearl Millet Workshop*. ICRISAT, Patancheru. 278 pp.

Leng, E.R. 1982. Status of sorghum production as compared to other cereals. Pages 25–32 in *Sorghum in the Eighties: Proceedings of the International Symposium on Sorghum*. November 2–7 1981. Patancheru, A.P. India. ICRISAT, Patancheru.

Lopez-Pereira, M.A., et al. 1990. New agricultural technologies in Honduras: an economic evaluation of new sorghum cultivars in southern Honduras. *Journal for Farming System Research and Extension* 1(2):81–103.

Matlon, P.J. 1990. Improving productivity in sorghum and pearl millet in semi-arid Africa. *Food Research Institute Studies* 22(1):1–43.

Quinby, J.R. 1971. *A Triumph of Research . . . Sorghum in Texas*. Texas A&M University Press, College Station. 28 pp.

Quinby, J.R. 1974. *Sorghum Improvement and the Genetics of Growth*. Texas A&M University Press. College Station, Texas. 108 pp.

Rooney, L.W., A.W. Kirleis, and D.S. Murty. 1986. Traditional foods from sorghum: their production, evaluation, and nutritional value. Pages 317–353 in *Advances in Cereal Science and Technology volume 3*. AACC, St. Paul, Minnesota.

Serna-Saldivar, S.O., C.M. McDonough, and L.W. Rooney. 1990. The millets. Pp. 271–300 in K.J. Lorenz and K. Kulp, eds., *Handbook of Cereal Science and Technology*. Dekker, New York.

Van de Venter, H.A. and W.H. Lock. 1992. An investigation of heat-shock and thermotolerance induction responses as indices of seed vigour in grain sorghum [*Sorghum bicolor* (L.) Moench.]. *Plant Varieties and Seeds* 5(1):13–18.

Wenzel, W.G. 1991. The inheritance of drought resistance characteristics in grain sorghum seedlings. *South African Journal of Plant and Soil* 8(4):169–171.

Zaongo, C.G.L., L.R. Hossner, and C.W. Wendt. 1994. Root distribution, water use, and nutrient uptake of millet and grain sorghum on West Africa soils. *Soil Science* 157(6):379–388.

SORGHUM, Subsistence Types

Carr, S.J. 1989. *Technology for Small-scale Farmers in Sub-Saharan Africa*. Technical Paper No. 109, World Bank, Washington, D.C. 106 pp.

DeWalt, B.R. 1985. Microcosmic and macrocosmic processes of agrarian change in southern Honduras: the cattle are eating the forests. Pages 165–185 in *Micro and Macro Levels of Analysis in Anthropology: Issues in Theory and Research*. Westview Press, Boulder, Colorado and London.

International Sorghum/Millet (INTSORMIL) Collaborative Research Support Program. 1990. *Annual Report 1990*. INTSORMIL, University of Nebraska, Lincoln. 246 pp.

INTSORMIL. 1991. *Annual Report 1991*. INTSORMIL, University of Nebraska, Lincoln. 256 pp.

Mukuru, S.Z. 1990. Traditional food grain processing methods in Africa. Pp. 216–221 in G. Ejeta, E.T. Mertz, L. Rooney, R. Schaffert, and J. Yohe, eds., *Sorghum Nutritional Quality: proceedings of an international conference held February 26–March 1, 1990*. Purdue University, West Lafayette, Indiana.

Rao, N.G.P. 1982. Transforming traditional sorghums in India. Pages 39–59 in *Sorghum in the Eighties: Proceedings of the International Symposium on Sorghum*. November 2–7 1981. ICRISAT, Patancheru.

SORGHUM, Commercial Types

Bennett, W.F., B.B. Tucker, and A.B. Maunder. 1990. *Modern Grain Sorghum Production*. Iowa State University Press, Ames. 169 pp.
DeWalt, B.R. and D. Barkin. 1987. Seeds of change: the effects of hybrid sorghum and agricultural modernization in Mexico. Pages 138–161 in H.R. Bernard and P.J. Pelto, eds., *Technology and Social Change*, 2nd ed. Waveland Press, Prospect Heights, Illinois. 393 pp.
Faure, J. 1988. Sorghum utilization in pasta production. Paper presented in the International Workshop on Policy and Potential Relating to Uses of Sorghum and Millets. June 8–12, 1988. SADCC/ICRISAT. Bulawayo, Zimbabwe.
ICRISAT. 1990. *Industrial Utilization of Sorghum: Summary of a Symposium on the Current Status and Potential of Industrial Uses of Sorghum in Nigeria*. December 4–6, 1989, Kano Nigeria. ICRISAT, Patancheru, andhra Pradesh, India. 59 pp.

SORGHUM, Specialty Types

Gupta, S.C., J.M.J. dewet, and J.R. Harlan. 1978. Morphology of *Saccharum -Sorghum* hybrid derivatives. *American Journal of Botany* 65:936–942. (sorghum relatives)
Gupta, S.C., J.R. Harlan, and J.M.J. dewet. 1978. Cytology and morphology of a tetraploid *Sorghum* population recovered from a *Saccharum = Sorghum* hybrid. *Crop Science* 18:879–883. (sorghum relatives)
Haggblade, S. and W.H. Holzapfel. 1989. Industrialization of Africa's indigenous beer brewing. Pages 191–283 in K.H. Steinkraus, ed., *Industrialization of Indigenous Fermented Foods*, Dekker, New York. (beer sorghums)
Hahn, D.H., L.W. Rooney, and C.F. Earp. 1984. Tannins and phenols of sorghum. *Cereal Foods World* 29(12):776–779. (tannin-free sorghums)
Jabri, A., R. Chaussat, M. Jullien, and Y. le Deunff. 1989. Callogenesis and somatic embryogenesis in leaf portions of three varieties of sorghum with and without tannins. *Agronomie* (France) 9(1):101–107. (tannin-free sorghums)
Novellie, L. 1982. Fermented beverages. In L.W. Rooney and D.S. Murty, eds., *Proceedings, International Grain Sorghum Quality Workshop*. ICRISAT, Patancheru.
Novellie, L. and P. De Schaepdrijver. 1986. Modern developments in traditional African beers. Pages 73–157 in M.R. Adams, ed., *Progress in Industrial Microbiology, Vol. 23*, Elsevier, Amsterdam. (beer sorghums)
Thorat, S.S., P.N. Satwadhar, D.N. Kulkarni, S.D. Choudhari, and U.M. Ingle. 1990. Varietal differences in popping quality of sorghum grains. *Journal of Maharashtra Agricultural Universities* 15(2):173–175. (popping sorghums)

SORGHUM, Fuel and Utility Types

Clegg, M.D, H.J. Gorz, J.W. Maranville, and F.A. Haskins. 1986. Evaluation of agronomic and energy traits of Wray sweet sorghum and the N39x Wray hybrid. *Energy Agriculture* 6:49–54.
Coleman, O.H. 1970. Syrup and sugar from sweet sorghum. Page 438 in J.S. Wall and W.M. Ross, eds., *Sorghum Production and Utilization*. Avi Publishing Company, Westport, Connecticut.
Cundiff, J.S. and D.H. Vaughan 1986. Sweet sorghum for ethanol industry for the Piedmont. Virginia Agricultural Experiment Station Bulletin 0096–6088, 86–9. Virginia Polytechnic Institute and State University, Blacksburg, Virginia.
Gascho, G.J., R.L. Nichols, and T.P. Gaines. 1984. Growing sweet sorghum as a source of fermentable sugars for energy. Research Bulletin 315. University of Georgia Experiment Station, Athens, Georgia.
Gas Research Institute. 1982. *SNG [synthetic natural gas] from land-based biomass: 1981 program final report*. Gas Research Institute, 8600 West Bryn Mawr Avenue, Chicago, IL 60631, USA. 111 pp.
Gibbons, W.R., C.A. Westby, and T.L. Dobbs. 1986. Intermediate-scale, semicontinuous solid-phase fermentation process for production of fuel ethanol from sweet sorghum. *Applied Environmental Microbiology* 51(1):115–122.
Hills, F.J., R.T. Lewellen, and I.O. Skoyen. 1990. Sweet sorghum cultivars for alcohol production. *California Agriculture* 44(1):14–16.
Kapocsi, I., J. Lezanyi, and B. Kevacs. 1983. Utilization of sweet sorghum for alcohol production. *Internationale Zeitschrift fuer Landwirtschaften* 5:441–443.

Kresovich, S., R.E. McGee, L. Panella, A.A. Reilley, and F.R. Miller. 1987. Application of cell and tissue culture techniques for the genetic improvement of sorghum, *Sorghum bicolor* (L.) Moench: progress and potential. *Advances in Agronomy* 41:147–170.

Kresovich, S. and D.M. Broadhead. 1988. Registration of 'Smith' sweet sorghum. *Crop Science* 28(1):195.

Kresovich, S., F.R. Miller, R.L. Monk, R.E. Dominy, and D.M. Broadhead. 1988. Registration of 'Grassl' sweet sorghum. *Crop Science* 28(1):194–195.

Lipinsky, E.S., D.R. Jackson, S. Kresovich, M.F. Arthur, and W.T. Lawhon. 1979. Carbohydrate Crops as a Renewable Resource for Fuels Production, Vol. 1: Agricultural research. Battelle Memorial Institute Columbus Laboratories, Columbus, Ohio 43201, USA

Lueschen, W.E., D.H. Putnam, B.K. Kanne, and T.R. Hoverstad. 1991. Agronomic practices for production of ethanol from sweet sorghum. *Journal of Production Agriculture* 4(4):619–625.

McClure, T.A., M.F. Arthur, S. Kresovich, and D.A. Scantland. 1980. Sorghums—viable biomass candidates for fuel alcohol production. Volume 1, pp. 123–130 in *Proceedings of the IV International Symposium on Alcohol Fuels Technology, Guaruja, São Paulo, Brasil, 5–8 Oct. 1980*. Instituto de Pesquisas Tecnológicas do Estado de São Paulo, São Paulo.

Miller, F.R. and G.G. McBee. 1993. Genetics and management of physiologic systems of sorghum for biomass production. Pages 41–49 in *Methane from Biomass: Science and Technology*. Biomass and Bioenergy, Vol. 5, No.1. Pergamon Press, Ltd., Oxford.

Morris, D.J. and I. Ahmed. 1992. *The carbohydrate economy: making chemicals and industrial materials from plant matter*. Institute for Local Self Reliance, Washington, DC.

Rajvanshi, A.K., R.M. Jorapur, and N. Nimbkar. 1989. *VITA News*, October. Pp. 7–8.

Ricaud, R. and A. Arceneaux. 1990. Sweet sorghum research on biomass and sugar production in 1990. Pp. 136–139 in *Report of Projects*, Louisiana Agricultural Experiment Station, Department of Agronomy, Louisiana State University, Baton Rouge, Louisiana.

Schaffert, R.E. 1988. Sweet sorghum substrate for industrial alcohol. Paper presented in the International Workshop on Policy and Potential Relating to Uses of Sorghum and Millets. June 8–12, 1988. SADCC/ICRISAT. Bulawayo, Zimbabwe.

Schaffert, R.E. and L.M. Gourley. 1982. Sorghum as an energy source. Pages 605–623 in *Sorghum in the Eighties: Proceedings of the International Symposium on Sorghum*. November 2–7 1981. ICRISAT, Patancheru.

Smith, B.A. 1982. Sweet sorghum. Page 611 in I.A. Wolf, ed., *CRC Handbook of Processing and Utilization in Agriculture. Volume 2*. CRC Press, Boca Raton, Florida.

Weitzel, T.T., J.S. Cundiff, and D.H. Vaughan. 1986. Improvement of juice expression after separation of sweet sorghum pith from rind leaf. American Society of Agricultural Engineers Microfiche Collection No. 86–6579. American Society of Agricultural Engineers, St. Joseph, Michigan.

TEF

Ashenafi, M. 1994. Microbial flora and some chemical properties of ersho, a starter for teff (*Eragrostis tef*) fermentation. *[World] Journal of Microbial Biotechnology* 10(1):69–73.

Berhe, T., L.A. Nelson, M.R. Morris, and J.W. Schmidt. 1989. Inheritance of phenotypic traits in tef: I - lemma color, II - seed color, III -panicle form. *Journal of Heredity* 80:62–70.

Besrat, A., A. Admasu, and M. Ogbai. 1980. Critical study of the iron content of tef (*Eragrostis tef*). *Ethiopian Medical Journal* 18:45–52.

Cheverton, M.R. and N.W. Galwey. 1989. Ethiopian t'ef: a cereal confined to its centre of variability. Pages 235–238 in G.E. Wickens, N. Haq, and P. Day, eds., *New Crops for Food and Industry*. Chapman and Hall, London and New York.

FAO. 1988. *Traditional Food Plants*. Food and Nutrition Paper 42. FAO, Rome.

Huffragel, H.P. 1961. *Agriculture in Ethiopia* FAO, Rome. 484 pp.

Jones, G. 1988. Endemic crops plants of Ethiopia I: t'ef (*Eragrostis tef*). *Walia* 11:37–43.

Ketema, S. 1988. Status of small millets in Ethiopia and Africa. Pp. 6–15 in *Small Millets: Recommendations for a Network*. Proceedings of the Small Millets Steering Committee Meeting, Addis Ababa, Ethiopia, 7–9 October 1987. International Development Research Centre, Ottawa.

Ketema, S. 1991. Germplasm evaluation and breeding work on teff (*Eragrostis tef*) in Ethiopia. Pp. 323–328 in Engels, J.M.M., J.G. Hawkes, and M. Worede, eds. 1991. *Plant Genetic Resources of Ethiopia*. Cambridge University Press, Cambridge, UK. 383 pp.

Lester, R.N. and E. Bekele. 1981. Amino acid composition of the cereal tef and related species of *Eragrostis* (Gramineae). *Cereal Chemistry* 58:113–115.

Mamo, T. and K.S. Killham. 1987. Effect of soil liming and vesicular-arbuscular-mycorrhizal inoculation on the growth and micronutrient content of the teff plant. *Plant and Soil* 102:257–259.

Mamo, T. and J.W. Parsons. 1987. Phosphorus-micronutrient interactions in teff (*Eragrostis tef*). *Tropical Agriculture* 64:309–312.
Mamo, T. and J.W. Parsons. 1987. Iron nutrition of *Eragrostis tef* (teff). *Tropical Agriculture* 64:313–317.
Marathee, J.P. Structure & characteristics of world millet economy. 1993. Pp. 159–178 in K.W. Riley, S.C. Gupta, A. Seetharam, and J.N. Mushonga, eds., *Advances in Small Millets*. Proceedings of an IDRC/ICRISAT conference. Oxford International Book House, New Delhi.
Osuji, P.O. and B. Capper. 1992. Effect of age on fattening and body condition of draught oxen fed teff straw (*Eragrostis teff*) based diets. *Tropical Animal Health and Production* 24(2):103–108.
Parker, M.L., M. Umeta, and R.M. Faulks. 1989. The contribution of flour components to the structure of injera, an Ethiopian fermented bread made from tef (*Eragrostis tef*). *Journal of Cereal Science* 10(2):93–104.
Riley, K.W., S.C. Gupta, A. Seetharam, J.N. Mushonga, eds. 1993. *Advances in Small Millets*. Proceedings of an IDRC/ICRISAT conference. Oxford International Book House, New Delhi.
Seetharam, A., K.W. Riley, and G. Harinarayana. 1989. *Small Millets in Global Agriculture*. IDRC, Ottawa, Canada. 392 pp.
Tadesse, E. 1969. *T'ef (Eragrostis tef): cultivation, usage, and some of the known disease and insect pests, Part I*. Experiment Station Bulletin No. 60. College of Agriculture, Haile Sellasie I University, Dire Dawa, Ethiopia. 56 pp.
Tadesse, E. 1975. T'ef (*Eragrostis tef*) cultivars: morphology and classification Part II. Experiment Station Bulletin No. 66. College of Agriculture, Addis Ababa University, Dire Dawa, Ethiopia. 73 pp.
Tadesse, D. 1993. Study of genetic variation of landraces of teff (*Eragrostis tef* (Zucc.) Trotter) in Ethiopia. *Genetic Resources and Crop Evolution* 40(2):101–104.
Tefera, H., S. Ketema, and T. Tesemma. 1990. Variability, heritability and genetic advance in tef (*Eragrostis tef* (Zucc.) Trotter) cultivars. *Tropical Agriculture* 67(4):317–320.
Twidwell, E.K., A. Boe, and D.P. Casper. 1991. Teff: a new annual forage grass for South Dakota. Extension Extra 8071. South Dakota Cooperative Extension Service, Brookings, South Dakota.
Umeta, M. and R.M. Faulks. 1988. The effect of fermentation on the carbohydrates in tef (*Eragrostis tef*). *Food Chemistry* 27:181–189.
Westphal, E. and J.M.C. Westphal-Stevels. 1975. *Agricultural Systems in Ethiopia*. Agricultural Research Reports 82b. Puduc, Wageningen. 278 pp.
Wolde-Gebriel, Z. 1988. Nutrition. In Zein Ahmed Zein and Helmut Kloos, eds., *The Ecology of Health and Disease in Ethiopia*. Ministry of Health, Addis Ababa.

OTHER CULTIVATED GRAINS

General

Harlan, J.R. 1986. *African Millets*. Food and Agriculture Organization of the United Nations (FAO), Rome
Harlan, J.R. 1989. Wild grass seed harvesting in the Sahara and Sub-Sahara of Africa. In *Foraging and Farming: the Evolution of Plant Exploitation*, D.R. Harris and G.C. Hillman, eds. Unwin-Hyman, London.

Emmer

Gasrataliev, G.S. 1982. Forms of *T. dicoccum* promising for southern Dagestan. *Bulletin of the N.I. Vavilov Institute of Plant Industry* 118:5–6.
Hakim, S., A.B. Damania, and M.Y. Moualla. 1992. Genetic variability in *Triticum dicoccum* Schubl. for use in breeding wheat for the dry areas. *FAO/IBPGR Plant Genetic Resources Newsletter* 88/89:11–15.
Mariam, G.H. and H. Mekbib. 1988. Agromorphological evaluation of *T. dicoccum*. *PGRC/E ILCA Germplasm Newsletter* 6–11.
Robinson, J. and B. Skovmand. 1992. Evaluation of emmer wheat and other Triticeae for resistance to Russian wheat aphid. *Genetic Resources and Crop Evolution* 39(3):159–163.

Barley

Endashaw Bekele. 1984. Relationships between morphological variance, gene diversity and flavonoid patterns in the land race populations of Ethiopian barley. *Hereditas* 100:271–294.

Engels, J.M.M., J.G. Hawkes, and M. Worede, eds. 1991. *Plant Genetic Resources of Ethiopia*. Cambridge University Press, Cambridge, UK. 383 pp.
Huffragel, H.P. 1961. *Agriculture in Ethiopia* FAO, Rome. 484 pp.
Munck, L. 1988. The importance of botanical research in breeding for nutritional quality characteristics in cereals. *Symbolae Botanicae Upsalienses* 28(3):69–78.
Nulugeta Negassa. 1985. Patterns of phenotypic diversity in an Ethiopian barley collection, and the Arsi-Bale Highland as a centre of Origin of barley. *Hereditas* 102:139–150.
Qualset, C.O. 1975. Sampling germplasm in a centre of diversity: an example of disease resistance in Ethiopian barley. Pages 81–96 in O.H. Frankel and J.G. Hawkes, eds., *Crop Genetic Resources for Today and Tomorrow*. Cambridge University Press, Cambridge.
Westphal, E. and J.M.C. Westphal-Stevels. 1975. *Agricultural Systems in Ethiopia*. Agricultural Research Reports 82b. Puduc, Wageningen. 278 pp.

Ethiopian oats

Baum, B.R. 1971. Taxonomic studies in *Avena abyssinica* and *Avena vaviloviana*, and some related species. *Canadian Journal of Botany* 49(12):2227–2232
Engels, J.M.M., J.G. Hawkes, and M. Worede, eds. 1991. *Plant Genetic Resources of Ethiopia*. Cambridge University Press, Cambridge, UK. 383 pp.
Ladizinsky, G. 1975. Oats in Ethiopia (*Avena barbata*, *Avena abyssinica*, cereal weeds). *Economic Botany* 29(3):238–241.

Guinea Millet

Harlan, J.R. 1986. *African Millets*. Food and Agriculture Organization of the United Nations (FAO), Rome

Kodo millet

De Wet, J.M.J., K.E. Prasada Rao, M.H. Mengesha, and D.E. Brink. 1983. Diversity in kodo millet, *Paspalum scrobiculatum*. *Economic Botany* 37(2):159–163.
Geervani, P. and B.O. Eggum. 1989. Effect of heating and fortification with lysine on protein quality of minor millets. *Plant Foods for Human Nutrition* 39(4):349–357.
Kapoor, P.N., S.P. Netke, and L.D. Bajpai. 1987. Kodo (*Paspalum scorbiculatum*) as a substitute for maize in chick diets. *Indian Journal of Animal Nutrition* 4(2):83–88.
Kaushik, S.K. and R.C. Gautam. 1985. Comparative performance of different millets at varying levels of nitrogen under dryland conditions. *Indian Journal of Agronomy* 30(4):509–511.
Ketema, S. 1988. Status of small millets in Ethiopia and Africa. Pp. 6–15 in *Small Millets: Recommendations for a Network*. Proceedings of the Small Millets Steering Committee Meeting, Addis Ababa, Ethiopia, 7–9 October 1987. International Development Research Centre, Ottawa.
Nayak, P. and S.K. Sen. 1991. Plant regeneration through somatic embryogenesis from suspension culture-derived protoplasts of *Paspalum scrobiculatum* L. *Plant Cell Reports* 10(6/7):362–365.
Sridhar, R. and G. Lakshminarayana. 1992. Lipid class contents and fatty acid composition of small millets: little (*Panicum sumatrense*), kodo (*Paspalum scrobiculatum*), and barnyard (*Echinocloa colona*). *Journal of Agricultural and Food Chemistry* 40(11):2131–2134.
Sudharshana, L., P.V. Monteiro, and G. Ramachandra. 1988. Studies on the proteins of kodo millet (*Paspalum scrobiculatum*). *Journal of the Science of Food and Agriculture* 42(4):315–323.

WILD GRAINS

Breman, H. and L. Diarra. 1989. Easy methods to follow the changes of vegetation in natural pastures in the Sahel. *Rapport CABO - Centre for Agrobiological Research (Netherlands)* No. 102. 40 pp.
Elberse, W.T. and H. Breman. 1990. Germination and establishment of Sahelian rangeland species. II. Effects of water availability. *Oecologia* 85(1):32–40. (*Eragrostis tremula*, kram-kram)
Harlan, J.R. 1989. Wild grass seed harvesting in the Sahara and Sub-Sahara of Africa. In D.R. Harris and G.C. Hillman, eds, *Foraging and Farming: the Evolution of Plant Exploitation*. Unwin-Hyman, London.

Kumar, A. 1976. Dry matter production and growth rates of three arid zone grasses in culture (*Dactyloctenium aegyptium*, *Cenchrus biflorus* and *Cenchrus ciliaris*). *Comparative Physiology and Ecology* 1(1):23–26. (Egyptian grass, kram-kram)

McKenzie, B. 1982. Resilience and stability of the grasslands of Transkei *Aristida*, *Themeda*, systems grazing, South Africa. *Proceedings of the Grasslands Society of South Africa* 17:21–24. (Drinn, *Themeda*)

Nicolaisen, J. 1963. *Ecology and Culture of the Pastoral Tuareg*. Copenhagen National Museum, Copenhagen.

Salih, O.M., A.M. Nour and D.B. Harper. 1992. Nutritional quality of uncultivated cereal grains utilised as famine foods in western Sudan as measured by chemical analysis. *Journal of the Science of Food and Agriculture* 58:417–424. (Egyptian grass, kram-kram, Shama millet, wadi rice)

Sharma, B.M. and A.O. Chivinge. 1982. Contribution to the ecology of *Dactyloctenium aegyptium* (L.) P. Beauv.: A nutritious fodder, Nigeria. *Journal of Range Management* 35(3):326–331. (Egyptian grass)

Siddiqui, K.A. 1987. Contribution of *Dactyloctenium aegyptium* (L.) Beauv. to bioreclamation of salt-affected soil. *Annals of Arid Zone* 26(4):301–303. (Egyptian grass)

United Nations Sudano-Sahelian Office (UNSO). 1990. *Lakes of Grass: Regenerating Bourgou in the Inner Delta of the Niger River*. Technical Publication No. 2, Spring 1990. UNSO, New York. 16 pp.

POTENTIAL BREAKTHROUGHS FOR GRAIN FARMERS (Appendix A)

Bruggers, R.L. and C.C.H. Elliott, eds. 1989. *Quelea quelea: Africa's Bird Pest*. Oxford University Press, Oxford, UK.

Butler, L.G., G. Ejeta, D. Hess, B. Siama, Y. Weerasuriya, and T. Cai. 1991. Some novel approaches to the *Striga* problem. Pp. 500–502 in J.K. Ransom, et al., eds., *Proceedings of the Fifth International Symposium on Parasitic Weeds*. Nairobi, Kenya, 24–30 May, 1991. Centro Internacional de Mejoramiento de Maíz y Trigo, Mexico City.

Mundy, P.J. and M.J.F. Jarvis, eds. 1989. *Africa's Feathered Locust*. Baobab Books, Harare, Zimbabwe. 166 pp.

POTENTIAL BREAKTHROUGHS IN GRAIN HANDLING (Appendix B)

Shankara, R., N.G. Malleshi, H. Krishnamurthy, M.N. Narayana, and H.S.R. Desikachar. 1985. Development of mini grain mill for dehusking and grinding of cereals. *Journal of Food Science Technology* 22:91.

POTENTIAL BREAKTHROUGHS IN CONVENIENCE FOODS (Appendix C)

Central Food Technology Research Institute (CFTRI). 1982. *Annual Report*. Mysore, Karnataka, India.

Coetzee, W.H.K. and I.S. Perold. 1958. Pre-cooked and enriched cereal products. *South African Journal of Agricultural Science* 1:327–333.

Desikachar, H.S.R. 1977. Processing of sorghum and millets for versatile food uses in India. In D.A.V. Dendy, ed., *Proceedings of a Symposium on Sorghum and Millets for Human Food, Vienna, 11–12 May, 1976*, Tropical Products Institute, London. 41 pp.

Hoseney, R.C., E.V. Marston and D.A.V. Dendy. 1981. Sorghum and millets. In Y. Pomeranz, ed., *Advances in Cereal Science and Technology, Vol IV*, AACC, St. Paul, Minnesota.

Hulse, J.H., E.M. Laing, and O.E. Pearson. 1980. Sorghum and millets: Their composition and nutritive value. IDRC. Ottawa, Academic Press, London.

Malleshi, N.G. and H.S.R. Desikachar. 1981. Varietal differences in puffing quality of ragi (*Eleusine coracana*) *Journal of Food Science and Technology* 18(1):30–32.

Malleshi, N.G. and H.S.R. Desikachar. 1982. Formulation of a weaning food with low hot-paste viscosity based on malted ragi (*Eleusine coracana*) and green gram (*Phaseolus radiatus*). *Journal of Food Science and Technology* 19(5):193–197.

Kumate, J. 1983. Relative Crispness and Oil Absorption Quality of *Sandige* (Extruded Dough) from Cereal Grains. M.Sc. Dissertation, University of Mysore. Mysore, India.

Malleshi, N.G. and H.S.R. Desikachar. 1986. Influence of malting conditions on quality of finger millet malt. *Journal of the Institute of Brewing* 92(1):81–83.

Malleshi, N.G. and H.S.R. Desikachar. 1986. Studies on comparative malting characteristics of some tropical cereals and millets. *Journal of the Institute of Brewing* 92(1):174–176.

Malleshi, N.G. and H.S.R. Desikachar. 1986. Nutritive value of malted millet flours. *Plant Foods for Human Nutrition* 36(3):191–196.

Perten, H. 1983. Practical experience in processing and use of millet and sorghum in Senegal and Sudan. *Cereal Foods World* 28:680–683.

Young, R., M. Haidara, L.W. Rooney, and R.D. Waniska. 1990. Parboiled sorghum: development of a novel decorticated product. *Journal of Cereal Science* 11:277–289. (Appendix C is based largely on this paper.)

POTENTIAL BREAKTHROUGHS IN CHILD NUTRITION (Appendix D)

Bang-Olsen, K., B. Stilling, and L. Munck. 1987. Breeding for high-lysine barley. *Barley Genetics* 5:865–870.

Bang-Olsen, K., B. Stilling, and L. Munck. 1991. The feasibility of high-lysine barley breeding: a summary. *Barley Genetics* 6:433–438.

Evans, D.J. and J.R.N. Taylor. 1990. Extraction and assay of proteolytic activities in sorghum malt. *Journal of the Institute of Brewing* 96(4):201–207.

Horn, C.H., J.C. Du Preez, and S.G. Kilian. 1992. Fermentation of grain sorghum starch by co-cultivation of *Schwanniomyces occidentalis* and *Saccharomyces cerevisiae*. *Bioresource Technology* 42(1):27–31.

Kumar, L.S., H.S Prakash, and H.S. Shetty. 1991. Influence of seed mycoflora and harvesting conditions on milling, popping, and malting qualities of sorghum (*Sorghum bicolor*). *Journal of the Science of Food and Agriculture* 55:617–625.

Kumar, L.S., M.A. Daodu, H.S. Shetty, and N.G. Malleshi. 1992. Seed mycoflora and malting characteristics of some sorghum cultivars. *Journal of Cereal Science* 15:203–209.

Malleshi, N.G. and H.S.R. Desikachar. 1982. Formulation of a weaning food with low hot-paste viscosity based on malted ragi (*Eleusine coracana*) and green gram (*Phaseolus radiatus*). *Journal of Food Science and Technology* 19(5):193–197.

Malleshi, N.G., M.A. Daodu, and A. Chandrasekhar. 1989. Development of weaning food formulations based on malting and roller drying of sorghum and cowpea. *International Journal of Food Science and Technology* 24:511–519.

Malleshi, N.G., H.S.R. Desikachar, and S. Venkat Rao. 1986. Protein quality evaluation of a weaning food based on malted ragi and green gram. *Plant Foods for Human Nutrition* 36(3):223–230.

Munck, L. 1988. The importance of botanical research in breeding for nutritional quality characteristics in cereals. *Symbolae Botanicae Upsalienses* 28(3):69–78.

Venkatnarayana, S., V. Screenivasmurthy, and B.A. Satyanarayana. 1979. Use of ragi in brewing. *Journal of Food Science Technology* 16:204.

Venkat Rao, S., S. Kurien, D.N. Swamy, V.A. Daniel, I.A.S. Murthy, N.G. Malleshi, and H.S.R. Desikachar. 1985. Clinical trials on a weaning food of low bulk based on ragi and green gram. Paper presented at the International Workshop on Weaning Foods, Iringa, Tanzania.

Appendix G
Research Contacts

AFRICAN RICE

African Countries

M. Agyen-Sampong, Mangrove Swamp Rice Research Station, West Africa Rice Development Association (WARDA), Private Mail Bag 678, Freetown, Sierra Leone
Joseph Amara, Njala University College, University of Sierra Leone, Private Mail Bag, Freetown, Sierra Leone
A. Franck Y. Attere, Department of Agronomy, University of Ibadan, Ibadan, Nigeria
Forson K. Ayensu, Plant Genetic Resources Unit, Crops Research Institute, Council for Scientific and Industrial Research (CSIR), PO Box 7, Bunso, Ghana
Jacob A. Ayuk-Takem, Institut de la Recherche Agronomique (IRA), Boîte Postal 2123, Yaoundé, Cameroon
Robert Cudjoe Aziawor, Grains Development Board, PO Box 343, Hohoe, Volta Region, Ghana
Osman Bah, Njala University College, University of Sierra Leone, Private Mail Bag, Freetown, Sierra Leone
J. Bozza, Institut de Recherches Agronomiques Tropicales et des cultures vivrières (IRAT), Boîte Postal 635, Bouaké, Côte d'Ivoire
Dana Burner, International Institute of Tropical Agriculture (IITA), Oyo Road, Private Mail Bag 5320, Ibadan, Nigeria (striga-germination promoters)
Saliou Diangar, Agronome/Programme Mil, Centre National de Recherches Agronomiques (CNRA), Institut Sénégalais de Recherches Agricoles (ISRA), Boîte Postal 53, Bambey, Senegal
Sahr N. Fomba, Mangrove Swamp Rice Research Station, WARDA, Private Mail Bag 678, Freetown, Sierra Leone
Kofi Goli, Institut des Savannes (IDESSA), Boîte Postal 633, Bouake, Côte d'Ivoire
Malcolm Jusu, Rokupr Rice Research Station, Private Mail Bag 736, Freetown, Sierra Leone
Serrie Kamara, Njala University College, University of Sierra Leone, Private Mail Bag, Freetown, Sierra Leone
Gueye Mamadou, West African Microbiological Research Centre (MIRCEN), CNRA, Boîte Postal 53, Bambey, Senegal
Kouamé Miezan, WARDA/ADRAO, Boîte Postal 2551, Bouake 01, Côte d'Ivoire
Sama Monde, Rokupr Rice Research Station, Private Mail Bag 736, Freetown, Sierra Leone
Helen Moss, 63 End Road, Linden Extension 2194, Randburg, South Africa
Folu M. Dania Ogbe, Department of Botany, University of Benin, Private Mail Bag 1154, Ugbowo Campus, Benin-City, Benin State, Nigeria
Rice Breeding Program, IITA, Oyo Road, Private Mail Bag 5320, Ibadan, Nigeria

B. Treca, Institut Français de Recherche Scientifique pour le Développement en Coopération, Office de la Recherche Scientifique et Technique Outre-Mer (ORSTOM), Boîte Postal 2528, Bamako, Mali

Patrice Vandenberghe, Appui au Développement de la Riziculture dans les Régions de Gao et de Tombouctou (A.R.G.T.), Boîte Postal 120, Bamako, Mali

West Africa Rice Development Association (WARDA), 01 Boîte Postal 2551, Bouaké, Côte d'Ivoire

WARDA, Abidjan Liaison Office, 01 Boîte Postal 4029, Abidjan 01, Côte d'Ivoire

WARDA, Monrovia Liaison Office, LBDI Building, Tubman Boulevard, Box 1019, Monrovia, Liberia

Sahel Irrigated Rice Research Station, WARDA, Boîte Postal 96, St. Louis, Côte d'Ivoire

Other Countries

J.P. Baudoin, Phytotechnie des Régions Chaudes, Faculté des Sciences Agronomiques de Gembloux, 2, Passage des Déportés, B-5800 Gembloux, Belgium

Gilles Bezançon, Laboratoire Ressources Génétiques et Amélioration des Plantes Tropicales (LRGAPT), Institut Français de Recherche Scientifique pour le Développement en Coopération de Montpellier, ORSTOM, Boîte Postal 5045, 34032 Montpellier Cédex, France

Lynne Brydon, Department of Sociology, University of Liverpool, PO Box 147, Liverpool L69 3BX, England

H.M. Burkill, Royal Botanic Gardens, Kew, Richmond, Surrey TW9 3AB, England (botany)

Matilde Causse, Institut Français de Recherche Scientifique pour le Développement en Coopération de Montpellier, ORSTOM, Boîte Postal 5045, 34032 Montpellier Cédex, France

G. Clement, Centre français du riz, Mas du Sonnailler, 13200 Arles, France

F. Cordesse, Institut Français de Recherche Scientifique pour le Développement en Coopération, Centre d'Etudes Phytosociologiques L. Emberger (CEPE), Centre National de la Recherche Scientifique (CNRS), ORSTOM, 1919 Route de Mende, Boîte Postal 5051, 34033 Montpellier-Cédex, France

Jeremy Davis, Plant Breeding International, PBI Cambridge Ltd., Maris Lane, Trumpington, Cambridge CB2 2LQ, England (forages)

M. Delseny, Laboratoire de Physiologie Végétale, U.A. 565, CNRS, Université de Perpignan, Avenue de Villeneuve, 66025 Perpignan-Cédex, France

Joseph DeVries, Department of Plant Breeding and Biometry, Emerson Hall, Room 255, Cornell University, Ithaca, New York 14851, USA

John Dudley, Plant Molecular Biology Laboratory, Agricultural Research Service (ARS), U.S. Department of Agriculture (USDA), Building 006 Room 118 BARC-West, Beltsville, Maryland 20705–2350, USA (biotechnology)

N.H. Fisher, Department of Chemistry, Louisiana State University, Baton Rouge, Louisiana 70803, USA (striga-germination promoters)

Alain Ghesquiere, Institut Français de Recherche Scientifique pour le Développement en Coopération de Montpellier, ORSTOM, Boîte Postal 5045, 34032 Montpellier Cédex, France

J.R. Harlan, 1016 North Hagan Street, New Orleans, Louisiana 70119, USA

Dale D. Harpstead, Department of Crop & Soil Sciences, Room 464, Plant and Soil Science Building, Michigan State University, East Lansing, Michigan 48824–1325, USA

Frank Nigel Hepper, The Herbarium, Royal Botanic Gardens, Kew, Richmond, Surrey, London TW9 3AE, England

David Hilling, Centre for Developing Areas Research, Department of Geography, Royal Holloway and Bedford New College, University of London, Egham Hill, Egham, Surrey TW20 0EX, England

Institut de Recherches Agronomiques Tropicales et des cultures vivrières (IRAT), Avenue du Val de Montferrand, Boîte Postal 5035, 34032 Montpellier Cédex, France

Institut Français de Recherche Scientifique pour le Développement en Coopération de Montpellier, ORSTOM, Boîte Postal 5045, 34032 Montpellier Cédex, France

International Plant Genetic Resources Institute (IPGRI, formerly IBPGR), Via delle Sette Chiese 142, 00145 Rome, Italy

M. Jacquot, Département du Centre de Coopération Internationale en Recherche Agronomique pour le Développement (CIRAD), IRAT, Boîte Postal 5035, 34032 Montpellier Cedex, France

Peter B. Kaufman, Department of Biology, University of Michigan, Ann Arbor, Michigan 48109–1048, USA

Clarissa T. Kimber, Department of Geography, Texas A&M University, College Station, Texas 77843–3147, USA

A. de Kochko, Department of Biology, Washington University, One Brookings Drive, Box 1137, St Louis, Missouri 63130, USA

J.M. Lock, Royal Botanic Gardens, Kew, Richmond, Surrey, London TW9 3AE, England

O.M. Lolo, Institut Français de Recherche Scientifique pour le Développement en Coopération, CEPE, CNRS, ORSTOM, 1919 Route de Mende, Boîte Postal 5051, 34033 Montpellier Cedex, France

J.P. Marathee, Crop and Grasslands Services, Plant Production and Protection Division, Food and Agricultural Organization of the United Nations (FAO), Via delle Terme di Caracalla, Rome 00100, Italy

Shegeta Masayoshi, The Center for African Area Studies, Kyoto University, 46 Shimoadachi-cho, Yoshida, Sakyo-ku, Kyoto 606, Japan

Hiroko Morishima, National Institute of Genetics, Mishima, 411, Japan

Nobuo Murata, Eco-Physiology Research Division, Tropical Agriculture Research Center (TARC), Ministry of Agriculture, Forestry and Fisheries, 1–2, Ohwashi, Tsukuba, Ibaraki 305, Japan

H.-I. Oka, National Institute of Genetics, Mishima, 411, Japan

Jean-Louis Pham, Laboratoire de Biologie et Genetique Evolutives, Universite Paris-Sud (UPS), Institut National de la Recherche Agronomique (INRA), Centre National de la Recherche Scientifique (CNRS), Gif sur Yvette, France

Paul Richards, Department of Anthropology, University College London, Gower Street, London WC1E 6BT, England

J. Neil Rutger, Jamie Whitter Delta States Research Center, ARS, USDA, PO Box 225, Stoneville, Mississippi 38776, USA (rice hybridization)

Gideon W. Schaeffer, Plant Molecular Biology Laboratory, ARS, USDA, Building 006 Room 118 BARC-West, Beltsville, Maryland 20705–2350, USA (biotechnology)

Gérard Second, CEPE, CNRS, ORSTOM, 1919 Route de Mende, Boîte Postal 5051, 34033 Montpellier Cedex, France

Francis T. Sharpe, Plant Molecular Biology Laboratory, ARS, USDA, Building 006 Room 118 BARC-West, Beltsville, Maryland 20705–2350, USA (biotechnology)

Steven D. Tanksley, Department of Plant Biology, 237 Plant Science Building, Cornell University, Ithaca, New York 14853, USA

Dat Van Tran, Plant Production and Protection Division, FAO, Via delle Terme di Caracalla, Rome 00100, Italy

G.E. Wickens, 50 Uxbridge Road, Hampton Hill, Middlesex TW12 3AD, England

FINGER MILLET

African Countries

A. Franck Y. Attere, Department of Agronomy, University of Ibadan, Ibadan, Nigeria

Jacob A. Ayuk-Takem, IRA, Boîte Postal 2123, Yaoundé, Cameroon

Stephen J. Carr, Christian Services Committee of Malawi, Private Bag 5, Zomba, Malawi

Abebe Demissie, Plant Germplasm Exploration & Collection, Plant Genetic Resources Centre/Ethiopia (PGRC/E), PO Box 30726, Addis Ababa, Ethiopia

S.C. Gupta, Regional Sorghum and Millets Improvement Program, Southern Africa Development Coordination Council (SADCC), International Crops Research Institute for the Semi-Arid Tropics (ICRISAT), PO Box 776, Bulawayo, Zimbabwe

Thomas V. Jacobs, Department of Botany, University of Transkei, Post Bag X1, Unitra, Umtata, Transkei, South Africa

Hilda Kigutha, Department of Home Economics, Egerton University, PO Box 536, Njoro, Kenya

J. Maud Kordylas, Arkloyd's Food Laboratory (A.F.L.), Boîte Postal 427, Douala, Cameroon

K. Anand Kumar, Pearl Millet Program, ICRISAT Sahelian Centre, Boîte Postal 12404, Niamey, Niger (INTSORMIL [International Sorghum/Millet Collaborative Research Support Program] collaborator)

I.M. Mharapara, Research and Specialist Services, Chiredzi Research Station, PO Box 97, Chiredzi, Zimbabwe

Sam Z. Mukuru, East Africa Regional Cereals and Legumes Centre (EARCAL), ICRISAT, PO Box 39036, Nairobi, Kenya (INTSORMIL collaborator)

Davidson K. Mwangi, Amaranth Seeds & Food Research & Development, Amaranth and Natural Foods, PO Box 376, Nanyuki, Central Province, Kenya

Figuhr Muza, Department of Research and Specialist Services, Private Bag 8100, Causeway, Harare, Zimbabwe

S.C. Nana-Sinkam, Joint ECA/FAO Agriculture Division, United Nations Economic Commission for Africa, PO Box 3001, Addis Ababa, Ethiopia

Nlandu ne Nsaku, Direction des Services Généraux Techniques, Institut de Recherche Agronomique et Zootechnique (IRAZ), Boîte Postal 91, Gitega, Burundi

J.C. Obiefuna, Department of Crop Production, Federal University of Technology, Owerri, Private Mail Bag 1526, Owerri, Imo State, Nigeria

Gregory Saxon, Caixa Postal 1152, Beira, Sofala, Mozambique (seeds and services)

Southern Africa Development Coordination Council (SADCC), Regional Sorghum and Millets Improvement Program, ICRISAT, PO Box 776, Bulawayo, Zimbabwe

Tharcisse Seminega, Départment de Biologie, Université Nationale du Rwanda, Boîte Postal 117, Butare, Rwanda (biotechnology and food industry)

A. Shakoor, The Dryland Farming Research and Development Project, Kenyan Ministry of Agriculture, FAO, PO Box 340, Katumani, Machakos, Kenya

P.S. Steyn, Division of Food Science and Technology, National Food Research Institute, CSIR, PO Box 395, Pretoria 0001, South Africa

J.H. Williams, Pearl Millet Improvement Program, ICRISAT Sahelian Centre, Boîte Postal 12404, Niamey, Niger (INTSORMIL collaborator)

Other Countries

David J. Andrews, Department of Agronomy, University of Nebraska, Lincoln, Nebraska 68583, USA (breeding)

S.C. Bal, Department of Soils and Agricultural Chemistry, Orissa University of Agriculture and Technology, Bhubaneswar 751 003, Orissa, India

K.V. Bondale, Millet Development Program, Ministry of Agriculture, 27 Eldams Road, Madras 600 018, India

Wayne Carlson, Maskal Forages, Inc., PO Box A, Caldwell, Idaho 83606, USA

William Critchley, Centre for Development Cooperation Services, Free University Amsterdam, van der Boechorststraat 7, 1081 BT Amsterdam, The Netherlands

Johannes M.M. Engels, Regional Office for South and Southeast Asia, IPGRI (IPBGR), c/o Pusa Campus, New Delhi 110 012, India

Charles A. Francis, Department of Agronomy, University of Nebraska, Lincoln, Lincoln, Nebraska 68583, USA

Zewdie Wolde Gebriel, Department of Human Nutrition, Wageningen Agricultural University, PO Box 8129, Wageningen 6700 EV, Netherlands

David Gibbon, School of Development Studies, Overseas Development Group, University of East Anglia, Norwich, Norfolk NR4 7TJ, England

Heiner E. Goldbach, Abta Agrarökologie, Lehrstuhl Biogeographie, Institut für Geowissenschaften der Universität Bayreuth, Postfach 101251, D-8580 Bayreuth, Germany

Le Dit Bokary Guindo, Special Program for African Agricultural Research, The World Bank, 1818 H Street, NW, Washington, DC 20433, USA

Khidir W. Hilu, Department of Biology, Virginia Polytechnic Institute and State University, Blacksburg, Virginia 24061, USA (molecular genetics)

S.C. Hiremath, Faculty of Science and Technology, Karnatak University, Dharwad 580 003, India

Leland R. House, Route 2 Box 136 A-1, Bakersville, North Carolina 28705, USA

R. Kulandaivelu, Agricultural Research Station, Bhavanisagar, Tamil Nadu, India

N.G. Malleshi, Cereal Science and Technology, Central Food Technological Research Institute, Cheluvamba Mansion, V.V. Mohalla PO, Mysore 570 013, India (processing, weaning foods)

J.P. Marathee, Crop and Grasslands Services, Plant Production and Protection Division, FAO, Via delle Terme di Caracalla, Rome 00100, Italy

Shegeta Masayoshi, The Center for African Area Studies, Kyoto University, 46 Shimoadachi-cho, Yoshida, Sakyo-ku, Kyoto 606, Japan

Alemu Mengistu, Department of Plant Pathology, University of Wisconsin, 1630 Linden Drive, Madison, Wisconsin 53706, USA

G.N. Mitra, Department of Soils and Agricultural Chemistry, Orissa University of Agriculture and Technology, Bhubaneswar 751 003, Orissa, India

Robert L. Myers, Alternative Crops and Products Project, Department of Agronomy, 210 Waters Hall, University of Missouri, Columbia, Missouri 65211, USA

R.C. Parida, Department of Soils and Agricultural Chemistry, Orissa University of Agriculture and Technology, Bhubaneswar 751 003, Orissa, India

Daniel H. Putnam, Department of Agronomy and Range Science, University of California, Davis, California 95616, USA

A. Appa Rao, Genetic Resources Unit, ICRISAT, Patancheru PO, Andhra Pradesh 502 324, India (germplasm)

Raman Rai, Department of Microbiology, Narendra Deva University of Agriculture and Technology, Narendranagar, Kumarganj, Faizabad 224 229, Uttar Pradesh, India (nitrogen fixation)

K.W. Riley, National Hill Crops Improvement Program, International Development Research Centre (IDRC), PO Box 1336, Kathmandu, Nepal

A. Seetharam, All India Coordinated Small Millets Improvement Project, G.K.V.K. Campus, University of Agricultural Sciences, Bangalore 560 065, India

K.V. Selvaraj, Agricultural Research Station, Bhavanisagar, Tamil Nadu, India

Y.M. Somasekhara, G.K.V.K. Campus, University of Agricultural Sciences, Bangalore 560 065, India

K. Vanangamudi, Agricultural Research Station, Bhavanisagar, Tamil Nadu, India

K.P.R. Vittal, Central Research Institute for Dryland Agriculture, Indian Council of Agricultural Research (ICAR), Santoshnagar, Hyderabad, Andhra Pradesh, India

C.E. West, Department of Human Nutrition, Wageningen Agricultural University, PO Box 8129, Wageningen 6700 EV, Netherlands

John Yohe, International Sorghum/Millet Collaborative Research Support Program (INTSORMIL CRSP), 54 Nebraska Center, University of Nebraska, Lincoln, Nebraska 68583, USA

Vincent Makumba Zake, c/o Department of Agronomy, Mississippi State University, Mississippi State, Mississippi 39762, USA (breeding)

FONIO (Acha)

Although fonio has been overlooked by most researchers, two special groups in Senegal have for some years been championing this crop's cause.

One, a program called "Fonio for the World," has been growing and distributing fonio seeds for research and development work. For information on availability of samples and terms under which they are supplied, write to Babacar N'Diaye, Conseiller Agricole chargé du Programme "Fonio pour le monde," Conseil Général de Diam-Diam, Koungheul, Senegal. The other program, concentrating on laboratory studies, is L'Institut des Sciences de l'Environnement of the Université Cheikh Anta Diop de Dakar, Boîte Postal 5005, Dakar-Fann, Senegal.

African Countries

Daniel K. Abbiw, Herbarium, Department of Botany, Faculty of Science, University of Ghana, Legon, Ghana

Diarra Aboubacak, Division Etude et Contrôle phytosanitaire, Service de la Protection des Végétaux, Boîte Postal 1560, Bamako, Mali

Samuel Agboire, National Cereals Research Institute, Private Mail Bag 8, Bida, Niger State, Nigeria

J.O. Akingbala, Department of Food Technology, University of Agriculture, Private Mail Bag 2240, Abeokuta, Ogun State, Nigeria

Association Malienne pour la Promotion des Jeunes, c/o Ministère de la Jeunesse et du Sport, Bamako, Mali

Forson K. Ayensu, Plant Genetic Resources Unit, Crops Research Institute, CSIR, PO Box 7, Bunso, Ghana

Robert Cudjoe Aziawor, Grains Development Board, PO Box 343, Hohoe, Volta Region, Ghana

S.O. Bennett-Lartey, Plant Genetic Resources Unit, Crops Research Institute, CSIR, PO Box 7, Bunso, Ghana

Carl W. Castleton, International Section, Animal and Plant Health Inspection Service (APHIS), U.S. Department of Agriculture, American Embassy -Abidjan, Boîte Postal 1712, Abidjan 01, Côte d'Ivoire

Saliou Diangar, Agronome/Programme Mil, CNRA, ISRA, Boîte Postal 53, Bambey, Senegal

Sahr N. Fomba, Mangrove Swamp Rice Research Station, WARDA, Private Mail Bag 678, Freetown, Sierra Leone

Ephraim O. Lucas, Department of Agronomy, University of Ibadan, Ibadan, Nigeria

Jon Kirby, Tamale Institute of Cross Cultural Studies, PO Box 42, Tamale, Northern Region, Ghana

Danladi Musa, Elwa Rural Development, Ltd., PO Box 63, Jos, Plateau State, Nigeria

S.C. Nana-Sinkam, Joint ECA/FAO Agriculture Division, United Nations Economic Commission for Africa, PO Box 3001, Addis Ababa, Ethiopia

Amadou Makhtar Ndiaye, Organisme de Recherches sur l'Alimentation et la Nutrition Africaines (ORANA), 39 avenue Pasteur, Boîte Postal 2089, Dakar, Senegal

Nyat Quat Ng, Genetic Resources Unit, IITA, Oyo Road, Private Mail Bag 5320, Ibadan, Nigeria

Emmanuel Ndu Onyedeke, Community Development, P.M.B.I. Mgbidi, Oru LGA, Imo State, Nigeria

O.B. Oyewole, Department of Food Science and Technology, University of Agriculture, P.M.B. 2240, Abeokuta, Ogun State, Nigeria

Z.J.L. Sanago, Division de Recherches sur les Systèmes de Production Rurale, Institut d'Economie Rurale, Sikasso, Mali

V.J. Temple, Department of Food and Nutrition, University of Jos, Jos, Plateau State, Nigeria

Mouhamadou Lamine Thiam, Biologie Végétale, Laboratoire de Microbiologie, Faculté Sciences et Techniques, Université Cheikh Anta Diop de Dakar, Boîte Postal 5005, Dakar-Fann, Senegal

Jane Toll, IPGRI (IBPGR), c/o ICRISAT Sahelian Centre, Boîte Postal 12404, Niamey, Niger

Jeanne Zoundjihekpon, Faculté des Sciences et Techniques, Université Nationale de Côte d'Ivoire, 22 Boîte Postal 582, Abidjan 22, Côte d'Ivoire

Other Countries

Kathleen M. Baker, Department of Geography, School of Oriental and African Studies, London University, Thornaugh Street, Russell Square, London WC1H 0XG, England

Robert Becker, U.S. Department of Agriculture, 800 Buchanan Street, Albany, California 94710, USA

Donald F. Beech, Division of Tropical Crops and Pastures, Commonwealth Scientific and Industrial Research Organisation (CSIRO), Mill Road, St. Lucia, Brisbane 4067, Queensland, Australia

H.M. Burkill, Royal Botanic Gardens, Kew, Richmond, Surrey TW9 3AB, England (botany)

Geoffrey P. Chapman, Wye College, University of London, Wye, Near Ashford, Kent TN25 5AH, England

James Duke, ARS, USDA, Room 133, Building 001, Beltsville, Maryland 20705, USA

P. Gosseye, Department of Agrosystems Research, Centrum voor Agrobiologisch Onderzoek, Centre for Agrobiological Research (CABO), Bornsesteeg 65, PO Box 14, Wageningen 6700 AA, Netherlands

Niels Hanssens, Kano State Agricultural and Rural Development Project, Hadejia Zone IV, Private Mail Bag 3130, Kano, Kano State, Nigeria, c/o Agroman, 34 New Cavendish Street, London W1M 7LH, England

Nazmul Haq, International Centre for UnderUtilized Crops, Andrews Building, Kings College London, Campden Hill Road, LI8 7AH, England

G. Harinarayana, All India Coordinated Pearl Millet Improvement Project, India Council of Agricultural Research (ICAR), College of Agriculture Campus, Shivajinagar, Pune 411 005, India

J.R. Harlan, 1016 North Hagan Street, New Orleans, Louisiana 70119, USA

Frank Nigel Hepper, The Herbarium, Royal Botanic Gardens, Kew, Richmond, Surrey, London TW9 3AE, England

Israel Afam Jideani, School of Science & Science Education, Abubakar Tafawa-Balewa University, Private Mail Bag 0248, Bauchi, Nigeria

International Centre for UnderUtilized Crops, Andrews Building, Kings College London, Campden Hill Road, LI8 7AH, England

Andrew Kidd, International Council for the Development of Underutilized Crops, Building 44, The University of Southhampton, Southampton SO9 5NH, England

Pius Michael Kyesmu, Wye College, University of London, Wye, Ashford, Kent TN25 5AH, England

Clare Madge, School of Geography, The University of Birmingham, PO Box 363, Egbaston, Birmingham B15 2TT, West Midlands, England

J.P. Marathee, Crop and Grasslands Services, Plant Production and Protection Division, FAO, Via delle Terme di Caracalla, Rome 00100, Italy

Shegeta Masayoshi, The Center for African Area Studies, Kyoto University, 46 Shimoadachi-cho, Yoshida, Sakyo-ku, Kyoto 606, Japan

Robert L. Myers, Alternative Crops and Products Project, Department of Agronomy, 210 Waters Hall, University of Missouri, Columbia, Missouri 65211, USA

Don Osborn, 4632 South Hagadorn Road #C-33, East Lansing, Michigan 48823, USA

Daniel H. Putnam, Department of Agronomy and Range Science, University of California, Davis, California 95616, USA

Paul Richards, Department of Anthropology, University College London, Gower Street, London WC1E 6BT, England

Barrie Sharpe, Department of Food Science, Kings College (Kensington Campus), University of London, London W8 7AH, England

B. Simpson, c/o Department of Resource Development, 323 Natural Resources Building, Michigan State University, East Lansing, Michigan 48824, USA

Margaret Steentoft, 7 The Purrocks, Petersfield, Hampshire GU32 2HU, England
E. Wagner, Crop and Grassland Production Service, Plant Production and Protection Division, FAO, Via delle Terme di Caracalla, Rome 00100, Italy

PEARL MILLET

African Countries

Adam Aboubacar, Cereal Chemistry, Institut National de Recherches Agronomiques au Niger (INRAN), Boîte Postal 429, Niamey, Niger (INTSORMIL collaborator)
Abdelmoneim Taha Ahmed, Economics and Statistical Section, Gezira Agricultural Research Station, Agricultural Research Corporation (ARC), PO Box 126, Wad Medani, Sudan (INTSORMIL collaborator)
O.C. Aworh, Department of Food Technology, Faculty of Technology, University of Ibadan, Ibadan, Nigeria
Sitt El Nafr Badi, Food Research Centre, Shambat Box 213, North Khartoum, Sudan (INTSORMIL collaborator)
Tareke Berhe, SAA - Global 2000, Kotoka International Airport, Post Office Private Mail Bag, Accra, Ghana (INTSORMIL collaborator)
Jacques Beyo, IRA, Boîte Postale 33, Maroua, Cameroon
Taye Bezuneh, STRC-SAFGRAD, Organization of African Unity (OAU), Boîte Postal 1783, Ouagadougou, Burkina Faso (INTSORMIL collaborator)
Ouendeba Botorou, PARA - Institut National de Recherches Agronomiques au Niger (INRAN), Boîte Postal 429, Niamey, Niger (INTSORMIL collaborator)
Stephen J. Carr, Christian Services Committee of Malawi, Private Bag 5, Zomba, Malawi
H.S. Chambo, Agricultural Research Institute (ARI) - Hombolo, PO Box 299, Dodoma, Tanzania (INTSORMIL collaborator)
Yacouba O. Doumbia, DRA/SRCVO - Institute Economic Rurale (IER), Sotuba Research Station, Boîte Postal 258, Bamako, Mali (INTSORMIL collaborator)
Amadou Fofana, CNRA, ISRA, Boîte Postal 53, Bambey, Senegal (INTSORMIL collaborator)
Walter Frölich, Sorghum and Millet Section, Nyankpala Agricultural Experiment Station (NAES), Crops Research Institute (CRI), PO Box 483, Tamale, Ghana
Lucas Gakale, Department of Agricultural Research, Private Bag 0033, Gaborone, Botswana (INTSORMIL collaborator)
S.C. Gupta, Regional Sorghum and Millets Improvement Program, SADCC, ICRISAT, PO Box 776, Bulawayo, Zimbabwe
M. Haidara, Institut Economie Rurale (IER), Boîte Postal 258, Bamako, Mali (INTSORMIL collaborator)
Dale Hess, ICRISAT Sahelian Centre, Boîte Postal 12404, Niamey, Niger (INTSORMIL collaborator)
Thomas V. Jacobs, Department of Botany, University of Transkei, Post Bag X1, Unitra, Umtata, Transkei, South Africa
Hilda Kigutha, Department of Home Economics, Egerton University, PO Box 536, Njoro, Kenya
J. Maud Kordylas, Arkloyd's Food Laboratory (A.F.L.), Boîte Postal 427, Douala, Cameroon
Joyce Lowe, Department of Botany and Microbiology, University of Ibadan, Ibadan, Nigeria
Fernando A.B. Marcelino, Instituto de Investigação Agronómica (IIA), Caixa Postal 406, Huambo, Angola
Demba M'Baye, CNRA, ISRA, Boîte Postal 53, Bambey, Senegal (INTSORMIL collaborator)
I.M. Mharapara, Research and Specialist Services, Chiredzi Research Station, PO Box 97, Chiredzi, Zimbabwe

Figuhr Muza, Department of Research and Specialist Services, Private Bag 8100, Causeway, Harare, Zimbabwe (breeding; INTSORMIL collaborator)

Oumar Niangado, Institute Economic Rurale (IER), Boîte Postal 258, Bamako, Mali (INTSORMIL collaborator)

Regional Sorghum and Millets Improvement Program, SADCC, ICRISAT, PO Box 776, Bulawayo, Zimbabwe

A. Shakoor, The Dryland Farming Research and Development Project, Kenyan Ministry of Agriculture, FAO, PO Box 340, Katumani, Machakos, Kenya

P. Soman, ICRISAT Sahelian Centre, Boîte Postal 12404, Niamey, Niger

P.S. Steyn, Division of Food Science and Technology, National Food Research Institute, CSIR, PO Box 395, Pretoria 0001, South Africa

Jens von Bargen, Nyankpala Agricultural Experiment Station (NAES), Crops Research Institute (CRI), PO Box 483, Tamale, Ghana

G.K. Weber, IITA, Private Mail Bag 5320, Ibadan, Nigeria

J.H. Williams, Pearl Millet Improvement Program, ICRISAT Sahelian Centre, Boîte Postal 12404, Niamey, Niger (INTSORMIL collaborator)

Ousmane Youm, ICRISAT Sahelian Centre, Boîte Postal 12404, Niamey, Niger (INTSORMIL collaborator)

Other Countries

David J. Andrews, Department of Agronomy, University of Nebraska, Lincoln, Nebraska 68583, USA (breeding; INTSORMIL collaborator)

John Axtell, Department of Agronomy, Purdue University, West Lafayette, Indiana 47907, USA (breeding; striga; INTSORMIL collaborator)

Glenn W. Burton, Forage and Turf Research, Georgia Coastal Plain Experiment Station, ARS, USDA, PO Box 748, Tifton, Georgia 31793, USA

Jeremy Davis, Plant Breeding International, PBI Cambridge Ltd., Maris Lane, Trumpington, Cambridge CB2 2LQ, England (forages)

Charles A. Francis, Department of Agronomy, University of Nebraska, Lincoln, Lincoln, Nebraska 68583, USA

Donald Fryrear, Big Spring Experiment Station, ARS, USDA, Box 909, Big Spring, Texas 79721, USA

Zewdie Wolde Gebriel, Department of Human Nutrition, Wageningen Agricultural University, PO Box 8129, Wageningen 6700 EV, Netherlands

P. Geervani, College of Home Science, Andhra Predesh Agricultural University, Rajendranagar, Hyderabad, Andhra Predesh 500 030, India

S.C. Geiger, College of Agriculture, Texas A&M University, College Station, Texas 77843, USA

Le Dit Bokary Guindo, Special Program for African Agricultural Research, The World Bank, 1818 H Street, NW, Washington, DC 20433, USA

Wayne W. Hanna, Department of Agronomy, Georgia Coastal Plain Experiment Station, ARS, USDA, PO Box 748, Tifton, Georgia 31793, USA

Tom Hash, ICRISAT, Patancheru PO, Andhra Pradesh 502 324, India (INTSORMIL collaborator)

G. M. Hill, Georgia Coastal Plain Experiment Station, College of Agriculture, Department of Animal Science, University of Georgia, PO Box 748, Tifton, Georgia 31793, USA

R.C. Hoseney, Department of Agronomy, Throckmorton Hall, Kansas State University, Manhattan, Kansas 66506, USA

Leland R. House, Route 2 Box 136 A-1, Bakersville, North Carolina 28705, USA

Catherine Howarth, Department of Environmental Biology, Welsh Plant Breeding Station, Institute for Grassland and Environmental Research, University College of Wales, PO Box 2, Old College, Aberystwyth, Dyfed SY23 3EB, Wales (modern millets)

Carl S. Hoveland, Department of Agronomy, College of Agriculture, University of Georgia, Athens, Georgia 30602, USA

R.L. Jensen, Department of Poultry Science, University of Geogia, Athens, Georgia 30602, USA

P. Kishore, Division of Entomology, Indian Agricultural Research Institute (IARI), New Delhi, India

K. Anand Kumar, Pearl Millet Program, ICRISAT Sahelian Centre, Boîte Postal 12404, Niamey, Niger (INTSORMIL collaborator)

James D. Maguire, Department of Agronomy and Soils, Washington State University, 201 Johnson Hall, Pullman, Washington 99164, USA

V. Mahalakshmi, Cereal Program, International Crops Research Institute for the Semi-Arid Tropics (ICRISAT), Patancheru PO, Andhra Pradesh 502 324, India

N.G. Malleshi, Cereal Science and Technology, Central Food Technological Research Institute, Cheluvamba Mansion, V.V. Mohalla PO, Mysore 570 013, India (processing, weaning foods)

J.P. Marathee, Crop and Grasslands Services, Plant Production and Protection Division, FAO, Via delle Terme di Caracalla, Rome 00100, Italy

Shegeta Masayoshi, The Center for African Area Studies, Kyoto University, 46 Shimoadachi-cho, Yoshida, Sakyo-ku, Kyoto 606, Japan

Paul L. Mask, 110 Extension Hall, Auburn University, Auburn, Alabama 36849–5633, USA

A.N. Misra, Department of Botany, Utkal University, PO Vani Vihar, Bhubaneswar 751 004, Orissa, India

N.A. Mnzava, The Asian Vegetable Research and Development Center, PO Box 42, Shanhua, Tainan, Taiwan 74199, Republic of China

Kenneth O. Rachie, 5434 Dynasty Drive, Pensacola, Florida 32504, USA

T.F. Rajewski, Department of Agronomy, University of Nebraska, Lincoln, Nebraska 68583, USA

A. Appa Rao, Genetic Resources Unit, ICRISAT, Patancheru PO, Andhra Pradesh 502 324, India (germplasm)

K.C. Reddy, Agency for International Development (AID) - Niamey, U.S. Department of State, Washington, DC 20523, USA

Shao Qiquan, Genetic Transformation Laboratory & Genetic Resources of Plants, Academia Sinica Institute of Genetics, De-Sheng-Men-Wai, Bei-Sha-Tan Building 917, Beijing 100012, China

R.L. Smith, Department of Agronomy, University of Georgia, Athens, Georgia 30602, USA

Robert J. Theodoratus, Department of Anthropology, Colorado State University, Fort Collins, Colorado 80523, USA (beer)

J.H. Topps, Division of Agricultural Chemistry and Biochemistry, School of Agriculture, University of Aberdeen, 581 King Street, Aberdeen AB2 4AQ, Scotland

S. Tostain, Institut Français de Recherche Scientifique pour le Développement en Coopération de Montpellier, Office de la Recherche Scientifique et Technique Outre-Mer (ORSTOM), Boîte Postal 5045, 34032 Montpellier Cédex, France

Rick J. Van Den Beldt, Forestry/Fuelwood Research and Development (F/FRED) Project, Winrock International Institute for Agricultural Development, Network Secretariat, PO Box 1038, Kasetsart Post Office, Bangkok 10903, Thailand

Paresh Verma, Institute of Agriculture and Natural Resources, Department of Agronomy, 205 KCRL, University of Nebraska, Lincoln, Nebraska 68583, USA

K.P.R. Vittal, Central Research Institute for Dryland Agriculture, Indian Council of Agricultural Research (ICAR), Santoshnagar, Hyderabad, Andhra Pradesh, India

C.E. West, Department of Human Nutrition, Wageningen Agricultural University, PO Box 8129, Wageningen 6700 EV, Netherlands

John Yohe, International Sorghum/Millet Collaborative Research Support Program (INTSORMIL CRSP), 54 Nebraska Center, University of Nebraska, Lincoln, Nebraska 68583, USA

SORGHUM

African Countries

Rashad A. Abo-Elenien, Field Crops Research Institute, Agricultural Research Center, Giza, Egypt (INTSORMIL collaborator)

Adam Aboubacar, Cereal Chemistry, Institut National de Recherches Agronomiques au Niger (INRAN), Boîte Postal 429, Niamey, Niger (INTSORMIL collaborator)

Moussa Adamou, Institut National de Recherches Agronomiques au Niger (INRAN), Boîte Postal 429, Niamey, Niger (INTSORMIL collaborator)

A. Adeyinka Adesiyun, Institute for Agricultural Research, Ahmadu Bello University, Samara, Zaria, Nigeria

Samuel Agboire, National Cereals Research Institute, Private Mail Bag 8, Bida, Niger State, Nigeria

Abdelmoneim Taha Ahmed, Economics and Statistical Section, Gezira Agricultural Research Station, Agricultural Research Corporation (ARC), PO Box 126, Wad Medani, Sudan (INTSORMIL collaborator)

Olupomi Ajayi, West African Sorghum Improvement Program (WASIP), ICRISAT, Private Mail Bag 3491, Kano, Nigeria

O.C. Aworh, Department of Food Technology, Faculty of Technology, University of Ibadan, Ibadan, Nigeria

Abdeljabar T. Babikher, Gezira Agricultural Research Station, Agricultural Research Corporation (ARC), Box 126, Wad Medani, Sudan (INTSORMIL collaborator)

Minbamba Bagayoko, Institute Economic Rurale (IER), Boîte Postal 258, Bamako, Mali (INTSORMIL collaborator)

M.A. Benhura, Department of Biochemistry, University of Zimbabwe, PO Box MP 176, Mount Pleasant, Harare, Zimbabwe (brewing)

Tareke Berhe, SAA - Global 2000, Kotoka International Airport, Post Office Private Mail Bag, Accra, Ghana (INTSORMIL collaborator)

Taye Bezuneh, STRC-SAFGRAD, Organization of African Unity (OAU), Boîte Postal 1783, Ouagadougou, Burkina Faso (INTSORMIL collaborator)

T.S. Brand, Elsenburg Agricultural Centre, Elsenburg, Cape Province, South Africa

Stephen J. Carr, Christian Services Committee of Malawi, Private Bag 5, Zomba, Malawi

J. Chantereau, CIRAD, West African Sorghum Improvement Program (WASIP), ICRISAT, Boîte Postal 320, Bamako, Mali

Edmund Chintu, Sorghum Research Office, Chitedze Research Station, PO Box 158, Lilongwe, Malawi (INTSORMIL collaborator)

Medson Chisi, Mutanda Research Station, Box 110312, Solwezi, Zambia (INTSORMIL collaborator)

Sidi Bekaye Coulibaly, SRCVO - Institute Economic Rurale (IER), PO Box 438, Bamako, Mali (INTSORMIL collaborator)

Sansan Da, Tropical Station Farako - BA, IRA, Boîte Postal 910, Bobo Dioulasso, Burkina Faso (INTSORMIL collaborator)

Alfredo A.F. Da Cunha, Breeding Research, Ministry of Agriculture, PO Box 527, Luanda, Angola (INTSORMIL collaborator)

Abera Debelo, Ethiopian Sorghum Improvement Program, Institute of Agricultural Research (IAR), PO Box 103, Nazreth, Ethiopia (INTSORMIL collaborator)

Siriba Dione, Institute Economic Rurale (IER), Boîte Postal 258, Bamako, Mali (INTSORMIL collaborator)

Mamourou Diourte, Institute Economic Rurale (IER), Sotuba, Boîte Postal 438, Bamako, Mali (INTSORMIL collaborator)

M.A. Daodu, Federal Institute of Industrial Research, Oshodi, Private Mail Bag 21023, Murtala Muhammed Airport, Ikeja, Lagos, Nigeria (weaning foods)

Yacouba O. Doumbia, DRA/SRCVO - Institute Economic Rurale (IER), Sotuba Research Station, Boîte Postal 258, Bamako, Mali (INTSORMIL collaborator)

Hamy El-Assuity, Department of Sugar Cane, Sorghum, Maize and Forage Crops, Agricultural Research Center, Giza, Egypt (INTSORMIL collaborator)

Ahmad Abu El Gassim, National Seed Administration, Ministry of Agriculture, PO Box 285, Khartoum, Sudan (INTSORMIL collaborator)

Osman O. El-Nagouly, National Sorghum Program, Agricultural Research Center, Giza, Egypt (INTSORMIL collaborator)

Osman Ibrahim El Obeid, Gezira Agricultural Research Station, Agricultural Research Corporation (ARC), Box 126, Wad Medani, Sudan (INTSORMIL collaborator)

Haroun El-Shafey, Department of Sugar Cane, Sorghum, Maize and Forage Crops, Agricultural Research Center, Giza, Egypt (INTSORMIL collaborator)

J. Peter Esele, Sorghum and Millet Unit, Uganda Agriculture and Forestry Research, PO Soroti, Serere, Uganda (INTSORMIL collaborator)

Saeed Farah, Gezira Agricultural Research Station, Agricultural Research Corporation (ARC), Box 126, Wad Medani, Sudan (INTSORMIL collaborator)

Lucas Gakale, Department of Agricultural Research, Private Bag 0033, Gaborone, Botswana (INTSORMIL collaborator)

Marcel Galiba, SAA - Global 2000, Kotoka International Airport, Post Office Private Mail Bag, Accra, Ghana (INTSORMIL collaborator)

M. Haidara, Institut Economie Rurale (IER), Boîte Postal 258, Bamako, Mali (INTSORMIL collaborator)

Dale Hess, ICRISAT Sahelian Centre, Boîte Postal 12404, Niamey, Niger (INTSORMIL collaborator)

C.H. Horn, University of the Orange Free State, PO Box 339, Bloemfontein 9300, Orange Free State, South Africa (fermentation)

Ben M. Kanyenji, National Dryland Farming Research Station Katumani, PO Box 340, Machakos, Kenya (INTSORMIL collaborator)

Helen Kasalu, Misamfu Regional Research Station, PO Box 410055, Kasama, Zambia (INTSORMIL collaborator)

Yilma Kebede, Pioneer Hybrid Seed Co., Private Bag BW 6237, Borrowdale, Harare, Zimbabwe (INTSORMIL collaborator)

Issoufou Kollo, Institut National de Recherches Agronomiques au Niger (INRAN), Boîte Postal 249, Niamey, Niger (INTSORMIL collaborator)

Mohale Mahanyele, National Sorghum Breweries Limited, PO Box 785067, Sandton 2146, South Africa (brewing; malting)

Charles Maliro, Sorghum Research Office, Chitedze Research Station, PO Box 158, Lilongwe, Malawi (INTSORMIL collaborator)

Anaclet Mansuetus, Agricultural Research Institute (ARI) - Ilonga, PO Box Ilonga, Ilosa, Tanzania (INTSORMIL collaborator)

Chris Manthe, Department of Agricultural Research, Private Bag 0033, Gaborone, Botswana (INTSORMIL collaborator)

Louis Mazhani, Department Agricultural Research, Private Bag 0033, Gaborone, Botswana (INTSORMIL collaborator)

C. Mburu, Western Agricultural Research Station, Kakamega, Kenya (INTSORMIL collaborator)

Emmanuel Monyo, SADCC, ICRISAT, PO 776, Bulawayo, Zimbabwe (INTSORMIL collaborator)

Lewis Mughogho, SADCC, ICRISAT, PO 776, Bulawayo, Zimbabwe (INTSORMIL collaborator)

Sam Z. Mukuru, RCAL, ICRISAT, PO Box 39063, Nairobi, Kenya (INTSORMIL collaborator)

D.S. Murty, West African Sorghum Improvement Program (WASIP), ICRISAT, Private Mail Bag 3491, Kano, Nigeria

Joseph N. Mushonga, Department of Research and Specialist Services, Crop Breeding Institute, Private Bag 8100, Causeway, Harare, Zimbabwe (breeding; INTSORMIL collaborator)

Mouhoussine Nacro, Organic Chemistry Laboratory, Chemistry Department, Ougadougou University, 01 PO Box 1955, Ougadougou 01, Burkina Faso (sorghum dyes and tanning)

Nlandu ne Nsaku, Direction des Services Généraux Techniques, IRAZ, Boîte Postal 91, Gitega, Burundi

Tunde Obilana, Sorghum and Millet Program, SADCC, ICRISAT, PO 776, Bulawayo, Zimbabwe (INTSORMIL collaborator)

El Hilu Omer, Gezira Agricultural Research Station, Agricultural Research Corporation (ARC), Box 126, Wad Medani, Sudan (INTSORMIL collaborator)

Caleb O. Othieno, Tea Research Foundation of Kenya, PO Box 820, Kericho, Kenya

Moussa Oumarou, Soil Laboratory, Institut National de Recherches Agronomiques au Niger (INRAN), Boîte Postal 429, Niamey, Niger (INTSORMIL collaborator)

Regional Sorghum and Millets Improvement Program, SADCC, ICRISAT, PO Box 776, Bulawayo, Zimbabwe

J.K. Rutto, Kenya Agricultural Research Institute, PO Box 57811, Nairobi, Kenya (INTSORMIL collaborator)

Peter Setimela, Department of Agricultural Research, PO Box 0033, Sebel, Botswana (INTSORMIL collaborator)

P.S. Steyn, Division of Food Science and Technology, National Food Research Institute, CSIR, PO Box 395, Pretoria 0001, South Africa

John R.N. Taylor, c/o Department of Food Science, University of Pretoria, Pretoria 0002, South Africa (brewing)

Pamela Thole, Zambia Seed Company, PO Box 35441, Lusaka, Zambia (INTSORMIL collaborator)

Abdoul Toure, DAR - Institute Economic Rurale (IER), Bamako, Mali (INTSORMIL collaborator)

Aboubacar Toure, Institute Economic Rurale (IER), Box 258, Bamako, Mali (INTSORMIL collaborator)

Moussa Traore, Ministere du Development Rural, Rue Mohamed V, Boîte Postal 61, Bamako, Mali (INTSORMIL collaborator)

Gilles Trouche, CNRA, ISRA, Boîte Postal 53, Bambey, Senegal (breeding; INTSORMIL collaborator)

H.A. Van de Venter, Faculty of Science, University of Pretoria, Pretoria 0002, South Africa (heat-shock sorghums)

Bhola Nath Verma, Mt. Makulu Research Station, Boîte Postal 7, Chilanga, Zambia (INTSORMIL collaborator)

W.G. Wenzel, Grain Crops Research Institute, Private Bag X1251, Potchefstroom 2520, South Africa

J.H. Williams, Pearl Millet Improvement Program, ICRISAT Sahelian Centre, Boîte Postal 12404, Niamey, Niger

V.M. Zake, Sorghum and Millet Unit, Uganda Agriculture and Forestry Research, PO Soroti, Serere, Uganda (INTSORMIL collaborator)

Other Countries

Irvin C. Anderson, Department of Agronomy, Iowa State University, Ames, Iowa 50011, USA (sweet sorghums)

David J. Andrews, Department of Agronomy, University of Nebraska, Lincoln, Nebraska 68583, USA (breeding; INTSORMIL collaborator)

John Axtell, Department of Agronomy, Purdue University, West Lafayette, Indiana 47907, USA (breeding; striga; INTSORMIL collaborator)

Sitt El Nafr Badi, Food Research Centre, Shambat Box 213, North Khartoum, Sudan (INTSORMIL collaborator)

Robert H. Baumann, Center for Energy Studies, Louisiana State University, Baton Rouge, Louisiana 70808, USA (sweet sorghums)

Dorothea Bedigian, Department of Biology, Washington University, Campus Box 1137, St. Louis, Missouri 63130, USA (ethnography)

Alessandro Bozzini, Ente Nazionale Energia Atomica, Viale Regina Margherita N.125, 00198, Rome, Italy (fuel, energy sorghums)

Paula Bramel-Cox, Crop, Soil, and Range Sciences, Department of Agronomy, Throckmorton Hall, Kansas State University, Manhattan, Kansas 66506–5501, USA (genetics and breeding; perennial sorghums)

Larry Butler, Department of Agronomy, Purdue University, West Lafayette, Indiana 47907, USA (striga)

David L. Carter, Soil and Water Management Research, ARS, USDA, Snake River Conservation Research Center, Route 1, Box 186, Kimberly, Idaho 83341, USA (soil reclamation)

Max D. Clegg, Department of Agronomy, Crop, Range, Soil, and Weed Sciences, University of Nebraska, Lincoln, Nebraska 68583, USA (sweet and fuel sorghums)

Elizabeth Colson, University of California at Berkeley (retired), 840 Arlington Boulevard, El Cerrito, California 94530, USA (ethnography)
Jeff Dahlberg, Sorghum Research Program, Department of Soil and Crop Sciences, Texas A&M University, College Station, Texas 77843, USA
Gene Dalton, Sorghum Research Program, Pioneer Hi-Bred International, Inc., 6800 Pioneer Parkway, PO Box 316, Johnston, Iowa 50131, USA
Division of Tropical Crops and Pastures, CSIRO, Mill Road, St. Lucia, Brisbane, Queensland 4067, Australia (sweet sorghums)
Ronny R. Duncan, Department of Agronomy, University of Georgia, Georgia Experiment Station, Griffin, Georgia 30223–1797, USA (editor of annual sorghum-research newsletter)
B.O. Eggum, Statens Husdyrbrugsfors☆g (National Institute of Animal Science), Postboks 39, 8830 Tjele, Denmark
Frank Einhellig, Department of Biology, University of South Dakota, Vermillion, South Dakota 57069, USA (sorghum for controlling weeds)
Gebisa Ejeta, Department of Agronomy, Purdue University, West Lafayette, Indiana 47907, USA (INTSORMIL collaborator)
Lindolfo Fernández, Las Playitas Experiment Station, Ministry of Natural Resources (MNR), Comayagua, Honduras
R.A. Frederiksen, Department of Plant Pathology and Microbiology, Texas A&M University, College Station, Texas 77843, USA
Gas Research Institute, 8600 West Bryn Mawr Avenue, Chicago, Illinois 60631, USA (sweet sweet-stalk sorghums for energy, and crop support)
Earl W. Gleaves, Department of Animal Science, Institute of Agriculture and Natural Resources, University of Nebraska, Lincoln, Nebraska 68583, USA
Francisco Gomez, Escuela Agrícola Panamericana, Apartado Postal 93, Tegucigalpa, Honduras
Patricio Gutierrez, Department of Agronomy, Escuela Agrícola Panamericana, Apartado Postal 93, Tegucigalpa, Honduras (INTSORMIL collaborator)
J.R. Harlan, 1016 North Hagan Street, New Orleans, Louisiana 70119, USA
F.J. Hills, Department of Agronomy and Range Science, University of California, Davis, California 95616, USA (sweet sorghums)
Leland R. House, Route 2 Box 136 A-1, Bakersville, North Carolina 28705, USA
Catherine Howarth, Department of Environmental Biology, Welsh Plant Breeding Station, Institute for Grassland and Environmental Research, University College of Wales, PO Box 2, Old College, Aberystwyth, Dyfed SY23 3EB, Wales, United Kingdom (heat-shock sorghum)
Kaiser Engineers, Inc., 1800 Harrison Street, PO Box 23210, Oakland, California 94623, USA (sweet sorghums)
Robert Kalton, Research Seeds, Inc., Answer Farm, Rural Route 2, Webster City, Iowa 50595, USA
Issoufou Kapran, Department of Agronomy, Purdue University, West Lafayette, Indiana 47907, USA (INTSORMIL collaborator)
Clarissa T. Kimber, Department of Geography, Texas A&M University, College Station, Texas 77843–3147, USA
Arthur Klatt, Division of Agriculture, 139 Agricultural Hall, Oklahoma State University, Stillwater, Oklahoma 74078–0500, USA
Steve Kresovich, Plant Genetic Resources Conservation Unit, ARS, USDA, Griffin, Georgia 30223–1797, USA (sweet, fuel sorghums)
Gerald R. Leather, Foreign Disease - Weed Science Research, ARS, USDA, Fort Detrick, Building 1301, Frederick, Maryland 21701, USA (allelopathy and sorghum for controlling weeds)
R.T. Lewellen, Department of Agronomy and Range Science, University of California, Davis, California 95616, USA (sweet sorghums)
David G. Lynn, Department of Chemistry, University of Chicago, Chicago, Illinois 60637, USA (striga)
N.G. Malleshi, Cereal Science and Technology, Central Food Technological Research Institute, Cheluvamba Mansion, V.V. Mohalla PO, Mysore 570 013, India (processing, weaning foods)

Paul L. Mask, 110 Extension Hall, Auburn University, Auburn, Alabama 36849–5633, USA (sweet-stalk sorghums)

A. Bruce Maunder, Sorghum and Tropical Maize Research, DeKalb Genetics Corporation, Route 2 Box 56, Lubbock, Texas 79415, USA

Dan H. Meckenstock, Programa Internacional de Sorgo y Mijo (INTSORMIL), Escuela Agrícola Panamericana, Apartado Postal 93, Tegucigalpa, Honduras

Humberto Mejia, Raul Valle Experiment Station, Ministry of Natural Resources (MNR), Olancho, Honduras (INTSORMIL collaborator)

Fred R. Miller, Department of Soil and Crop Sciences, Texas A&M University, College Station, Texas 77843 (breeding and crop support)

Kenneth J. Moore, Wheat, Sorghum, and Forage Research Unit, ARS, USDA, 344 Keim Hall, University of Nebraska East Campus, Lincoln, Nebraska 68583, USA (utility sorghums)

Charles F. Murphy, Grain Crops, Plant Sciences, National Program Staff, Agricultural Research Service (ARS), U.S. Department of Agricuture (USDA), Building 005, BARC-West, Beltsville, Maryland 20705, USA

George R. Newkome, Center for Energy Studies, Louisiana State University, Baton Rouge, Louisiana 70808, USA (sweet sorghums; fuel sorghums)

M.H. Nguyen, Hawkesbury Agricultural College, Richmond, N.S.W. 2753, Australia (sweet sorghums)

Evelyn Oviedo, La Lujosa Experiment Station, Ministry of Natural Resources (MNR), Choluteca, Honduras (INTSORMIL collaborator)

Alejandro Palma, Department of Agronomy, Escuela Agrícola Panamericana, Apartado Postal 93, Tegucigalpa, Honduras (INTSORMIL collaborator)

Gary Peterson, Texas A&M Agricultural Experiment Station, Lubbock, Texas 79401, USA (INTSORMIL collaborator)

Hector Portillo, Department of Agronomy, Escuela Agrícola Panamericana, Apartado Postal 93, Tegucigalpa, Honduras (INTSORMIL collaborator)

Martin L. Price, Educational Concerns for Hunger Organization (ECHO), 17430 Durrance Road, North Fort Myers, Florida 33917, USA (fuel and specialty sorghums)

Alan Putnam, Department of Agronomy, Michigan State University, East Lansing, Michigan 48824, USA (sorghum for controlling weeds)

Kenneth O. Rachie, 5434 Dynasty Drive, Pensacola, Florida 32504, USA

A.K. Rajvanshi, Nimbkar Agricultural Research Institute, PO Box 23, Phaltan-Lonand Road, Phaltan 415 523, Satata District, Maharashtra, India (sweet sorghums; alcohol fuels)

K.V. Ramaiah, ICRISAT, Patancheru PO, Andhra Pradesh 502 324, India

Korivi Eswara Prasada Rao, Genetic Resources Program, ICRISAT, Patancheru PO, Andhra Pradesh 502 324, India

James Rasmussen, Mount Marty College, 1105 West 8th, Yankton, South Dakota 57078, USA (sorghum for controlling weeds)

James L. Riopel, Department of Biology, University of Virginia, Charlottesville, Virginia 29428, USA (striga)

Charles W. Robbins, Soil and Water Management Research, ARS, USDA, 3793 North 3600 East, Kimberly, Idaho 83341, USA (sorghum/sudangrass hybrids; sodic soils)

Lloyd W. Rooney, Cereal Quality Lab, Department of Soil and Crop Sciences, Texas A&M University, College Station, Texas 77843, USA (food science)

Darrell Rosenow, Texas A&M Agricultural Experiment Station, Research and Extension Center, Lubbock, Texas 79401, USA (INTSORMIL collaborator)

Edgar Salguero, Instituto de Ciencia y Tecnología Agrícolas (ICTA), Jutiapa, Guatemala (INTSORMIL collaborator)

Manuel Santos, Centro Nacional de Tecnología Agropecuaria (CENTA), San Andres, El Salvador (INTSORMIL collaborator)

R.E. Schaffert, Centro Nacional de Pesquisa de Milho e Sorgo (CNPMS), Empresa Brasileira de Pesquisa Agropecuária (EMBRAPA), Caixa Postal 151, Sete Lagoas, Minas Gerais 35700, Brazil (breeding; sweet sorghums)

I.O. Skoyen, Department of Agronomy and Range Science, University of California, Davis, California 95616, USA (sweet sorghums)

Bluebell R. Standal, Department of Food Science and Nutrition, College of Tropical

Agriculture and Human Resources, University of Hawaii, 3190 Maile Way, Honolulu, Hawaii 96822, USA (nutrition; specialty sorghums)
Keith Steinkraus, Department of Food Science, Cornell University, Ithaca, New York 14456, USA (fermented foods)
J.W. Stenhouse, ICRISAT, Patancheru PO, Andhra Pradesh 502 324, India (INTSORMIL collaborator, breeding)
Robert J. Theodoratus, Department of Anthropology, Colorado State University, Fort Collins, Colorado 80523, USA (beer)
K.P.R. Vittal, Central Research Institute for Dryland Agriculture, Indian Council of Agricultural Research (ICAR), Santoshnagar, Hyderabad, Andhra Pradesh, India
Fred M. Wrighton, Center for Energy Studies, Louisiana State University, Baton Rouge, Louisiana 70808, USA (sweet sorghums)
John Yohe, International Sorghum/Millet Collaborative Research Support Program (INTSORMIL CRSP), 54 Nebraska Center, University of Nebraska, Lincoln, Nebraska 68583, USA
J.R. Zanini, Departamento de Agricultura, Universidade Estadual Paulista "Júlio de Mesquita Filho" (UNESP), Ilha Solteira, São Paulo, Brazil (sweet sorghums)

TEF

African Countries

Endashaw Bekele, The National Herbarium, Addis Ababa University, PO Box 3434, Addis Ababa, Ethiopia
Abebe Demissie, Plant Germplasm Exploration & Collection, PGRC/E, PO Box 30726, Addis Ababa, Ethiopia
Susan Burnell Edwards, The National Herbarium, Addis Ababa University, PO Box 3434, Addis Ababa, Ethiopia
Tewolde Berhan Gebre Egziabher, c/o The National Herbarium, Addis Ababa University, PO Box 3434, Addis Ababa, Ethiopia
Kifle Gozeguze, Regional Soil and Water Conservation Department, c/o Ministry of Agriculture, PO Box 62347, Addis Ababa, Ethiopia
Tantigegn Kerede Kassa, Zonal Team in Soil Conservation, Bahrder, Ethiopia
Abebe Kirub, Information Services, Institute of Agricultural Research, PO Box 2003, Addis Ababa, Ethiopia
Helmut Kreiensiek, Agriculture and Soil Conservation, German AgroAction -FSAP, PO Box 6, Maseru 100, Lesotho
Mahmoud Ahmed Mahmoud, Arab Organization for Agricultural Development (AOAD), PO Box 474, Khartoum, Sudan
Dejene Makonnen, Alemaya University of Agriculture, PO Box 138, Alemaya, Ethiopia
P.C.J. Maree, Department of Agronomy and Pastures, University of Stellenbosch, Victoria Street, Stellenbosch 7600, Cape Province, South Africa
Mateos Megiso, Soil Conservation Department, Ministry of Agriculture, PO Box 62347, Addis Ababa, Ethiopia
Gebru Teka Mehereta, Natural Resources Department, Ministry of Agriculture, PO Box 62347, Addis Ababa, Ethiopia
Solomon Mengistu, International Livestock Centre for Africa (ILCA), PO Box 30709, Nairobi, Kenya
Getachew Beyene Misker, Community Forestry Department, Ministry of Agriculture, PO Box 62347, Addis Ababa, Ethiopia
Helen Moss, 63 End Road, Linden Extension 2194, Randburg, South Africa
P.O. Osuji, International Livestock Centre for Africa (ILCA), P.O. Box 5689, Addis Ababa, Ethiopia (forage)
Norman F.G. Rethman, Department of Plant Production, University of Pretoria, Pretoria 0002, South Africa

Adebacho Watchiso, Community Forestry Department, Ministry of Agriculture, PO Box 62347, Addis Ababa, Ethiopia

J.J.P. Van Wyk, Research Institute for Reclamation Ecology, Potchefstroom 2520, South Africa (soil reclamation)

Other Countries

Donald F. Beech, Division of Tropical Crops and Pastures, CSIRO, Mill Road, St. Lucia, Brisbane 4067, Queensland, Australia

K.V. Bondale, Millet Development Program, Ministry of Agriculture, 27 Eldams Road, Madras 600 018, India

Wayne Carlson, Maskal Forages, Inc., PO Box A, Caldwell, Idaho 83606, USA

Geoffrey P. Chapman, Wye College, University of London, Wye, Near Ashford, Kent TN25 5AH, England

M. Cheverton, Wye College, University of London, Wye, Ashford, Kent TN25 5AH, England

R.H. Ellis, Department of Agriculture, University of Reading, Reading, Berkshire, RG6 2AH, England

Johannes M.M. Engels, Regional Office for South and Southeast Asia, IPGRI (IBPGR), c/o Pusa Campus, New Delhi 110 012, India

Don F. Gaff, Department of Ecology and E.B., Monash University, Wellington Road, Clayton, Victoria 3168, Australia

Heiner E. Goldbach, Abta Agrarökologie, Lehrstuhl Biogeographie, Institut für Geowissenschaften der Universität Bayreuth, Postfach 101251, D-8580 Bayreuth, Germany

Pamela M. Goode, Environmental Resources Unit, University of Salford, Salford M5 4WT, England

Le Dit Bokary Guindo, Special Program for African Agricultural Research, The World Bank, 1818 H Street, NW, Washington, DC 20433, USA

David Hilling, Centre for Developing Areas Research, Department of Geography, Royal Holloway and Bedford New College, University of London, Egham Hill, Egham, Surrey TW20 0EX, England

B.M. Glyn Jones, Biology Department, Huntersdale New College, Egham, Surrey TW20 0EX, England

J.P. Marathee, Crop and Grasslands Services, Plant Production and Protection Division, FAO, Via delle Terme di Caracalla, Rome 00100, Italy

Melak H. Mengesha, Genetic Resources Unit, ICRISAT, Patancheru PO, Andhra Pradesh 502 324, India

Frederick G. Meyer, c/o Herbarium, U.S. National Arboretum, 3501 New York Avenue, NE, Washington, DC 20002, USA

Redwood City Seed Company, PO Box 361, Redwood City, California 94064, USA

Michael Stocking, Soils and Land Use Development Studies, Overseas Development Group, University of East Anglia, Norwich, Norfolk NR4 7TJ, England

D. Theodoropoulos, J.L. Hudson, Seedsman, PO Box 1058, Redwood City, California 94064, USA

H.D. Tindall, 8 Church Avenue, Ampthill, Bedford MK45 2PN, England

J.H. Topps, Division of Agricultural Chemistry and Biochemistry, School of Agriculture, University of Aberdeen, 581 King Street, Aberdeen AB2 4AQ, Scotland

E.K. Twidwell, South Dakota Cooperative Extension Service, South Dakota State University, College Station, Brookings, South Dakota 57007, USA (forage)

C.E. West, Department of Human Nutrition, Wageningen Agricultural University, PO Box 8129, Wageningen 6700 EV, Netherlands

Josien M.C. Westphal-Stevels, Department of Plant Taxonomy, Wageningen Agricultural University, Gen. Foulkesweg 37, PO Box 8010, Wageningen 6700 ED, Netherlands

Erica F. Wheeler, Centre for Human Nutrition, The London School of Hygiene and Tropical Medicine, 2 Taviton Street, London WC1H 0BT, England

M. Zewdie, Department of Agriculture, University of Reading, Reading, Berkshire, RG6 2AH, England

CULTIVATED AND WILD GRAINS

African Countries

Samuel Agboire, National Cereals Research Institute, Private Mail Bag 8, Bida, Niger State, Nigeria

Abebe Demissie, Plant Germplasm Exploration & Collection, PGRC/E, PO Box 30726, Addis Ababa, Ethiopia (emmer, barley)

Susan Burnell Edwards, The National Herbarium, Addis Ababa University, PO Box 3434, Addis Ababa, Ethiopia

Seyfu Ketema, Tef Improvement Programme, Holetta Research Station, Institute of Agricultural Research (IAR), PO Box 2003, Addis Ababa, Ethiopia

H. Gebre Mariam, Institute of Agricultural Research, PO Box 2003, Addis Ababa, Ethiopia

Bede Okigbo, Institute for Natural Resources, United Nations University (UNU), c/o United Nations Development Programme (UNDP), PO Box 1423, Accra, Ghana

Other Countries

Bernard R. Baum, Central Experimental Farm, Centre for Land & Biological Resources Research, Ottawa K1A 0C6, Ontario, Canada (Ethiopian oats)

K.V. Bondale, Millet Development Program, Ministry of Agriculture, 27 Eldams Road, Madras 600 018, India (kodo millet)

Geoffrey P. Chapman, Wye College, University of London, Wye, Near Ashford, Kent TN25 5AH, England (buffel grass; wild grains)

A.B. Damania, Genetic Resources Unit, International Center for Agricultural Research in the Dry Areas (ICARDA), PO Box 5466, Aleppo, Syria (emmer)

Frances Cook, Economic and Conservation Section (ECOS), Royal Botanic Gardens, Kew, Richmond, Surrey TW9 3AB, England (botany)

Johannes M.M. Engels, Regional Office for South and Southeast Asia, IPGRI (IBPGR), c/o Pusa Campus, New Delhi 110 012, India (barley)

Jeffrey Gritzner, Public Policy Research Institute, Department of Geography, University of Montana, Missoula, Montana 59812–1018, USA

S. Hakim, Faculty of Agriculture, Tishreen University, Lattakia, Syria (emmer)

Wayne W. Hanna, Department of Agronomy, Georgia Coastal Plain Experiment Station, ARS, USDA, PO Box 748, Tifton, Georgia 31793, USA

J.R. Harlan, 1016 North Hagan Street, New Orleans, Louisiana 70119, USA (wild grains, barley)

David B. Harper, Food and Agricultural Chemistry Department, The Queen's University of Belfast, Newforge Lane, Belfast BT9 5PX, United Kingdom (wild grains)

Arthur Klatt, Division of Agriculture, 139 Agricultural Hall, Oklahoma State University, Stillwater, Oklahoma 74078–0500, USA

J.P. Marathee, Crop and Grasslands Services, Plant Production and Protection Division, FAO, Via delle Terme di Caracalla, Rome 00100, Italy

M.Y. Moualla, Faculty of Agriculture, Tishreen University, Lattakia, Syria (emmer)

John Compton Tothill, Division of Tropical Crops and Forages, CSIRO, 306 Carmody Road, St. Lucia 4069, Brisbane, Queensland, Australia (taxonomy)

John Yohe, International Sorghum/Millet Collaborative Research Support Program (INTSORMIL CRSP), 54 Nebraska Center, University of Nebraska, Lincoln, Nebraska 68583, USA

Appendix H
Note on Nutritional Charts

In the earlier chapters we have included tables of nutritional information, as well as charts that show how this information compares with that of a standard cereal such as maize or rice. They appear on the following pages.

Crop	Page
African rice	27
finger millet	44, 45
fonio	64
pearl millet	86, 87
sorghum	134, 135
tef	222, 223
kram-kram	263
shama millet	268
Egyptian grass	269
wadi rice	270

These tables and charts should be taken only as rough indications of the lost crop's merits, not the definitive word. Some species in this book are so neglected that their nutritional components have been reported merely once or twice. It is thus probable that the figures we have used are not representative of average samples, let alone especially nutritious forms. Moreover, natural variation can occur in the nutritional content of grain from any particular species as a result of nongenetic factors such as climate and the availability of nutrients in the soil. It could be, therefore, that even better types will be discovered and developed.

The bar graphs provide what we think is a simple, but visually powerful, representation of the relative nutritional merits of two foods. With them nutritional figures between two foods (or between a food and a recommended daily allowance) can be compared almost instantly. This technique, in which the relative merits can be seen at a glance, was devised specifically for this project, but comparable approaches could be employed equally well in Africa.*

The maize and rice values against which the African grains are compared in the bar graphs are taken from U.S. Department of Agriculture tables. The actual figures (converted to a dry-weight basis) are given below.

Component	Maize	Rice
Food energy (Kc)	408	406
Protein (g)	10.5	8.1
Carbohydrate (g)	83	90
Fat (g)	5.3	0.7
Fiber (g)	3.2	0.3
Ash (g)	1.3	0.7
Thiamin (mg)	0.43	0.08
Riboflavin (mg)	0.22	0.06
Niacin (mg)	4.1	1.8
Vitamin B6 (mg)	0.58	0.02
Folate (μg)	0.0	9.1
Pantothenic acid (mg)	0.47	1.15
Calcium (mg)	8	32
Copper (mg)	0.35	0.25
Iron (mg)	3.0	0.9
Magnesium (mg)	142	130
Manganese (mg)	0.55	1.1
Phosphorus (mg)	234	130
Potassium (mg)	320	130
Sodium (mg)	39	6
Zinc (mg)	2.5	1.2

In each of the essential-amino-acid bar graphs, the figures were compared on the basis of the amounts occurring in the protein of each grain (that is, grams per 100 grams of protein). In the other bar graphs, all nutrients were compared on a dry-weight basis so as to eliminate the distortions of different (and varying) amounts of moisture. Digestibility and other metabolic factors were not factored into the calculations. For vitamin A, the values for Retinol Equivalents were derived using standard formulas to convert literature figures given for carotenoids, ß-carotene, or International Units.

* The bar graphs were plotted electronically, so their resolution exceeds the standard error of the data (which is at minimum 10 percent). Duplicate data were discarded, and ranges were treated as separate values.

Amino Acid	Maize	Rice
Cystine	1.8	2.0
Isoleucine	3.6	4.3
Leucine	12.3	8.3
Lysine	2.8	3.6
Methionine	2.1	2.4
Phenylalanine	4.9	5.3
Threonine	3.8	3.6
Tryptophan	0.7	1.2
Tyrosine	4.1	3.3
Valine	5.1	6.1
Total	41.1	38.1

Grams per 100 g protein.

In most of the charts in the chapters we have compared the native grains in their whole-grain form with whole-grain rice and maize. A more realistic comparison might have been against polished rice and maize meal (in which the germ has been removed). This is the form in which rice and maize are normally consumed, whereas the native grains—pearl millet, fonio, finger millet, tef, and (in most cases at least) sorghum—are eaten as whole grains. Comparing nutritive values for the forms in which each is actually eaten creates an even more graphic picture of the nutritional superiority of the native grains.

APPENDIX I

Lost Crops of Africa Series

This is the first in a series of books highlighting the promise to be found in food plants native to Africa. The second and third volumes in the series are now being prepared for publication, and the fourth, fifth, and sixth are in the planning stage. Following are lists of the plants now being considered.

Volume 2: Cultivated Fruits

Balanites (Desert Date)	*Balanites aegyptiaca*
Baobab	*Adansonia digitata*
Butterfruit (Africado)	*Dacryodes edulis*
Carissa	*Carissa* spp., esp. *C. macrocarpa*
Horned Melon	*Cucumis metuliferus*
Kei Apple	*Dovyalis caffra*
Marula	*Sclerocarya caffra*
Melon	*Cucumis melo*
Tamarind	*Tamarindus indica*
Watermelon	*Citrullus lanatus*
Ziziphus	*Ziziphus mauritiana*

Volume 3: Wild Fruits

African Medlars	*Vangueria madagascariensis*
Aizen	*Boscia* spp.
Chocolate Berries	*Vitex* spp.
Custard Apples	*Annona senegalensis*
Figs	*Ficus* spp.
Gemsbok Cucumber	*Acanthosicyos naudinianus*
Gingerbread Plums	*Parinari* spp.
Grapes	*Vitis* spp.
Icacina (False Yam)	*Icacina oliviformis*
Imbe (African Mangosteen)	*Garcinia livingstonei*
Milkwoods	*Mimusops* spp.

Monkey Apple	*Anisophyllea laurina*
Monkey Orange	*Strychnos* spp.
Nara	*Acanthosicyos horrida*
Raisin Trees	*Grewia* spp.
Rubber Fruits	*Landolphia* spp.
Sour Plum	*Ximenia* spp.
Star Apples	*Chrysophyllum* spp.
Sugar Plums	*Uapaca* spp.
Sweet Detar	*Detarium senegalense*
Tree Grapes	*Lannea* spp.
Tree Strawberry	*Nauclea* spp.
Velvet Tamarind	*Dialium guineense*
Water Berry	*Syzygium guineense*
Wild Plum	*Pappea capensis*

Volume 4: Vegetables

African Eggplant	*Solanum macrocarpon*
Amaranths	*Amaranthus* spp.
Bitterleaf	*Vernonia amygdalina*
Bitter Melon	*Momordica* spp.
Baobab	*Adansonia digitata*
Bologi	*Crassocephalum biafrae*
Bungu	*Ceratotheca sesamoides*
Bur Gherkin	*Cucumis* spp.
Celosia	*Celosia* spp.
Cleome	*Cleome gynandra*
Crotalaria	*Crotalaria* spp.
Dayflowers	*Commelina* spp.
Edible Flowers	Various species
Edible Mushrooms	Various species
Edible Trees	Various species
Egusi-ito	*Cucumeropsis mannii*
Enset	*Ensete ventricosum*
Ethiopian Mustard	*Brassica carinata*
Fluted Pumpkin	*Telfairia occidentalis*
Garden Cress	*Lepidium* spp.
Gherkins	*Cucumis* spp.
Horned Melon (Kiwano)	*Cucumis metuliferus*
Jilo	*Solanum gilo*
Mock Tomato	*Solanum aethiopicum*
Okra	*Abelmoschus esculentus*
Ogunmo	*Solanum melanocerasum*

Oyster Nut	*Telfairia pedata*
Spirulina	*Spirulina* spp.
Water Leaf	*Talinum* spp.

Volume 5: Legumes

Bambara Groundnut	*Vigna subterranea*
Cowpea	*Vigna unguiculata*
Grass Pea	*Lathyrus* spp.
Guar	*Cyamopsis tetragonoloba*
Groundbean	*Macrotyloma geocarpa*
Lablab	*Lablab purpureus*
Locust Beans	*Parkia* spp.
Marama Bean	*Bauhinia esculenta*
Pigeon Pea	*Cajanus cajan*
Sword Bean	*Canavalia* spp.
Velvet Tamarind	*Dialium* spp.

Volume 6: Roots and Tubers

African Yam Bean	*Sphenostylis* spp.
Anchote	*Coccinia* spp.
Guinea Yam	*Dioscorea* x *cayenensis*
Potato Yam	*Dioscorea esculenta*
Other Yams	*Dioscorea* spp.
Hausa Potato	*Solenostemon rotundifolius*
Sudan Potato	*Solenostemon parviflorus*
Livingstone Potato	*Plectranthus esculentus*
Wing bean Roots	*Psophocarpus* spp.
Tiger Nut (Chufa)	*Cyperus esculentus*
Vigna Roots	*Vigna* spp., especially *V. vexillata*

We hope that this set of reports will alert everyone to the wealth of foods that are Africa's own heritage. We also hope to continue the series with volumes on nuts, oilseeds, spices, beverage plants, and others. Collectively, the resulting wealth of knowledge and guidance might well lead to a "second front" in the war on hunger in what is now the most hunger-ravaged part of the world.

We would very much like to hear from readers who would like to contribute to these future volumes. Send your name and the crop in which you're interested to:

Noel D. Vietmeyer, FO 2060
National Academy of Sciences
2101 Constitution Avenue, N.W.
Washington, DC 20418, USA

Fax: (202) 334-2660
Email: nvietmey @ nas. edu

Above all, we'd like to appeal for photographs. Locating pictures for this book on grains has been a monumental headache; finding interesting shots for the future volumes will likely be even harder.

INDEX OF FOODS

INDEX OF PLANTS

The BOSTID Innovation Program

Since its inception in 1970, BOSTID has had a small project to evaluate innovations that could help the Third World. Formerly known as the Advisory Committee on Technology Innovation (ACTI), this small program has been identifying unconventional developments in science and technology that might help solve specific developing-country problems. In a sense, it acts as an "innovation scout"—providing information on options that should be tested or incorporated into activities in Africa, Asia, and Latin America.

So far, the BOSTID innovation program has published about 40 reports, covering, among other things, underexploited crops, trees, and animal resources, as well as energy production and use. Each book is produced by a committee of scientists and technologists (including both skeptics and proponents), with scores (often hundreds) of researchers contributing their knowledge and recommendations through correspondence and meetings.

These reports are aimed at providing reliable and balanced information, much of it not readily available elsewhere and some of it never before recorded. In its two decades of existence, this program has distributed more than 500,000 copies of its reports. Among other things, it has introduced to the world grossly neglected plant species such as jojoba, guayule, leucaena, mangium, amaranth, and the winged bean.

BOSTID's innovation books, although often quite detailed, are designed to be easy to read and understand. They are produced in an attractive, eye-catching format, their text and language carefully crafted to reach a readership that is uninitiated in the given field. In addition, most are illustrated in a way that helps readers deduce their message from the pictures and captions, and most have brief, carefully selected bibliographies, as well as lists of research contacts that lead readers to further information.

By and large, these books aim to catalyze actions within the Third World, but they usually also have utility in the United States, Europe, Japan, and other industrialized nations.

So far, the BOSTID innovation project on underexploited Third-World resources (Noel Vietmeyer, Director and Scientific Editor) has produced the following reports.

Ferrocement: Applications in Developing Countries (1973). 104 pp.

Mosquito Control: Perspectives for Developing Countries (1973). 76 pp.

Some Prospects for Aquatic Weed Management in Guyana (1974). 52 pp.

Roofing in Developing Countries: Research for New Technologies (1974). 84 pp.
An International Centre for Manatee Research (1974). 38 pp.
More Water for Arid Lands (1974). 165 pp.
Products from Jojoba (1975). 38 pp.
Underexploited Tropical Plants (1975). 199 pp.
The Winged Bean (1975). 51 pp.
Natural Products for Sri Lanka's Future (1975). 53 pp.
Making Aquatic Weeds Useful (1976). 183 pp.
Guayule: An Alternative Source of Natural Rubber (1977). 92 pp.
Aquatic Weed Management: Some Prospects for the Sudan (1976). 57 pp.
Ferrocement: A Versatile Construction Material (1976). 106 pp.
More Water for Arid Lands (French edition, 1977). 164 pp.
Leucaena: Promising Forage and Tree Crop for the Tropics (1977). 123 pp.
Natural Products for Trinidad and the Caribbean (1979). 50 pp.
Tropical Legumes (1979). 342 pp.
Firewood Crops: Shrub and Tree Species for Energy Production (volume 1, 1980). 249 pp.
Water Buffalo: New Prospects for an Underutilized Animal (1981). 126 pp.
Sowing Forests from the Air (1981). 71 pp.
Producer Gas: Another Fuel for Motor Transport (1983). 109 pp.
Producer Gas Bibliography (1983). 50 pp.
The Winged Bean: A High-Protein Crop for the Humid Tropics (1981). 58 pp.
Mangium and Other Fast-Growing Acacias (1983). 72 pp.
Calliandra: A Versatile Tree for the Humid Tropics (1983). 60 pp.
Butterfly Farming in Papua New Guinea (1983). 42 pp.
Crocodiles as a Resource for the Tropics (1983). 69 pp.
Little-Known Asian Animals With Promising Economic Future (1983). 145 pp.
Firewood Crops: Shrub and Tree Species for Energy Production, Volume 2 (1983). 103 pp.
Casuarinas: Nitrogen-Fixing Trees for Adverse Sites (1984). 128 pp.
Amaranth: Modern Prospects for an Ancient Crop (1984). 90 pp.
Leucaena: Promising Forage and Tree Crop (Second edition, 1984). 110 pp.
Jojoba: A New Crop for Arid Lands (1985). 112 pp.
Quality-Protein Maize (1988). 112 pp.
Triticale: A Promising Addition to the World's Cereal Grains (1989). 113 pp.

Lost Crops of the Incas: Little-Known Plants of the Andes with Promise for Worldwide Cultivation (1989). 427 pp.
Microlivestock: Little-Known Small Animals with a Promising Economic Future (1991). 468 pp.
Neem: A Tree for Solving Global Problems (1992). 151 pp.
Vetiver: A Thin Green Line Against Erosion (1993).
Lost Crops of Africa: Volume 1 - Grains (1995).
Lost Crops of Africa: Volume 2 - Cultivated Fruits (1995).
Foods of the Future: Tropical Fruits (1995)

Board on Science and Technology for International Development

BOSTID Publications

BOSTID manages programs with developing countries on behalf of the U.S. National Research Council. Reports published by BOSTID are sponsored in most instances by the U.S. Agency for International Development. They are intended for distribution to readers in developing countries who are affiliated with governmental, educational, or research institutions, and who have professional interest in the subject areas treated by the reports.

BOSTID Publications and Information Services (FO-2060Z)
National Research Council
2101 Constitution Avenue, N.W.
Washington, D.C. 20418 USA

Energy

33. **Alcohol Fuels: Options for Developing Countries.** 1983, 128 pp. Examines the potential for the production and utilization of alcohol fuels in developing countries. Includes information on various tropical crops and their conversion to alcohols through both traditional and novel processes. ISBN 0–309–04160–0.

36. **Producer Gas: Another Fuel for Motor Transport.** 1983, 112 pp. During World War II Europe and Asia used wood, charcoal, and coal to fuel over a million gasoline and diesel vehicles. However, the technology has since been virtually forgotten. This report reviews producer gas and its modern potential. ISBN 0–309–04161–9.

56. **The Diffusion of Biomass Energy Technologies in Developing Countries.** 1984, 120 pp. Examines economic, cultural, and political factors that affect the introduction of biomass-based energy technologies in developing countries. It includes information on the opportunities for these technologies as well as conclusions and recommendations for their application. ISBN 0–309–04253–4.

Technology Options

14. **More Water for Arid Lands: Promising Technologies and Research Opportunities.** 1974, 153 pp. Outlines little-known but promising technologies to supply and conserve water in arid areas. ISBN 0–309–04151–1.

34. **Priorities in Biotechnology Research for International Development: Proceedings of a Workshop.** 1982, 261 pp. Report of a workshop

organized to examine opportunities for biotechnology research in six areas: 1) vaccines, 2) animal production, 3) monoclonal antibodies, 4) energy, 5) biological nitrogen fixation, and 6) plant cell and tissue culture. ISBN 0–309–04256–9.

61. **Fisheries Technologies for Developing Countries.** 1987, 167 pp. Identifies newer technologies in boat building, fishing gear and methods, coastal mariculture, artificial reefs and fish aggregating devices, and processing and preservation of the catch. The emphasis is on practices suitable for artisanal fisheries. ISBN 0–309–04260–7.

73. **Applications of Biotechnology to Traditional Fermented Foods**. 1992, 207 pp. Microbial fermentations have been used to produce or preserve foods and beverages for thousands of years. New techniques in biotechnology allow better understanding of these transformations so that safer, more nutritious products can be obtained. This report examines new developments in traditional fermented foods. ISBN 0–309–04685–8.

Plants

47. **Amaranth: Modern Prospects for an Ancient Crop.** 1983, 81 pp. Before the time of Cortez, grain amaranths were staple foods of the Aztec and Inca. Today this nutritious food has a bright future. The report discusses vegetable amaranths also. ISBN 0–309–04171–6.

53. **Jojoba: New Crop for Arid Lands.** 1985, 102 pp. In the last 10 years, the domestication of jojoba, a little-known North American desert shrub, has been all but completed. This report describes the plant and its promise to provide a unique vegetable oil and many likely industrial uses. ISBN 0–309–04251–8.

63. **Quality-Protein Maize.** 1988, 130 pp. Identifies the promise of a nutritious new form of the planet's third largest food crop. Includes chapters on the importance of maize, malnutrition and protein quality, experiences with quality-protein maize (QPM), QPM's potential uses in feed and food, nutritional qualities, genetics, research needs, and limitations. ISBN 0–309–04262–3.

64. **Triticale: A Promising Addition to the World's Cereal Grains.** 1988, 105 pp. Outlines the recent transformation of triticale, a hybrid between wheat and rye, into a food crop with much potential for many marginal lands. The report discusses triticale's history, nutritional quality,

breeding, agronomy, food and feed uses, research needs, and limitations. ISBN 0–309–04263–1.

70. **Saline Agriculture: Salt-Tolerant Plants for Developing Countries.** 1989, 150 pp. The purpose of this report is to create greater awareness of salt-tolerant plants and the special needs they may fill in developing countries. Examples of the production of food, fodder, fuel, and other products are included. Salt-tolerant plants can use land and water unsuitable for conventional crops and can harness saline resources that are generally neglected or considered as impediments to, rather than opportunities for, development. ISBN 0–309–04266–6.

74. **Vetiver Grass: A Thin Green Line Against Erosion.** 1992, 000 pp. Vetiver is a little-known grass that seems to offer a practical solution for controlling soil loss. Hedges of this deeply rooted species catch and hold back sediments. The stiff foliage acts as a filter that also slows runoff and keeps moisture on site, allowing crops to thrive when neighboring ones are desiccated. In numerous tropical locations, vetiver hedges have restrained erodible soils for decades and the grass—which is pantropical—has shown little evidence of weediness. ISBN 0–309–04269–0.

77. **Lost Crops of Africa: Volume 1, Grains.** (1995) Many people predict that Africa in the near future will not be able to feed its projected population. What is being overlooked, however, is that Africa has more than 2,000 native food plants, few of which are receiving research or recognition. This book describes the potentials of finger millet, fonio, pearl millet, sorghum, tef, and other cereal grains that are native to Africa.

Innovations in Tropical Forestry

35. **Sowing Forests from the Air.** 1981, 64 pp. Describes experiences with establishing forests by sowing tree seed from aircraft. Suggests testing and development of the techniques for possible use where forest destruction now outpaces reforestation. ISBN 0–309–04257–7.

41. **Mangium and Other Fast-Growing Acacias for the Humid Tropics.** 1983, 63 pp. Highlights 10 acacia species that are native to the tropical rain forest of Australasia. That they could become valuable forestry resources elsewhere is suggested by the exceptional performance of *Acacia mangium* in Malaysia. ISBN 0–309–04165–1.

42. **Calliandra: A Versatile Small Tree for the Humid Tropics.** 1983, 56

pp. This Latin American shrub is being widely planted by villagers and government agencies in Indonesia to provide firewood, prevent erosion, provide honey, and feed livestock. ISBN 0–309–04166-X.

43. **Casuarinas: Nitrogen-Fixing Trees for Adverse Sites.** 1983, 118 pp. These robust, nitrogen-fixing, Australasian trees could become valuable resources for planting on harsh eroding land to provide fuel and other products. Eighteen species for tropical lowlands and highlands, temperate zones, and semiarid regions are highlighted. ISBN 0–309–04167–8.

52. **Leucaena: Promising Forage and Tree Crop for the Tropics.** 1984 (2nd edition), 100 pp. Describes a multipurpose tree crop of potential value for much of the humid lowland tropics. Leucaena is one of the fastest growing and most useful trees for the tropics. ISBN 0–309–04250-X.

71. **Neem: A Tree for Solving Global Problems.** 1992, 151 pp. The neem tree is potentially one of the most valuable of all trees. It shows promise for pest control, reforestation, and improving human health. Safe and effective pesticides can be produced from seeds. Neem can grow in arid and humid tropics and is a fast-growing source of fuelwood. ISBN 0–309–04686–6.

Managing Tropical Animal Resources

32. **The Water Buffalo: New Prospects for an Underutilized Animal.** 1981, 188 pp. The water buffalo is performing notably well in recent trials in such unexpected places as the United States, Australia, and Brazil. Report discusses the animal's promise, particularly emphasizing its potential for use outside Asia. ISBN 0–309–04159–7.

44. **Butterfly Farming in Papua New Guinea.** 1983, 36 pp. Indigenous butterflies are being reared in Papua New Guinea villages in a formal government program that both provides a cash income in remote rural areas and contributes to the conservation of wildlife and tropical forests. ISBN 0–309–04168–6.

45. **Crocodiles as a Resource for the Tropics.** 1983, 60 pp. In most parts of the tropics, crocodilian populations are being decimated, but programs in Papua New Guinea and a few other countries demonstrate that, with care, the animals can be raised for profit while protecting the wild populations. ISBN 0–309–04169–4.

46. **Little-Known Asian Animals with a Promising Economic Future.** 1983, 133 pp. Describes banteng, madura, mithan, yak, kouprey, babirusa, javan warty pig, and other obscure but possibly globally useful wild and domesticated animals that are indigenous to Asia. ISBN 0–309–04170–8.

68. **Microlivestock: Little-Known Small Animals with a Promising Economic Future.** 1990, 449 pp. Discusses the promise of small breeds and species of livestock for Third World villages. Identifies more than 40 species, including miniature breeds of cattle, sheep, goats, and pigs; eight types of poultry; rabbits; guinea pigs and other rodents; dwarf deer and antelope; iguanas; and bees. ISBN 0–309–04265–8.

Health

49. **Opportunities for the Control of Dracunculiasis.** 1983, 65 pp. Dracunculiasis is a parasitic disease that temporarily disables many people in remote, rural areas in Africa, India, and the Middle East. Contains the findings and recommendations of distinguished scientists who were brought together to discuss dracunculiasis as an international health problem. ISBN 0–309–04172–4.

55. **Manpower Needs and Career Opportunities in the Field Aspects of Vector Biology.** 1983, 53 pp. Recommends ways to develop and train the manpower necessary to ensure that experts will be available in the future to understand the complex ecological relationships of vectors with human hosts and pathogens that cause such diseases as malaria, dengue fever, filariasis, and schistosomiasis. ISBN 0–309–04252–6.

60. **U.S. Capacity to Address Tropical Infectious Diseases.** 1987, 225 pp. Addresses U.S. manpower and institutional capabilities in both the public and private sectors to address tropical infectious disease problems. ISBN 0–309–04259–3.

Resource Management

50. **Environmental Change in the West African Sahel.** 1984, 96 pp. Identifies measures to help restore critical ecological processes and thereby increase sustainable production in dryland farming, irrigated agriculture, forestry and fuelwood, and animal husbandry. Provides baseline information for the formulation of environmentally sound projects. ISBN 0–309–04173–2.

51. **Agroforestry in the West African Sahel.** 1984, 86 pp. Provides

development planners with information regarding traditional agroforestry systems—their relevance to the modern Sahel, their design, social and institutional considerations, problems encountered in the practice of agroforestry, and criteria for the selection of appropriate plant species to be used. ISBN 0–309–04174–0.

72. **Conserving Biodiversity: A Research Agenda for Development Agencies.** 1992, 127 pp. Reviews the threat of loss of biodiversity and its context within the development process and suggests an agenda for development agencies. ISBN 0–309–04683–1.

Forthcoming Books from BOSTID

Lost Crops of Africa: Volume 2, Cultivated Fruits. (1996) This report details the underexploited promise of about a dozen native African fruits. Included are such species as baobab, butter fruit, horned melon, marula, and watermelon.

Foods of the Future: Volume 1, Promising Tropical Fruits. (1996) Of the more than 3,000 edible tropical fruits, only four—banana, pineapple, mango, and papaya—are major resources. This report will describe many more that could have significance for increasing worldwide food supplies, improving nutrition, adding new flavors to world cuisine, and boosting economic advancement.

ART CREDITS

Page	
16, 37	Duncan Vaughan, courtesy International Rice Research Institute, Los Baños, the Philippines
57	E. Westphal
75	Folu M. Dania Gobe
160	U.S. Department of Commerce, Bureau of the Census
214	James Bruce (this drawing appears in the famous explorer's report of his historic penetration of Ethiopia in 1770)
219	Taylor Horst
229	Glenn Oakley (Photograph shows Wayne Carlson)
286	Victor Englebert

Drawings on pages 38, 144, 158, 236, and 250 are by Sara Joelle Hall and are reproduced (with permission) from *Food from the Veld* by F.W. Fox and M.E. Norwood Young, Delta Books (Pty) Ltd., South Africa.

Drawings on pages 143, 157, 175, 193, and 212 are by Vantage Art, Inc., and are reproduced (with permission) from the *Fieldbook of Natural History*, Second Edition. E. Laurence Palmer and H. Seymour Fowler. Copyright 1975 by McGraw-Hill, Inc.

Cover Design by David Bennett

ORDER FORM

Please indicate on the labels below the names of colleagues, government institutions, universities, and others who could use a copy of Lost Crops of Africa: Volume 1, Grains *in their work.* The addresses of public libraries, magazines, and newspapers would be also welcome. Copies are available for classroom use throughout Africa.

Please return this form to: African Grains Report, FO 2060
National Academy of Sciences
2101 Constitution Avenue, N.W.
Washington, D.C. 20418, USA

77 77

77 77

77 77